MW01484569

MACHINERY VIBRATION AND ROTORDYNAMICS

MACHINERY VIBRATION
AND ROTORDYNAMICS

John Vance, Fouad Zeidan, Brian Murphy

JOHN WILEY & SONS, INC.

This book is printed on acid-free paper. ∞

Copyright © 2010 by John Wiley & Sons, Inc. All rights reserved.

Published by John Wiley & Sons, Inc., Hoboken, New Jersey
Published simultaneously in Canada

No part of this publication may be reproduced, stored in a retrieval system, or transmitted in any form or by any means, electronic, mechanical, photocopying, recording, scanning, or otherwise, except as permitted under Section 107 or 108 of the 1976 United States Copyright Act, without either the prior written permission of the Publisher, or authorization through payment of the appropriate per-copy fee to the Copyright Clearance Center, 222 Rosewood Drive, Danvers, MA 01923, (978) 750-8400, fax (978) 646-8600, or on the web at www.copyright.com. Requests to the Publisher for permission should be addressed to the Permissions Department, John Wiley & Sons, Inc., 111 River Street, Hoboken, NJ 07030, (201) 748-6011, fax (201) 748-6008, or online at www.wiley.com/go/permissions.

Limit of Liability/Disclaimer of Warranty: While the publisher and the author have used their best efforts in preparing this book, they make no representations or warranties with respect to the accuracy or completeness of the contents of this book and specifically disclaim any implied warranties of merchantability or fitness for a particular purpose. No warranty may be created or extended by sales representatives or written sales materials. The advice and strategies contained herein may not be suitable for your situation. You should consult with a professional where appropriate. Neither the publisher nor the author shall be liable for any loss of profit or any other commercial damages, including but not limited to special, incidental, consequential, or other damages.

For general information about our other products and services, please contact our Customer Care Department within the United States at (800) 762-2974, outside the United States at (317) 572-3993 or fax (317) 572-4002.

Wiley also publishes its books in a variety of electronic formats. Some content that appears in print may not be available in electronic books. For more information about Wiley products, visit our web site at www.wiley.com.

Library of Congress Cataloging-in-Publication Data:

Vance, John M.
  Machinery vibration and rotordynamics / John Vance, Brian Murphy, Fouad Zeidan.
        p. cm.
  Includes bibliographical references and index.
  ISBN 978-0-471-46213-2 (cloth)
  1. Rotors–Dynamics. 2. Rotors–Vibration. 3. Machinery–Vibration. 4.
Turbomachines–Dynamics. I. Murphy, Brian, 1956- II. Zeidan, Fouad. III. Title.
  TJ177.V36 2010
  621.8′11—dc22   2009045963

Printed in the United States of America

10 9 8 7 6 5 4 3

The first author gratefully dedicates his part in this book to his loving wife Louise, who made the book possible by her unselfish support of the task and devotion to her husband while it was being written.

John M. Vance

CONTENTS

PREFACE

This book follows the first author's book *Rotordynamics of Turbomachinery* in its practical approach and style. Much of the material in that book has been updated and extended with new information, new examples, and a few corrections that reflect what has been learned since then. Of particular interest and significance are the new chapters (4, 5, and 6) on bearings, seals, and computer modeling contributed by the co-authors Dr. Fouad Zeidan and Dr. Brian Murphy. Dr. Zeidan is the president of two companies that design and manufacture high performance bearings and seals. These products often require the design and modeling of the complete rotor-bearing system to ensure reliable operation and compatibility. Dr. Murphy is the author of XLRotor™, one of the most widely used computer programs for rotordynamic analysis. Chapters 1 and 7 are also completely new. Chapter 1 describes the classical analytical techniques used by engineers for troubleshooting vibration problems. Chapter 7 gives a history of the most important rotordynamics analysis and experiments since 1869.

The authors have noted (with some surprise) for many years that the subject material of this book is not taught in most engineering colleges, even though rotating machines are probably the most common application of mechanical engineering. The book is organized so that the first three or four chapters could be used as a text for a senior or graduate college elective course. These chapters have exercises at the end that can be assigned to the students, which will greatly enhance their understanding of the chapter material. The later chapters will serve the same students well after graduation as reference source material with examples of analysis and test results for real machines, bearings, and seals. But for the majority of engineers assigned to troubleshoot a rotating machine, or to design it for reliability, and having no relevant technical background, this entire book can be the substitute for the course they never had.

It is the author's hope that this book will make a significant contribution to the improvement of rotating machines for the service of mankind in the years to come.

John M. Vance
Fouad Y. Zeidan
Brian T. Murphy

1

FUNDAMENTALS OF MACHINE VIBRATION AND CLASSICAL SOLUTIONS

This chapter is focused on practical applications of mechanical vibrations theory. The reader may want to supplement the chapter with one of the vibration textbooks in the reference list at the end of the chapter if he has no background in the theory.

THE MAIN SOURCES OF VIBRATION IN MACHINERY

The most common sources of vibration in machinery are related to the inertia of moving parts in the machine. Some parts have a reciprocating motion, accelerating back and forth. In such a case Newton's laws require a force to accelerate the mass and also require that the force be reacted to the frame of the machine. The forces are usually periodic and therefore produce periodic displacements observed as vibration. For example, the piston motion in the slider-crank mechanism of Fig. 1-1 has a fundamental frequency equal to the crankshaft speed but also has higher frequencies (harmonics). The dominant harmonic is twice crankshaft speed (2nd harmonic). Figure 1-2a shows the displacement of the piston. It looks almost like a sine wave but it is slightly distorted by higher-order harmonics due to the nonlinear kinematics of the mechanism. Fig. 1-2b shows the acceleration of the piston, where the 2nd harmonic is amplified since the acceleration amplitude is frequency-squared times the displacement amplitude.

Even without reciprocating parts, most machines have rotating shafts and wheels that cannot be perfectly balanced, so according to Newton's laws, there must be a rotating force vector at the bearing supports of each

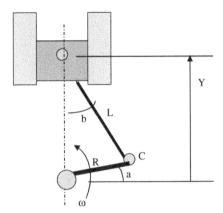

Figure 1-1 Slider-crank mechanism.

rotor to produce the centripetal acceleration of the mass center. Most of these force vectors are rotating and therefore produce a rotating displacement vector (all real machine parts are elastic) that can be observed as an orbit if two orthogonal vibration transducers are employed. Each of the transducers will produce a time trace similar to Fig. 1-2a or 1-2b. Harmonics and resulting distortion similar to Fig. 1-2a and 1-2b can be produced by shaft misalignment or by nonlinearity of the bearing stiffness. The fundamental frequency of the X and Y (orthogonal) vibration vectors is shaft speed ω, so the fundamental vibration is $x(t) = X \cos(\omega t)$ and $y(t) = Y \sin(\omega t)$. This type of vibration is referred to as *forced response* or *synchronous response to unbalance*. The vibration amplitude can become very large if the excitation frequency (rotor speed for example) becomes close to one of the natural frequencies of the machine structure. This is called a *resonance* or a *critical speed*, but it is not an unstable motion since the amplitude does not grow with time (unless there is no damping).

Another type of machine vibration problem, less common but more difficult to deal with, can come from the characteristic natural vibration frequencies (eigenvalues) of the machine structure and its supports, even if no imbalance or excitation is present. Natural frequencies die out in static structures due to the energy dissipated by damping, but in rotating machines they can grow larger with time. This is known as *self-excited instability* or *rotordynamic instability*. It is an innate potential characteristic of some rotating machines, especially when fluid pressures are present (e.g., bearings, impellers, turbine wheels, or seals).

Every real structure has an infinite number of natural frequencies, but many machinery vibration problems involve just one of these frequencies. That is why the simple single degree of freedom (SDOF) model (with just one natural frequency) presented in vibration textbooks [1–3] can be

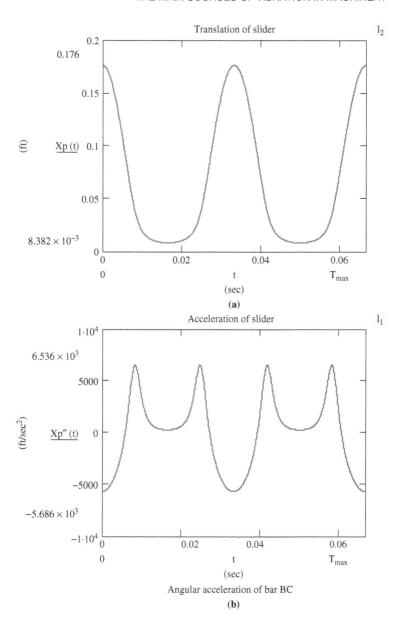

Figure 1-2 (a) Displacement of the piston, and (b) acceleration of the piston.

useful for analyzing vibration in machines. In fact, a SDOF model, consisting of one rigid mass, one spring, and one damper can be constructed to represent the vibration characteristics of any real machine in the neighborhood of a particular natural frequency of interest. This is called a *modal model*. To make physical sense out of complex machinery vibration data, or from realistic computer simulations of machinery vibration, the details of the SDOF mathematical model, its variations, and its solutions must be burned indelibly into the mind of the vibration engineer.

THE SINGLE DEGREE OF FREEDOM (SDOF) MODEL

The SDOF model as seen in most vibration textbooks is shown in Fig. 1-3. Here it will be referred to as system A. The stiffness, damping, and mass are k, c, and m, respectively. The undamped natural frequency is given by

$$\omega_n = \sqrt{\frac{k}{m}} \quad \text{rad/sec} \tag{1-1}$$

The circular frequency ω_n can be converted to hertz (Hz) (cycles/sec) as $f_n = \omega_n/2\pi$, or to revolutions per minute (rpm) as $N = 60 f_n$.

With a sinusoidal force applied to the mass, the differential equation of motion

$$m\ddot{x} + c\dot{x} + kx = F \sin(\omega t) \tag{1-2}$$

has a solution made up of two parts: (1) the particular solution for x that gives $F \sin(\omega t)$ on the right-hand side, and (2) the homogeneous solution for x that gives zero on the right-hand side. The sum of the two solutions, of course, gives $F \sin(\omega t)$, which satisfies the equality sign. The two solutions represent the two types of machine vibration described in the previous section, that is, forced response and characteristic (free) vibration. The particular solution for forced response is

$$x_p(t) = F \sin(\omega t + \phi) / \sqrt{\left(k - m\omega^2\right)^2 + (c\omega)^2} \tag{1-3}$$

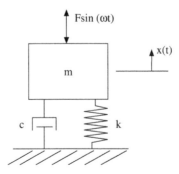

Figure 1-3   Single degree of freedom vibration model (system A).

Notice that the frequency ω of the forced vibration response is the same as the frequency of the excitation. The angle ϕ gives the time ϕ/ω by which the response x lags the excitation force F. For analyzing a vibration problem it is important to understand how k, c, and m influence the response amplitude. They have different effects depending on the frequency ratio ω/ω_n, as we shall see in the section to follow. Looking at Eq. 1-3 we can see that the amplitude X of the forced vibration response is

$$X = F \Big/ \sqrt{\left(k - m\omega^2\right)^2 + (c\omega)^2} \tag{1-4}$$

which depends on k, c, m, ω, and F. Notice that the denominator gets small when the exciting frequency ω is ω_n (Eq. 1-1) unless the damping coefficient c is large. A plot of Eq. 1-4 is shown in Fig. 1-7. It is called the *Bode amplitude plot* or the *frequency response plot* for system A.

The homogeneous part of the solution (for free vibration) with $F = 0$ is given by

$$x_h(t) = Ae^{st} \tag{1-5}$$

where s is a complex number, $s = \lambda + i\omega_d$. s is called the *eigenvalue*. Using the law of exponents, Eq. 1-5 can be rewritten as

$$x_h(t) = Ae^{\lambda t}e^{i\omega_d t} \tag{1-6}$$

where

$$e^{i\omega_d t} = \cos(\omega_d t) + i \sin(\omega_d t) \tag{1-7}$$

Equation 1-5 or 1-6 satisfies the differential Eq. 1-2 with $F = 0$ provided that the real part of the eigenvalue is $\lambda = -c/2m$ and the imaginary part is the square root of $\omega_d^2 = k/m - (c/2m)^2$. The amplitude A in Eq. 1-5 is of little interest here since it is determined only by the initial condition that instigates the free vibration. In rotating machinery, the differential equations are more complicated but still are of the same class as (1-2) and have the same form of homogeneous solution as (1-5). The imaginary part of s, ω_d, is the damped natural frequency. Notice that it becomes equal to ω_n, Eq. 1-1, when the damping coefficient $c = 0$.

The real part λ of the eigenvalue s determines how fast the free vibration dies out. It is often converted into a *damping ratio* $\zeta = c/c_{cr}$, where the critical damping $c_{cr} = 2m\omega_n$. Critical damping is the amount required to prevent free vibration (and no more). The conversion equation is $\zeta = -\lambda/\omega_n$. Figure 1-4a shows free vibration with $\zeta = 0.05$ (5% of critical damping); Fig. 1-4b shows the same system with $\zeta = 0.25$ (25% of critical damping). If a free vibration is graphed like Fig. 1-4, the damping can be expressed as the natural logarithm of

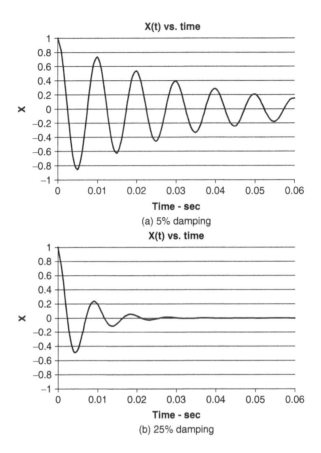

Figure 1-4   (a) Free vibration with 0.05 damping ratio; (b) free vibration with 0.25 damping ratio.

the ratio of successive amplitudes X_n/X_{n+1}. The logarithmic decrement $\delta = \ln(X_n/X_{n+1}) = 2\pi\zeta/(1+\zeta^2)^{1/2}$. The inverse expression is often useful: $\zeta = \delta/[(2\pi)^2 + \delta^2]^{1/2}$.

The algebraic sign of the real part of the eigenvalue λ is the mathematical test for vibration stability, i.e., whether the free vibration of frequency ω_d will die out or, in the unstable case, will grow with time. For example, in the simple system of Fig. 1-3, λ becomes positive if the damping c is negative. Negative damping is possible in mechanical systems, especially when fluid pressures are acting.

USING SIMPLE MODELS FOR ANALYSIS AND DIAGNOSTICS

Techniques and methods for solving vibration problems can often be developed by using the simple one degree of freedom model even though the real system is more complicated. The main purpose of the model is to provide an understanding of the type of problem being encountered so that the most effective type of "fix" can be identified. Sometimes a simple model can even yield useful approximations for the optimum parametric values, such as stiffness and damping to be employed. In contrast to the large and detailed finite element models being promoted by some for all diagnostic vibration analysis, this approach suggests that the engineer should first use the simplest possible model that contains the relevant physical characteristics and resort to the more detailed models only when the simple models do not yield sufficient guidance for modifications to the design or when improved accuracy is desired.

In addition to system A of Fig. 1-3, two more single degree of freedom models are shown in Figs. 1-5 and 1-6. All three of these systems have a single natural frequency determined by their modal mass and stiffness, but there are subtle differences between the three models that are related to the type of excitation.

The constant amplitude exciting force F in system A is generally unrealistic. Inertia forces in rotating machinery are proportional to speed squared. Model C in Fig. 1-6 has an unbalanced rotor so that the exciting force $F = m\omega^2 u$, where u is the offset of the center of rotor mass m from the axis of rotation. Note that the mass m is the rotating mass, not the total mass, so m on the left side of differential equation (1-2) must be replaced by the total mass M unless the nonrotating mass is negligible.

In some cases the excitation is a vibration displacement at the base, rather than a force. This is represented by system B in Fig. 1-5.

These small differences in the models produce different frequency response curves. The differences are useful in diagnosing problems and determining solutions. Obviously, to use these differences, the engineer must have a complete and thorough knowledge of the three models and their responses. The three systems illustrated in Figs. 1-3, 1-5, and 1-6 and their mathematical analyses are described in most vibration textbooks [1–3]. In some cases the damping should be included in the most realistic way possible, i.e., as viscous, Coulomb, hysteretic, or aerodynamic damping. However, if the damping is other than viscous, it may usually be

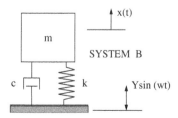

Figure 1-5 SDOF model with base excitation.

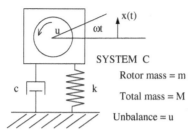

Figure 1-6   SDOF model with rotating unbalance.

represented by an equivalent viscous damping coefficient that varies with frequency [1, page 73]. For purely steel structures, it is usually less than 5% of the critical value. System B may have its predominant damping either (1) between the vibrating base and the modal mass, or (2) from the mass to ground. It is important to recognize the difference and set up the model correctly.

The frequency response curves for systems A, B, and C are plots of the amplitude of forced vibration versus the frequency. The response amplitude for system A is computed from Eq. 1-4 at each frequency, using appropriate values for k, c, m, and F. Figure 1-7 shows the response curve for system A with parameter values from Table 1-1. For plotting the curve, frequency ω (rad/sec) has been converted to rpm (cpm). X_static in the table is F/k, the displacement at zero frequency, which is the deflection of the spring under a static force F. Resonance is the undamped natural

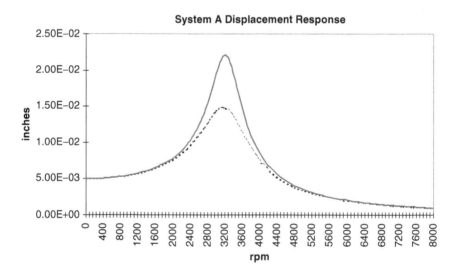

Figure 1-7   Forced response of system A (constant amplitude excitation force F).

Table 1-1   System A values for Fig. 1-7

	Data	Units
Input		
Mass	100	lb
Kstiff	30,000	lb/in
Cdamp	20	lb-sec/in
Force	150	lb
Freqstart	0	rpm
Freqstop	8000	rpm
Npoints	101	use 101
Output		
Resonance	3251.252	rpm
Zeta	0.11349	none
X_static	5.00E-03	in

frequency ω_n converted to cpm. Zeta is the critical damping ratio, i.e., the percentage of critical damping divided by 100. The solid curve in Fig. 1-7 has all the parametric values of Table 1-1.

The dashed curve in Fig. 1-7 has all the values of Table 1-1 except that the damping coefficient c has been increased from 20 lb-sec/in. (in the solid curve) to 30 lb-sec/in. The main effect of the increased damping is to reduce the vibration amplitude at the critical speed. It has very little effect at frequencies away from the critical speed. The critical speed (where the peak vibration occurs) is 3200 rpm for the solid curve and about 3150 rpm for the dashed curve. These are both slightly below the undamped natural frequency of 3251 cpm. Thus, damping tends to lower the critical speed. (This effect is reversed in system C (below) when the constant shaking force F is replaced with a rotating unbalance force $m\omega^2 u$). In Fig. 1-7, notice that the response amplitude X ($= 5$ mils at zero frequency) becomes large near the natural frequency, and approaches zero at very high frequencies. Figure 1-8 shows how the vibration X (the dashed curve) lags the force F with a phase angle ϕ (see Eq. 1-3). Figure 1-9 shows how the phase angle varies with frequency. More damping (the dashed curve) makes the phase angle change more gradually as the excitation frequency passes through ω_n. The phase angle is 90 degrees at the undamped natural frequency ω_n, regardless of the amount of damping. This fact is useful in determining the value of ω_n, since the phase angle can be measured but ω_n cannot be measured.

Graphs like Figs. 1-7 and 1-9 are often referred to as the frequency response curves, or Bode plots. If the parameter values (k, c, m) are changed, then the response curves will look similar but will have different

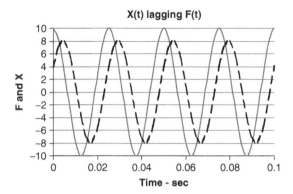

Figure 1-8 X (dashed) lagging force (solid).

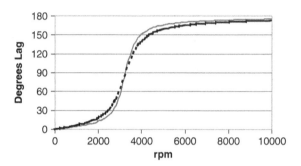

Figure 1-9   Phase lag response of system A.

values of response amplitude and phase. Increasing the damping generally brings the peak amplitude down but has a negligible effect at frequencies away from the natural frequency.

The necessity to plot many different curves for different values of F, k, and m is avoided by plotting the curve with dimensionless ratios as shown in Fig. 1-10. The abscissa in Fig. 1-10 is frequency ratio ω/ω_n; the ordinate Xk/F is X/X_static (the ratio of vibration amplitude to static displacement under the force F).

The frequency response of system B (Fig. 1-5, base vibration excitation) is given by

$$X = Y\sqrt{\frac{k^2 + (\omega c)^2}{\left(k - m\omega^2\right)^2 + (c\omega)^2}} \tag{1-8}$$

Figure 1-11 shows the response amplitude X calculated with the parametric values of Table 1-2. In the table, X_Base is the displacement amplitude Y of the vibrating support. Notice that damping in system B (the dashed

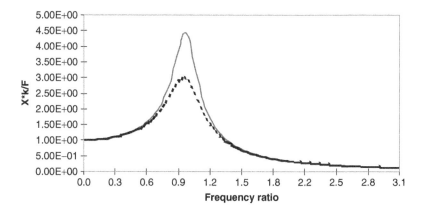

Figure 1-10   Dimensionless response of system A.

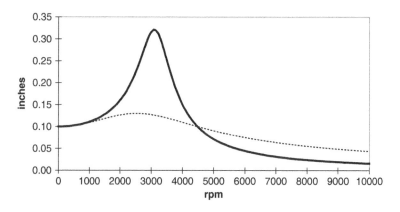

Figure 1-11   Response to base excitation of system B.

curve) actually increases the vibration response at high frequencies. Solving the differential equation for system B [1, page 66] shows that the crossover frequency is 1.4 times the undamped natural frequency. All the curves with different damping values cross at this frequency, and the amplitude there is the same as X_Base. The frequency range above this is called the *isolation range*, since the response there is reduced below what would be obtained with a hard support. A vibrating system with a fixed excitation frequency can be put into the isolation range by softening the spring Kstiff between the vibrating base and the mass.

The frequency response of system C (Fig. 1-6) is given by Eq. 1-9, where u is the unbalance (C.G. offset of the rotor), m is the rotor mass, and M is the total mass:

$$X = m\omega^2 u / \sqrt{\left(k - M\omega^2\right)^2 + (c\omega)^2} \qquad (1\text{-}9)$$

Table 1-2 System B parameters for Fig. 1-11

	Data	Units
Input		
Mass	0.35	lb
Kstiff	100	lb/in
Cdamp	0.1	lb-sec/in
Cdamp2	0.4	lb-sec/in
X_Base	0.1	in
Freqstart	0	rpm
Freqstop	10,000	rpm
Npoints	101	use 101
Output		
Resonance	3172.897	rpm
Zeta	0.166132	none
Zeta2	0.66453	none

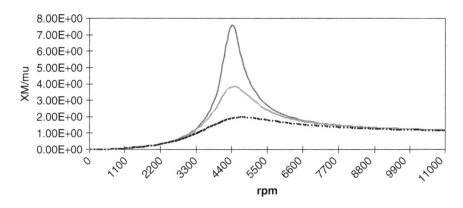

Figure 1-12 Response with an unbalanced rotor for three damping values.

The ratio X/u is often used and is sometimes called the *magnification factor*. The response calculated with the parametric values of Table 1-3 is shown in Fig. 1-12 with dimensionless amplitude XM/mu. In the table, m = Rotrmass and M = Rotrmass + Housmass. Note that XM/mu is approximately X/u in this case, since m/M = 0.98 (the housing mass is negligible). Figure 1-12 shows that system C response starts out at zero and damping in system C reduces the peak amplitude of vibration response and *raises* the critical speed. At very high frequencies the vibration amplitude approaches a limiting value determined by the amount of unbalance. Increasing the housing mass will reduce this value.

Table 1-3 System C parameters for
Fig. 1-12

	Data	Units
Input		
Rotrmass	50	lb
Housmass	1	lb
Kstiff	28,000	lb/in
Cdamp	8	lb-sec/in
Cdamp2	16	lb-sec/in
Cdamp3	32	lb-sec/in
Unbalance	0.0015	in
Freqstart	0	rpm
Freqstop	11000	rpm
Npoint	101	use 101
Output		
Resonance	4398.29	rpm
Zeta	0.065798	none
Zeta2	0.131597	none
Zeta3	0.263193	none
Totalmass	51	lb
Massratio	0.980392	none

SIX TECHNIQUES FOR SOLVING VIBRATION PROBLEMS WITH FORCED EXCITATION

When vibration measurements from the real system are compared and identified with the theoretical response from the appropriate model (A, B, or C) one of the following techniques for reducing the vibration will often become apparent.

1. *Identify and reduce the excitation source.* This most obvious solution is also the one least likely to be possible in systems of type A or type B, but it should be investigated first. In rotating machinery (system C), this technique is implemented by balancing the rotating parts. Balancing will be effective only when the vibration frequency is equal to the speed of a rotating part or its integer harmonics, and this fact is the corollary of a diagnostic rule: *Frequency components in a measured spectrum that are synchronous with a rotating speed or one of its harmonics are often caused by rotating imbalance.* In a reciprocating machine (Fig. 1-1), balancing the 2nd harmonic often requires a separate unbalanced *balance*

shaft rotating at twice crankshaft speed to cancel out the inertia forces.

2. *Tune the natural frequency to a value further away from the frequency of excitation to avoid resonance.* A study of the frequency response curves for any of the systems A, B, or C reveals that the vibratory excitation is highly magnified at frequencies near the natural frequency. This magnification factor R, or Q factor as it is sometimes called, can typically range from 5 to 50 or more depending on the amount of damping. The excitation frequency can seldom be changed, but the natural frequency can sometimes be easily changed by changing the modal stiffness. This is one place where intelligent construction of the analytical model becomes important, since the modal stiffness may be made up of several real stiffnesses in parallel or in series. In parallel combinations the very low stiffnesses have little effect in determining the modal stiffness, while in series combinations the very high stiffnesses have little effect. The tuning method is effective only when the excitation frequency is constant or when it only varies over a narrow range.

3. *Isolate the modal mass from the vibratory excitation by making the modal stiffness very low.* Notice that all the response curves show a very low response to the vibratory excitation at frequencies much higher than the natural frequency (far to the right on the response curves). Once again, the excitation frequency usually cannot be changed but the natural frequency can be brought far down by a very soft modal stiffness, thus placing the system response far to the right of resonance on the response curve. This method is particularly effective in systems of type B. A typical application is isolating an electronics box from a vibrating vehicle frame.

4. *Add damping to the system.* Damping is added by incorporating mechanisms that dissipate vibratory energy into heat. When they work, damping mechanisms produce forces that act in opposition to the vibratory velocity. Contrary to popular belief, however, adding damping indiscriminately does not always reduce vibration. Damping does work well whenever operation is near resonance (and this is the operating condition most likely to cause a problem). At frequencies away from resonance damping has very little effect, except to increase the forces transmitted to ground at high frequencies far above resonance. In a system B application where isolation is used, damping added between the modal mass and the vibrating support will actually increase the vibration of the mass at high frequencies. In a system C (rotating machinery) application with rolling element bearings, adding damping to the bearing supports will increase the

dynamic bearing loads and shorten bearing life for operation at high supercritical speeds [4, page 14].

5. *Add a vibration absorber.* A vibration absorber is a separate spring–mass assembly, which is added to the original system to "absorb" the vibration. This method works well only under a strict set of conditions: (a) the excitation frequency must be constant and resonant (i.e., equal to a natural frequency of the system), (b) the absorber spring–mass assembly must be tuned to a natural frequency equal to the resonant frequency of the original system, (c) the absorber mass should be at least 20 percent of the modal mass of the original system, and (d) the absorber spring–mass assembly should not have much damping. Under all of these conditions the modal mass of the original system will stand still while the absorber mass vibrates with a large amplitude. Since the absorber adds a degree of freedom to the analytical model, it follows that mathematical analysis of absorber performance requires at least a two degree of freedom model with two differential equations to solve simultaneously [2, page 293].

6. *Stiffen the system.* This method is listed last because it is valid only for systems of type A, but often is mistakenly suggested for all type systems. On the dimensionless response curves for system A (Fig. 1-10), notice on the vertical amplitude axis that the vibration amplitude X is determined by multiplying the graph value by F/k. Thus, the vibration amplitude can be made smaller at any frequency by raising the stiffness k. Once again, this applies only to systems of type A in which there is a constant amplitude force excitation that does not vary with frequency.

SOME EXAMPLES WITH FORCED EXCITATION

Illustrative Example 1

Problem: Figure 1-13 shows a car towing a trailer. This car/trailer system has a vibration problem in the direction of travel, which occurs only during braking. The car has a warped front brake disk, which produces a vibratory braking torque and braking force P. The trailer hitch is flexible in the direction of travel such as might be produced by installing the hitch ball directly onto a lightly constructed rear bumper. During braking the vibration frequency decreases as the front wheel speed decreases. At some particular speed the excitation frequency becomes equal to the natural frequency of the car/trailer system and the amplitude becomes very large. The car and trailer move longitudinally as rigid bodies (out of phase) in the vibratory motion.

Analysis: Let the car displacement be X_1 and the trailer be X_2 (relative to the displacement produced by travel speed). There are two degrees of freedom (dof), but the system can be reduced to 1 dof because there is no spring to ground. The two differential equations (one for each dof) can be combined by subtracting one from the other (because the first mode has zero frequency). $X = X_1 - X_2$ is a *modal coordinate*. This produces the system A differential equation (1-2), where m_e is the modal or equivalent mass and P_e is the modal or equivalent force as follows:

$$m_e \ddot{X} + KX = P_e \qquad (1\text{-}10)$$

where

$$m_e = \frac{m_1 m_2}{m_1 + m_2} \qquad (1\text{-}11)$$

$$P_e = \frac{m_2 P}{m_1 + m_2} \qquad (1\text{-}12)$$

Figure 1-14 is the dimensionless response curve with the modal parameters and with a small amount of damping added to keep the amplitudes positive.

See Problem 5-6 in [1] for the torsional analogy to this problem. The coordinates X_1 and X_2 describe the displacement of the car and trailer, respectively, as rigid bodies. This model has two degrees of freedom, but since neither the car nor the trailer has spring connections to ground, only one degree of freedom is relevant. Any movement of the system in which the car and trailer move in unison is irrelevant since they cannot vibrate together in phase; hence, the vibration coordinate of interest is the *relative* displacement $X = X_1 - X_2$. Mathematically this is the modal coordinate of the second mode, as the first mode has zero natural frequency. In the model of Fig. 1-13 the vibratory braking torque has been translated into a vibratory braking force with amplitude P and frequency ω equal to the rotational speed of the front wheel. The two differential equations in X_1 and X_2 have been subtracted one from another to produce the single differential equation in x (this is possible only when there are no springs to ground). Inspection of the resulting differential equation in X shows that

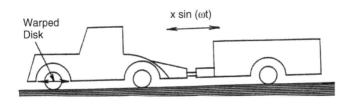

Figure 1-13   Car and trailer with flexible bumper/hitch.

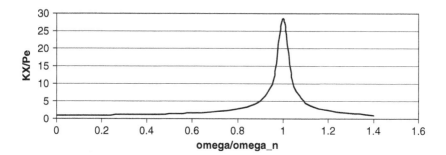

Figure 1-14   Dimensionless response of the car and trailer to the warped brake disk.

the modal mass m_e is $m_1 m_2 (m_1 + m_2)$ and the equivalent excitation force P_e is $P m_2 / (m_1 + m_2)$. The modal stiffness is simply the hitch connection stiffness K. The system is of Type A since *the differential equation has exactly the same form as the system A equation 1–2*. This is true because the equivalent excitation is a *constant force* $F = P_e$. Notice that the modal mass can be easily calculated from the weights of the car and trailer, and the modal stiffness K can be measured directly by applying static forces to the hitch or by measuring the resonant frequency and calculating $K = \omega^2 m_e$. The numerical magnitude of P need not be known to arrive at useful solutions as will be seen below.

Solution: Consider the six different methods described above for reducing vibration. The first method, reducing the source, could be implemented by replacing the warped brake disk and would be the ideal solution. If for some reason this cannot be done, consider the remaining methods. Tuning or absorption will not work because the excitation frequency is variable. Isolation will not work because the excitation frequency goes all the way down to zero. Damping would help at frequencies near resonance, but requires the addition of an expensive damping element to the flexible hitch connection at the rear bumper. Method 6, stiffening the system, can be implemented by stiffening the rear bumper or hitch connection and would be the best approach if the warped brake disk cannot be corrected or replaced. On the dimensionless response curve, note the effect of K on the value of X at every frequency. Stiffening the system in this case will reduce the response at all frequencies.

Illustrative Example 2

Problem: It is desired to mount an electronics package onto a vibrating surface with assurance that the electronics will survive. This generally requires that testing be done to define the limits of the vibratory environment that could damage the electronics in the package and also that the vibration amplitude of the mounting surface be known as a function of frequency (preferably from testing). In this example an electronic box weighing 0.35 lb is to be supported on a bracket that is welded to a vibrating bulkhead (Fig. 1-15). The excitation is rotating unbalance. A rubber mounting pad is to be designed as a vibration *isolator*. The vibration limits specified by the electronics manufacturer are shown in Fig. 1-16. The bulkhead vibration measured with an accelerometer is shown in Fig. 1-17.

Analysis: This type of problem is almost always addressed with method 3 (isolation) and is modeled by system B. It is helpful to plot the system B response curve in terms of dimensionless parameters as shown in Fig. 1-18. The frequency ratio is the ratio of the exciting frequency ω to the undamped natural frequency ω_n.

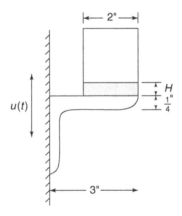

Figure 1-15   Electronic box installation.

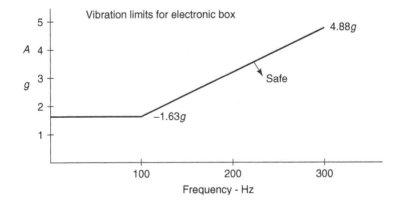

Figure 1-16   Vibration limits for the electronics.

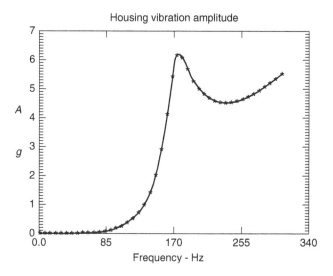

Figure 1-17 Measured bulkhead vibration, *g*'s.

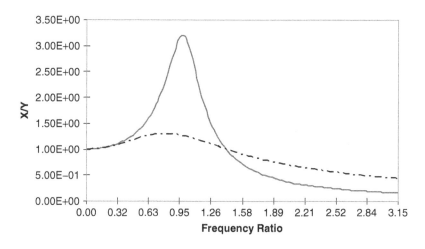

Figure 1-18 Dimensionless system B response (transmissibility).

The dimensionless *transmissibility* (ordinate) in Fig. 1-18 is the ratio of the vibratory amplitude of the electronics box to the vibratory amplitude of the mounting surface. There are two different curves for two different damping factors. The dashed curve line is for the higher damping. Inspection of the response curve shows that the best place to be on the response curve is far to the right at very high-frequency ratios, and with less damping. This can be accomplished if the mounting stiffness can be made soft enough and with relatively small damping. It is helpful in quantifying a solution to fit a curve to the measured vibration of the mounting surface, which is shown in Fig. 1-17. The problem may now be stated mathematically as follows: At every frequency in the operating range the product of the transmissibility and the amplitude in g's at the mounting surface must be less than the vibration limit shown in Fig. 1-16. The function $\tau = X/Y$ that generates the transmissibility curve in Fig. 1-18 is the dimensionless form of Eq. 1-8. The function is

$$\tau = \sqrt{\frac{1 + \eta^2}{\left(1 - r^2\right)^2 + \eta^2}} \tag{1-13}$$

where η is a damping factor and r is the frequency ratio ω/ω_n. For viscous damping the damping factor is

$$\eta = 2\xi r \tag{1-14}$$

where ξ is the ratio of the viscous damping coefficient to the critical value and r is the frequency ratio. For hysteretic damping, which elastomeric materials exhibit, the damping factor η is simply the loss factor of the elastomeric material (generally published by the manufacturer with other material properties).

Solution: The transmissibility function can be multiplied by the curve-fitted function for A_b (Fig. 1-17) and compared with the vibration limits in Fig. 1-16. This process is well suited for computer coding and it is found that the modal stiffness k should be 320 lb/in or less to keep the vibration amplitude of the electronic box below the specified limits at all frequencies. The solution is found to be insensitive to variations in the loss factor for typical rubber materials. The analysis shows that the damping should be small. If the electronic box is mounted on a bracket as shown in Fig. 1-15, then the bracket becomes part of the modal stiffness for the system. The metal bracket, however, is much stiffer than 320 lb/in, so a rubber pad will probably be needed to get the support stiffness down low enough. A single pad, or several pads, made of butadiene compound can be sized so that $AE/t = 320$ lb/in, where A is the total contact area of the pads, E is the elastic modulus, and t is the pad thickness. Then the bracket and the rubber in series will have $k < 320$ lb/in and a composite loss factor less than the rubber alone.

Illustrative Example 3

Problem: A typical beach house structure is shown in Fig. 1-19. The house is built on tall piers (without the cross braces shown with question marks).

At the beach, the piers are set into soft damp sand and this gives the structure a significant amount of damping to attenuate lateral vibration. When
such a house is built at locations with a hard rock foundation, the damping
is much less, about 1 percent of the critical value (Q factor $= 50$). There
are typically two modes of lateral vibration in which the house vibrates as
a rigid body with the piers acting as cantilever beams to produce lateral
stiffness. A rectangular house plan produces orthotropic stiffness (with the
higher stiffness in the longer direction) and consequently the two modes.
The two natural frequencies are about 3 and 5 Hz.

Consider the response of the structure to running a washing machine
with a vertical rotating axis (the tub). The unbalanced tub spins up from
zero to a speed much higher than the natural frequency of the house
on piers and then coasts back down to zero, thus producing resonance
twice in each of the two modes during each spin cycle. The amplitude at
resonance is about 1/4″, which is enough to rattle dishes.

Analysis: It is tempting to simply rely on experience and recall that
stiffening the system in the car/trailer problem reduced the response at
all frequencies. It would thus seem that cross braces should be added to
the pier support structure as shown in Fig. 1-19. However, this problem
is of type C (rotating excitation) instead of type A. Look at the dimensionless group of variables on the vertical axis of the system C response
curve (Fig. 1-12) and notice that the stiffness k does not directly appear.

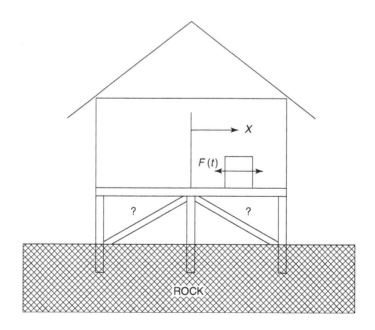

Figure 1-19 Beach house on piers.

To get the vibration amplitude from the curve, one multiplies the graphed value by mu/M, that is, the rotor mass times the unbalance divided by the total mass. In system C the excitation is *centrifugal force*, which increases as the square of rotating frequency. Increasing the stiffness of such a system raises resonance to a higher frequency where the excitation force is higher. Figure 1-20 shows the effect of adding stiffness to the supports. The assumed parameters are shown in Table 1-4. Notice that the housing mass in this case is much larger than the rotor mass. Although X/u is plotted here, the basic dimensionless group is XM/mu, so the large housing mass reduces the peak vibration to be less than the unbalance. But it is still unacceptable and adding stiffness makes it worse.

Now consider the other methods described above to reduce the vibration. Method 1, reducing the source, might be implemented by replacing the washing machine (which already has an automatic balancer). If method 1 is not practical, we move to method 2 or 3, tuning or isolation, neither of which work because the frequency of excitation is variable and starts at zero. Method 5, absorption, would require an absorber mass 1/5 the mass of the house and would introduce additional natural frequencies in the operating range. This leaves only method 4, damping, which is the parameter we lost by moving away from the beach and cementing the piers into hard rock.

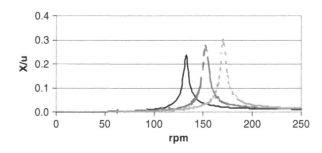

Figure 1-20   Stiffness raises the vibration amplitude in system C.

Table 1-4   Parameter values for the beach house

Rotrmass	100	lb
Housmass	12,000	lb
Kstiff	6,000	lb/in
Kstiff2	8,000	lb/in
Kstiff3	10,000	lb/in
Cdamp	15	lb-sec/in
Unbalance	1	in

Solution: Figure 1-21 illustrates the tremendous reduction in resonant vibration that can be obtained by increasing the damping. The two lower curves have 30 and 60 lb-sec/in of viscous damping, repectively.

To determine the existing damping, the fraction ξ of critical damping (*damping ratio*) can be obtained by measuring the logarithmic decay δ of free vibration and calculating

$$\xi = \frac{\delta}{\sqrt{(2\pi)^2 + \delta^2}} \tag{1-15}$$

The effective viscous modal damping coefficient is then given by

$$C = 2m\omega_n\xi \tag{1-16}$$

where m is the mass of the house (lb-sec^2/in, not lb) and ω_n is the natural frequency in radians/sec ($= 2\pi$ times 2.25 Hz or about 14 rad/sec).

Since the damping must act on motion of the house relative to the ground (i.e., absolute motion), there are practical problems associated with installing it. A single damping element at one point would probably flex the house structure and possibly fail at the attachment point. A number of steel cables from the tops of piers out to ground anchors, with viscous shock absorbers (*dashpots*), could likely be made to work but might be a visual detraction from the house. In this case the sum of all the damping coefficients of the added dashpots acting in the same direction should be 2 or 3 times the value of the existing C calculated from formula 1-16.

In actual practice, this problem was solved by replacing the washing machine with one that has a much better automatic balancing mechanism, with the rotating tub on much softer supports.

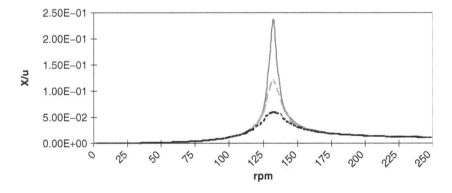

Figure 1-21 Effect of damping on the beach house vibration response.

Illustrative Example 4

Problem: A power turbine rotor in a turboprop aircraft engine has large vibration on start-up when the lube oil is hot. The rotor is mounted on squirrel cage bearing supports with stiffness much lower than the ball bearings themselves. (This is common practice in aircraft turbine engines). Figure 1-22 shows the rotor–bearing assembly mounted on pedestals in the Turbomachinery Laboratory at Texas A&M University. Figure 1-23 shows the measured vibration response with oil at three temperatures ranging from 94 to 204°F (operating temperature). The squeeze film damper becomes less effective due to the loss of viscosity at higher temperatures, which almost doubles the peak vibration response.

It is desired to minimize the amplitude of response at the critical speed, independent of temperature. The rotor speed of aircraft engines is highly variable and it is impossible to avoid passing through some of the lower critical speeds on start-up.

Analysis: It is often tempting to do the analysis with intuition, which suggests stiffening the bearing supports—perhaps even mounting the ball bearings solidly in the engine housing. Recall once again, however, that this approach moves the resonance to higher speeds where the force of the unbalanced rotor mass is higher by the square of rotor speed.

Figure 1-22 Power turbine rotor.

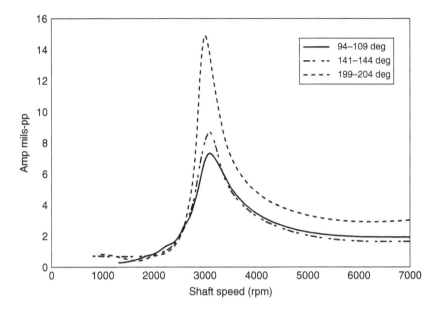

Figure 1-23 Response to unbalance at three oil temperatures.

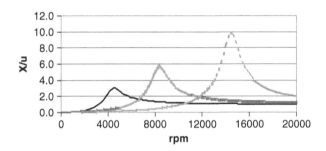

Figure 1-24 Effect of increased support stiffness on rotor response.

It would also move the critical speed up into the operating speed range. Figure 1-24 shows graphically the effect of stiffening the supports and shows that the amplitude of vibration actually *increases* with support stiffness. Other undesirable things happen as well to high-speed rotors when support stiffness is raised, as will be shown in following chapters on rotordynamics. Method 1, balancing the rotor, is often used in cases like this. It works, but tedious precision balancing is required, and the state of balance usually tends to degrade with operation time. Since we want to reduce the amplitude at the critical speed (near resonance), Fig. 1-12 shows that damping is the preferred approach. Like most aircraft turbine engines, this system already has squeeze film bearing dampers for this purpose. But the damping coefficient should be independent of temperature, and squeeze film dampers depend on the viscosity of the lube oil. Figure 1-25 shows the computed effect of damping on the unbalance response of this rotor at its first critical speed.

Solution: Experiments in the Turbomachinery Laboratory at Texas A&M University have shown that bearing supports made of woven wire mesh can provide the same effective damping as a squeeze film damper at operating temperature, since their damping is independent of temperature. Figure 1-26 shows the response of this power turbine on metal mesh supports, measured at three temperatures up to 210°F. More about this new type of bearing damper is presented in Chapter 5.

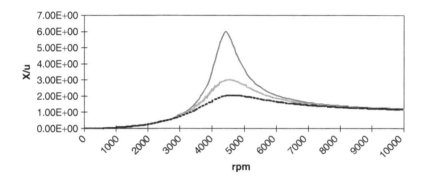

Figure 1-25   Effect of bearing support damping on the power turbine.

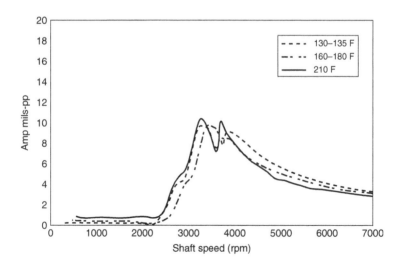

Figure 1-26   Power turbine response on metal mesh at three temperatures.

SOME OBSERVATIONS ABOUT MODELING

The last example above raises a question about the adequacy of a single degree of freedom model to represent the power turbine rotor-bearing system. In this case it was found that the CG of the rotor was directly above the outboard bearing support. This suggested that the first critical speed would be a mode with very little pitch (i.e., a cylindrical whirl mode, not conical) and therefore with little gyroscopic effect. The rotor was also modeled with XLROTOR using 17 stations with 68 degrees of freedom as shown in Fig. 1-27. The computed response to unbalance in Fig. 1-28 is identical to the response computed from the one degree of freedom (dof) model. The judgments required in constructing an appropriate one-dof model for the power turbine must be based on some knowledge about rotordynamics. This material is presented in following chapters.

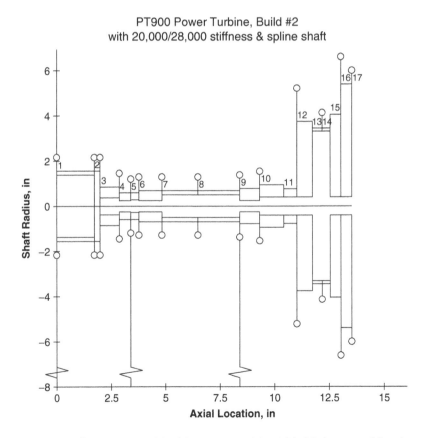

Figure 1-27   Computer model of the power turbine with 68 degrees of freedom.

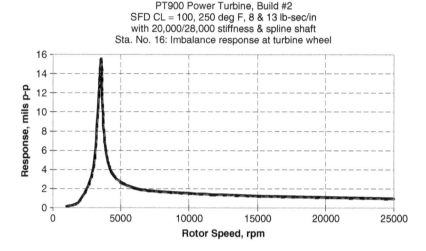

Figure 1-28 Response curve computed from the Fig. 1-27 model.

UNSTABLE VIBRATION

None of the examples presented above involve unstable vibration as represented by the homogeneous solutions (1-5) and (1-6) to the differential equation. Furthermore, all six of the presented techniques for solving vibration problems are based on the particular solution, i.e., response to excitation. Unstable vibration in nonrotating structures is rare. Those few cases usually involve fluid flow across a transverse member that sheds Von Karman vortices. The fluid pressure on the transverse member acts in phase with the vibratory velocity of the member, thus producing a negative damping coefficient. Since the real part of the eigenvalue is $\lambda = -c/2m$ in Eq. 1-6, negative c produces a growing exponential function and the vibration amplitude grows without limit. The frequency of unstable vibration is always the natural frequency ω_d of the system, independent of any external exciting frequency.

Figure 1-29 shows a simple apparatus (from Den Hartog [5]) in which negative damping can be generated. The flow of air around the beam of semicircular cross section produces a pressure distribution, which pushes the beam in the same direction as its instantaneous velocity (for motion in a vertical plane), as shown in Fig. 1-30. The differential equation for the vertical translational displacement Y of free vibration is

$$m\ddot{Y} - c\dot{Y} + kY = 0 \qquad (1\text{-}17)$$

where m is the mass of the beam, c is the (negative) damping coefficient, and k is the total effective stiffness of all the springs. The solution of Eq. 1-17 is

$$Y(T) = Ae^{st} \qquad (1\text{-}18)$$

where A is a constant, and the values of s that satisfy Eq. 1-17 are the complex conjugate eigenvalues. They are

$$s = \frac{c}{2m} \pm i\sqrt{\frac{k}{m} - \left(\frac{c}{2m}\right)^2} \qquad (1\text{-}19)$$

The positive real part of the eigenvalue indicates that the natural frequency of the beam will be unstable, with an amplitude that grows exponentially with time. In practice, the motion will become bounded at some finite amplitude large enough to render the linear equation (1-17) no longer valid. Note that every term in Eq. 1-17 contains the coordinate Y or one of its derivatives. This makes the equation *homogeneous*, which is a general property of the type of equations used to predict instabilities.

Practical examples of this phenomenon do exist. The oscillating eddies of the air are called *Von Karman vortices*. The beam could be a long pipe in a heat exchanger, a vertical smokestack, an electrical transmission

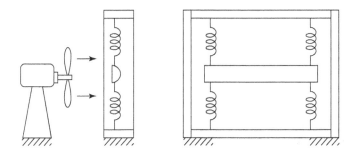

Figure 1-29 Apparatus to demonstrate unstable vibration. From Den Hartog [5].

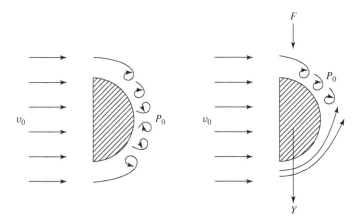

Figure 1-30 Negative damping produced by aerodynamic flow separation. From Den Martog [5].

wire, or a guy wire. However, unstable vibration is much more common in rotating machinery than in structures and can be very destructive. In rotating machinery it is called *rotordynamic instability*. It is generally caused by cross-coupled stiffness instead of negative direct damping. It will be analyzed and discussed in chapters to follow.

REFERENCES

[1] Thomson, W. T. *Theory of Vibration with Applications*, 4th ed. Englewood Cliffs, NJ: Prentice Hall, 1993.

[2] Steidel, R. F., Jr. *An Introduction to Mechanical Vibrations*. New York: Wiley, 1989.

[3] Den Hartog, J. P. *Mechanical Vibrations*, 4th ed. New York: McGraw-Hill, 1956.

[4] Vance, J. M. *Rotordynamics of Turbomachinery*. New York: Wiley, 1988.

[5] Den Hartog, J. P. *Mechanical Vibration*, 4th ed. New York: McGraw-Hill, 1956, p. 301.

EXERCISES

1-1. Use Excel or some other application to plot the force F and displacement X versus time t for a synchronous vibration at machine speed 3000 rpm with no higher harmonics. The peak force and amplitude values are 10 lb and 8 mils (0.008″), respectively. Let the displacement lag the force by 90°. Show that the acceleration amplitude is 790 times larger than the displacement and is equivalent to 2.04g.

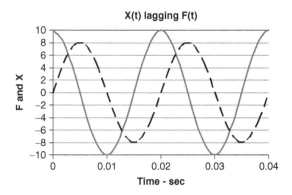

Ex. Figure 1-1

1-2. Use Excel or some other application to plot the total displacement amplitude $(x_1 + x_2)$ and the total acceleration amplitude $(a_1 + a_2)$ versus time for a machine vibration containing synchronous component $x_1(t)$ and a second harmonic $x_2(t)$ with half the amplitude. The fundamental amplitude is 10 mils ($0.010''$). The machine speed is 6000 rpm. For measurements, note that a displacement transducer is preferable to an accelerometer here if the primary interest is synchronous vibration.

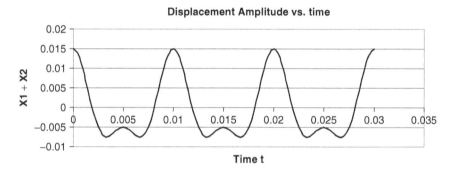

Ex. Figure 1-2a

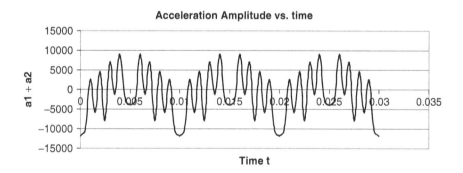

Ex. Figure 1-2b

1-3. Substitute Eq. 1-3 into Eq. 1-2 to show that the particular solution satisfies the differential equation.

1-4. Substitute Eq. 1-5 into Eq. 1-2 to show that the homogeneous solution satisfies the differential equation with $F = 0$. Show that the real part of the eigenvalue $\lambda = -c/2m$ and the imaginary part is the square root of $\omega_d^2 = k/m - (c/2m)^2$.

1-5. Referring to Eq. 1-6, the solution to the homogeneous differential equation for free vibration with no damping is $Ae^{i\omega t}$, since $\lambda = 0$.

Show that the solution can also can be written as $A_1 \cos(\omega_n t) + A_2 \sin(\omega_n t)$, where A_1 and A_2 are real numbers and $\omega_n = (k/m)^{1/2}$, provided that A is an arbitrary complex number.

1-6. Show that eq. 1-6 with nonzero damping can be expressed as $x_h(t) = e^{-\frac{c}{2m}t}[A_1 \cos(\omega_d t) + A_2 \sin(\omega_d t)]$. Assuming initial conditions to give $A_2 = 0$, take the ratio of successive amplitudes X_n/X_{n+1} to show that the logarithmic decrement $\delta = \ln(X_n/X_{n+1}) = 2\pi\zeta/(1+\zeta^2)^{1/2}$, where $\zeta = c/2m\omega_n$.
   Hint: Note that the period of the damped vibration is $2\pi/\omega_d$.

1-7. The tuning method on page 14 states that intelligent construction of the analytical model is important, since the modal stiffness may be made up of several real stiffnesses in parallel or in series. In parallel combinations the very low stiffnesses have little effect in determining the modal stiffness, while in series combinations the very high stiffnesses have little effect.

   a. Show that the effective stiffness of k_1 and k_2 in parallel is practically k_1 if $k_1 = 100k_2$.
   b. Show that the effective stiffness of k_1 and k_2 in series is practically k_2 if $k_1 = 100k_2$.

1-8. Derive the dimensionless form of Eq. 1-3 for the purpose of plotting Fig. 1-14.

1-9. Referring to Illustrative Example 1 and Fig. 1-13:

   a. Derive the differential equation in X_1 for the car, with P_e as the excitation force.
   b. Derive the differential equation in X_2 for the trailer.
   c. Divide each equation by the mass and subtract the X_2 equation from the X_1 equation.
   d. Do the math to obtain Eq. 1-10.
   e. Use Excel or some other application to plot Fig. 1-14 in dimensional variables (X versus ω). Assume a speed range 70 mph down to zero with a tire diameter $D = 30''$ and brake excitation force $P = 100$ lb. Assume resonance occurs at 50 mph. Assume the car weighs 3000 lb and the trailer weighs 1000 lb. Include small damping $C = 2$ lb-sec/in using Eq. 1-4 so that the amplitude curve is always positive. Vary the stiffness K and note how it changes the response curve.

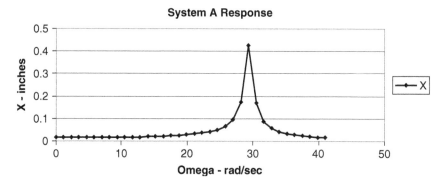

System A Response

Ex. Figure 1-9

1-10. See Illustrative Example 2, where it is suggested that "it is helpful in quantifying a solution to fit a curve to the measured vibration of the mounting surface, which is shown in Fig. 1-17." Develop a mathematical function that will approximate the data in Fig. 1-17. *Solution:* This can be done with existing curve-fit software, but a more instructive approach is to realize that the excitation is likely due to some rotating unbalance since the data begin at the origin, which is unique to system C. The reason that the data do not look like Fig. 1-12 for system C is that the data are acceleration, not displacement. Multiplication by ω^2 and division by acceleration of gravity g converts Eq. 1-9 for system C to acceleration in g's. Assuming a small housing mass and dividing numerator and denominator by $M = m$ yields

$$\frac{\omega^2 X}{g} = \frac{\omega^4 u/g}{\sqrt{(\omega_n^2 - \omega^2)^2 + (2\xi\omega_n\omega)^2}} \qquad (1\text{-}20)$$

The broad-banded peak response suggests a damping ratio ζ about equal to 0.085. The peak acceleration in the data is 6.2g or 2396 in/sec². This allows a calculation of the unbalance u = 0.00035″ to match the peak acceleration. The critical speed is seen to be about 170 Hz, so $\omega_n = 1068$ rad/sec. The angular velocity ω must be converted to hertz $= \omega/(2\pi)$ for the graph. With these values, Eq. 1-20 produces the graph shown here.

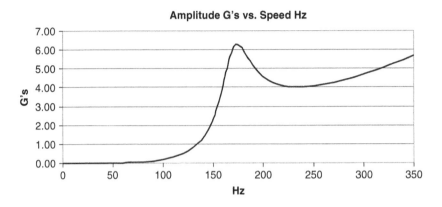

Ex. Figure 1-10

2

TORSIONAL VIBRATION

Torsional vibration is oscillatory twisting of the shafts in a rotor assembly. When superposed on the running speed, it is evidenced as an oscillatory variation of speed. The frequency can be externally forced, or can be an eigenvalue (natural frequency of the torsional system). Resonance will occur if a forcing frequency coincides with a natural frequency (see Chapter 1).

There is a rich well-documented history of torsional vibration problems and solutions associated with diesel engines driving electric generators or marine propellers. Calculation of the excitation frequencies and equivalent inertias for reciprocating machines is a major part of the required analysis in those cases (see [1, 2]). With a few exceptions the applications in this chapter will be turbomachinery drive trains, and details of the reciprocating models will not be given.

Individual turbomachinery rotors are generally stiff enough in torsion to put their natural frequencies of torsional vibration above the range of most torsional excitations. An exception is the torsional excitation in some steam turbines with long rotors, which can fatigue blades. When turbomachines are connected together by shaft couplings, however, each of the individual rotors can act as a single massive inertia. The torsional stiffness values of the couplings and connecting shafts are then often low enough to bring natural frequencies in torsion down into the range of excitation frequencies. This may also occur when a turbomachine is connected to some other type of rotating machine (e.g., an electric motor or a reciprocating engine). Such drive trains often include gearboxes as well. It is the non-turbomachinery components in such drive trains that

Figure 2-1 Turbomachines connected by shaft couplings. Courtesy of GE Energy.*

Figure 2-2 Shaft coupling.

usually provide the torsional excitation. The excitation frequencies are generally quite low, often below 20 Hz.

An example is illustrated in Fig. 2-1, which shows a large electric motor driving turbomachines through a gearbox. The components in industrial trains are often quite large. The motor can be several thousand horsepower.

Figure 2-2 shows a typical shaft coupling that has two flex halves with bolted flanges connected by a spacer shaft or tube. The two flex halves are designed to allow angular compliance and misalignment. The long spacer shaft allows radial (lateral) misalignment, and also produces a softer (lower) torsional stiffness of the coupling.

TORSIONAL VIBRATION INDICATORS

Rotating machinery trains are not commonly instrumented to measure torsional vibration unless a costly torsional-related problem has become obvious. Furthermore, torsional vibration seldom produces noise or vibration of the stationary frame. Quite often, the first indicator of a torsional vibration problem is a broken shaft coupling or a broken shaft. In the latter case, the failure usually begins in the vicinity of a stress concentration and propagates at 45 degrees to the shaft axis due to metal fatigue.

*System 1 is a registered trademark of Bently Nevada, Inc.

In machines with gears, torsional vibration can produce a high noise level if the gears become periodically unloaded due to the fluctuations of torque. In these cases, rapid gear wear and deterioration of tooth surfaces (pitting of the pitch line) may occur. Periodic torque reversals, coupled with backlash, can break the gear teeth.

Torsional and Lateral Vibration—The Key Differences

1. Lateral vibration is easily detected through standard instrumentation or through vibrations transmitted to housings and foundations. Also, large amplitudes of lateral vibration are often noticed due to rubbing of rotating seals and process wheels. On the other hand, instrumentation for torsional vibration is not usually installed, and large amplitudes can occur silently and without much effect on housings and foundations.
2. Natural frequencies of lateral vibration are influenced by rotating speed, whereas natural frequencies of torsional vibration are independent of rotating speed and can be measured with the machine at rest if excitation can be provided.
3. Lateral vibration in rotating machines can become unstable; this is very rare for torsional vibration in machines without speed control feedback.
4. The most common excitation of lateral vibration is synchronous ($1\times$) from rotor imbalance. Rotor imbalance has no effect on torsional vibration, except indirectly in machines with gears where the lateral vibration produces dynamic torque.
5. Analysis of lateral vibration can usually be performed on each body in the train separately, whereas analysis of torsional vibration must include all rotors in the train. In many cases of torsional analysis, each rotor in the train can be treated as rigid.

OBJECTIVES OF TORSIONAL VIBRATION ANALYSIS

The typical engineering objectives of torsional vibrations analysis are as follows:

1. Predict the natural frequencies.
2. Evaluate the effect on the natural frequencies and vibration amplitudes due to changing one or more design parameters or components in the train (i.e., *sensitivity analysis*).

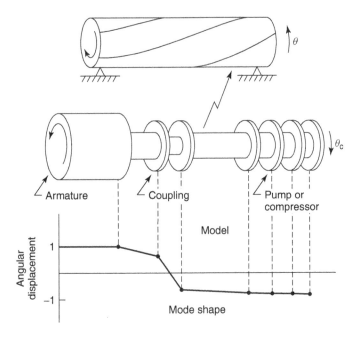

Figure 2-3   Torsional mode shape for a motor driving a compressor. From [1].

3. Compute the vibration amplitudes and peak torques under steady-state torsional excitation.
4. Compute the dynamic torques and gear tooth loads under transient conditions (e.g., during a machine start-up).
5. Evaluate the torsional stability of drive trains with automatic speed control.

Objective 1 can often be accomplished with useful accuracy using analysis with simplified models. The remaining objectives usually require a computer simulation (see Chapter 4).

Figure 2-3, from Eshleman [1], shows the mode shape of a typical train consisting of an electric motor driving a pump or compressor. The motor is oscillating against (out of phase with) the compressor. Notice that there is very little twist in the motor rotor and the compressor rotor, so they can be modeled as rigid.

SIMPLIFIED MODELS

I_{P1} and I_{P2} in Fig. 2-4 are the polar mass moments of inertia (in–lb – sec^2) of two machine rotors connected by a coupling and shafts that have

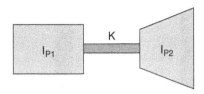

Figure 2-4   A two-inertia simplified model.

a combined torsional stiffness K (in-lb/rad). Ignoring damping, which is often negligible in torsional vibration, the fundamental natural frequency (in Hz) is

$$f_n = \frac{1}{2\pi}\sqrt{K\frac{I_{P1}+I_{P2}}{I_{P1}\cdot I_{P2}}} \tag{2-1}$$

Trains with gears can be reduced to an equivalent system on one shaft, called the *reference shaft*, at the speed of the reference shaft. Figure 2-5 shows two rotors connected through a gearbox. Choose the driver I_{P1} as the reference shaft at speed N_1.

Torsional stiffness K_2 and inertia I_{P2} are at a different speed N_2, so they must be multiplied by the gear ratio G_R squared. Assume here that I_{P1} and I_{P2} are much larger than the gear inertia, so we still have a two-inertia model for practical purposes. Figure 2-6 shows the equivalent model referred to shaft 1. K_1 and K_2 are in series, so the effective stiffness is $K_{\mathrm{eff}} = K_1 K_{2\mathrm{eff}}/(K_1 + K_{2\mathrm{eff}})$, where $K_{2\mathrm{eff}} = G_R^2 K_2$. The effective inertia of I_{P2} on the reference shaft is $I_{2\mathrm{eff}} = G_R^2 I_{P2}$. The fundamental

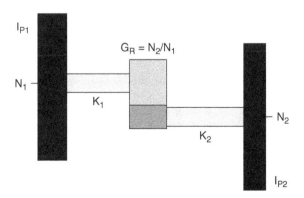

Figure 2-5   A two-inertia model with a gearbox.

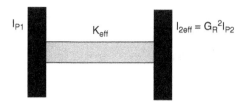

Figure 2-6   An equivalent model for Fig. 2-5.

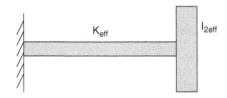

Figure 2-7   Single degree of freedom torsional model.

natural frequency of torsional vibration can be calculated from these numbers as

$$f_n = \frac{1}{2\pi}\sqrt{K_{\text{eff}}\frac{I_{P1} + I_{2\text{eff}}}{I_{P1} \cdot I_{2\text{eff}}}} \tag{2-2}$$

Further simplification is possible if one of the two inertias is much larger than the other, as the oscillations of the larger inertia become negligible. The model then becomes practically one degree of freedom as shown in Fig. 2-7 where I_{P1} is much larger than $I_{2\text{eff}}$. Equation 2-3 gives the natural frequency:

$$f_n = \frac{1}{2\pi}\sqrt{\frac{K_{\text{eff}}}{I_{2\text{eff}}}} \tag{2-3}$$

Table 2-1 shows the errors associated with the assumption that $I_{P1} \gg I_{2\text{eff}}$, comparing Eq. 2-3 with Eq. 2-1. One can see that I_{P1} must be more than ten times $I_{2\text{eff}}$ to keep the error <5%.

In some cases the polar inertia of the coupling hubs and the spacer torsional stiffness may have a significant effect on the torsional model, and subsequently the value for the first natural frequency. This is often the case when converting from a rigid or relatively rigid coupling, to a dry flexible coupling. These two limiting factors (rigid vs flexible coupling) will be illustrated in the following example.

Table 2-1

$K_{eff} = 20E6,$ $I_{2eff} = 1000$	Equation 2–1 gives (Hz)	Equation 2–3 gives (Hz)	% Error
$I_{P1} = 2I_{2eff}$	27.57	22.51	18.35
$I_{P1} = 10I_{2eff}$	23.61	22.51	4.65
$I_{P1} = 100I_{2eff}$	22.62	22.51	0.5

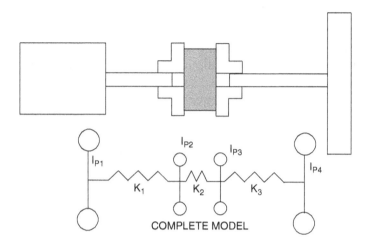

Figure 2-8   Including the coupling components.

The schematic in Figure 2-8 shows a typical driver (motor or turbine) driving an overhung fan or blower through a power transmission coupling. This complete physical model is represented by the inertias and torsional springs shown below the schematic. The complete model consists of four degrees of freedom; four inertias (inertia of the driving machine, inertia of the driven machine, and the two coupling hubs inertias), and three torsional springs (driving shaft, driven shaft, and the coupling spacer shaft or tube). This model suggests the use of a computer program or a lengthy equation to solve for the natural frequencies, but evaluation of the relative values for the inertias and the torsional springs in this model, can allow simplification where natural frequencies can be calculated by hand without loss in accuracy.

In the case of a flexible coupling model where the coupling spacer is much more torsionally flexible than the shafts in the driving and the driven machine, the softer spring dominates. This is a very realistic assumption in many real machines that get retrofitted with flexible couplings. In this case the driving and driven shafts are assumed rigid relative to the coupling spacer. Since in this case there is no relative movement between the inertia of each of the hubs and the inertia of the driving and driven machine, the inertia of the coupling hubs are lumped with the machines inertias as shown in the schematic of Figure 2-9. This simplification allows the use of equation 2-1 to solve for the natural frequency.

In contrast, the simplified model with the rigid coupling is shown in Figure 2-10. Since the coupling is rigid there is no relative motion between the two coupling hubs, so their inertias are lumped together. This simplified model has three inertias and the solution to the natural frequency can be solved using equation 2-5 on page 44.

The two simplified models discussed for Fig. 2-8 are tabulated with numerical values in the Exercise section at the end of the chapter. The solution to the first natural frequency using the simplified models are listed in Table 2-2 for reference. It is apparent that the simplified models provide a very good approximation in comparison with the full model for both the flexible coupling and the rigid coupling cases. It is also relevant to note that the difference in natural frequency between the rigid and the flexible coupling case is significant. The flexibility of the coupling allows the designer to tune the natural frequency of the train, a tool often used

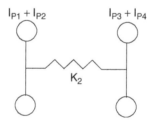

Figure 2-9   A two-inertia model when K_2 is soft.

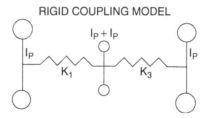

Figure 2-10   Three inertia system when K_2 is very stiff.

Table 2-2

Coupling Type	Simplified Model	Complete Mode 1st Nat. Freq.	Complete Mode 2nd Nat. Freq.
Flexible Coupling	9.81 Hz	9.05 Hz	359 Hz
Rigid Coupling	25.6 Hz	23.6 Hz	318 Hz

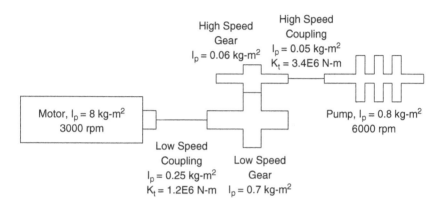

Figure 2-11 Model of an induction motor, 2:1 gearbox, and multistage pump.

to move the natural frequency away from interference with an excitation source.

Figure 2-11 shows an example torsional analysis model of an oil refinery pump. The system consists of a two pole induction motor running at 3000 rpm, a 2-to-1 speed increasing gearbox, and a multistage centrifugal pump. The low-speed and high-speed couplings are flexible disk pack couplings with the properties given in the figure. The fundamental natural frequency of this system can be approximated using an equivalent two degrees of freedom (dof) system, shown in Fig. 2-12, and Eq. 2-1. The two coupling stiffnesses have been combined in series, and the two inertias are the total inertias of the low-speed and high-speed shaft lines. Since this equivalent model has 2-dof, it has two natural frequencies: one frequency is zero (i.e., rigid body mode where the two inertias move together), and the other is 103.9 Hz where the two inertias move relative to each other:

$$ f_n = \frac{1}{2\pi} \sqrt{\frac{(1.1027 \times 10^6)(8.95 + 3.64)}{(8.95)*(3.64)}} = 103.9 \, \text{Hz} \qquad (2\text{-}4) $$

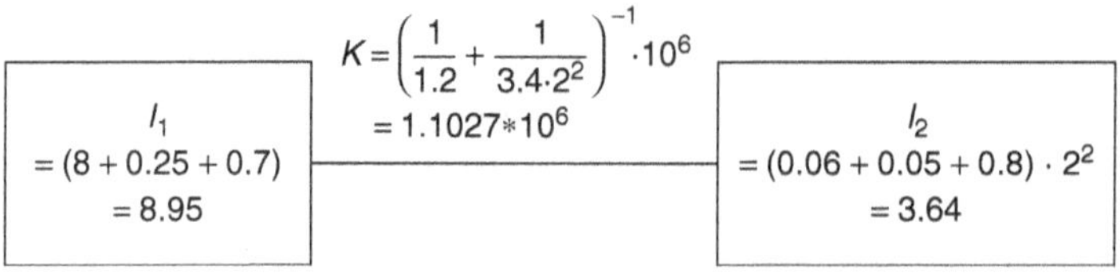

$$K = \left(\frac{1}{1.2} + \frac{1}{3.4 \cdot 2^2}\right)^{-1} \cdot 10^6$$

$$= 1.1027 * 10^6$$

I_1		I_2
$= (8 + 0.25 + 0.7)$		$= (0.06 + 0.05 + 0.8) \cdot 2^2$
$= 8.95$		$= 3.64$

Figure 2-12 Equivalent 2-dof system for the model shown in Fig. 2-10.

I_1	$K_1 = 1.2 \cdot 10^6$	I_2	$K_2 = (3.4 \cdot 10^6) \cdot 2^2$	I_3
$= (8 + 0.125)$		$= 0.125 + 0.7 + (0.06 + 0.025) \cdot 2^2$		$= (0.025 + 0.8) \cdot 2^2$
$= 8.125$		$= 1.165$		$= 3.3$

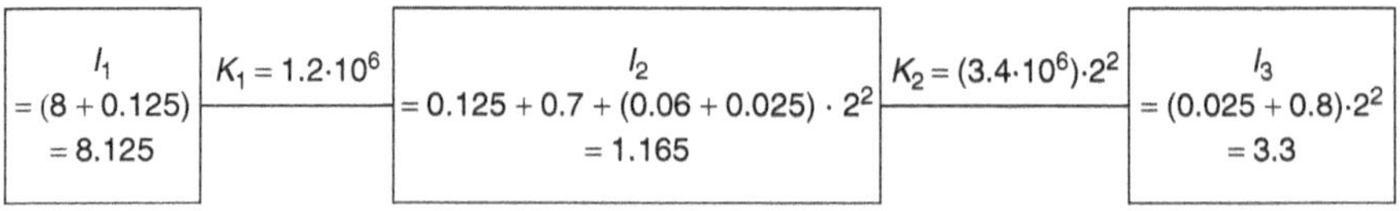

Figure 2-13 Equivalent 3-dof system for the model shown in Fig. 2-10.

As an additional level of refinement, an equivalent three-inertia system can be created as illustrated in Fig. 2-13. As the figure shows, the three inertias are the motor, the gearbox, and the pump. The inertias of the two couplings have been split in half and added to their respective shafts.

The three degrees of freedom of this system imply that it has three natural frequencies. One of these is zero, and the other two are calculated using the following expression:

$$f_n = \frac{1}{2^{1.5}\pi} \sqrt{\frac{K_1 + K_2}{I_2} + \frac{K_1}{I_1} + \frac{K_2}{I_3} \pm \sqrt{\frac{\left(\frac{K_1 + K_2}{I_2} + \frac{K_1}{I_1} + \frac{K_2}{I_3}\right)^2 - 4K_1 K_2}{\left(\frac{1}{I_1 I_2} + \frac{1}{I_1 I_3} + \frac{1}{I_2 I_3}\right)}}}$$

$$f_n = 100.26\,\text{Hz} \quad \text{and} \quad 647.98\,\text{Hz}$$

(2-5)

The first natural frequency is seen to be close, within 1 percent of 2 times motor speed and 1 times pump speed.[1] So there are two possible sources of torsional excitation that could excite large amplitude response of the first torsional mode. The typical course of action would be to change one or both of the couplings to try to get at least 10 percent separation between the torsional natural frequency and the excitation frequency.

---

[1]There is no 1X excitation here, but API requires that it be avoided.

But before doing that, the model should be checked for inadequacies. Pump shafts can be long and flexible, and this may need to be accounted for in the analysis. Including this will lower the first natural frequency. The motor shaft also may have significant compliance. An exercise at the end of this chapter addresses these issues.

COMPUTER MODELS

For models with more than two degrees of freedom (more than two inertias), a computer code such as XLRotor, as described in Chapter 4, will be desirable. Figure 2-14 shows the lumped parameter model with N inertias connected by $N - 1$ massless torsional springs. The simplified cases above can often be used to check that results from the computer code are in the ballpark.

For the model with N degrees of freedom (N inertias), the lumped parameter (or finite element) approach leads to N second-order differential equations to be solved. The homogeneous equations with no excitation can be solved for the natural frequencies. The nonhomogeneous equations can be solved for the response to torsional excitation.

The five-inertia system in Fig. 2-15 will be used to demonstrate the derivation of system equations both with and without flexibility of the gear teeth. The derivation is intended to explain how a computer program such as XLRotor works. The software user does not need to perform this himself. The dof at each inertia is defined using a strict right-hand notation. This means that if the gear ratio is entered as a positive number,

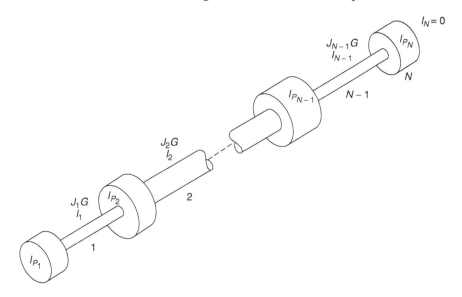

Figure 2-14 Lumped parameter torsional model.

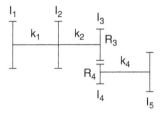

Figure 2-15 Five-inertia geared torsional system.

then the two shafts rotate in the same direction. To have them rotate in opposite directions, the gear ratio must be negative. First, we look at the case without gear tooth flexibility, where dof θ_4 will be eliminated in favor of θ_3.

Kinetic Energy Expression

$$\text{KE} = \tfrac{1}{2}\left(I_1\dot\theta_1^2 + I_2\dot\theta_2^2 + I_3\dot\theta_3^2 + I_4\dot\theta_4^2 + I_5\dot\theta_5^2\right) \tag{2-6}$$

Let

$$\theta_4 = n\theta_3 \qquad \theta_5 = n\hat\theta_5 \tag{2-7}$$

where

$$n = \frac{R_3}{R_4} \tag{2-8}$$

is the gear ratio, which can be positive or negative. Substituting 2-7 into 2-6 will effectively convert the high-speed shaft to an equivalent shaft spinning at the same speed as stations 1, 2, 3:

$$\text{KE} = \tfrac{1}{2}\left[I_1\dot\theta_1^2 + I_2\dot\theta_2^2 + (I_3 + n^2 I_4)\dot\theta_3^2 + I_5 n^2 \dot{\hat\theta}_5^2\right] \tag{2-9}$$

Potential Energy

$$\text{PE} = \tfrac{1}{2}\left[k_1(\theta_2 - \theta_1)^2 + k_2(\theta_3 - \theta_2)^2 + k_4(\theta_5 - \theta_4)^2\right] \tag{2-10}$$

$$\text{PE} = \tfrac{1}{2}\left[k_1(\theta_2 - \theta_1)^2 + k_2(\theta_3 - \theta_2)^2 + n^2 k_4(\hat\theta_5 - \theta_3)^2\right] \tag{2-11}$$

Externally applied torques, if any are handled as

$$\delta W = T_1\delta\theta_1 + T_2\delta\theta_2 + T_3\delta\theta_3 + T_4(n\delta\theta_3) + T_5(n\delta\hat\theta_5)$$
$$\delta W = T_1\delta\theta_1 + T_2\delta\theta_2 + (T_3 + nT_4)\delta\theta_3 + nT_5\delta\hat\theta_5 \tag{2-12}$$

By a applying the method of Lagrangian mechanics in taking partial derivatives of KE, PE, and δW, the following system of equations is obtained:

$$
\begin{bmatrix} I_1 & & & \\ & I_2 & & \\ & & I_3 + n^2 I_4 & \\ & & & n^2 I_5 \end{bmatrix} \begin{Bmatrix} \ddot{\theta}_1 \\ \ddot{\theta}_2 \\ \ddot{\theta}_3 \\ \ddot{\hat{\theta}}_5 \end{Bmatrix}
$$

$$
+ \begin{bmatrix} k_1 & -k_1 & & \\ -k_1 & k_1 + k_2 & -k_2 & \\ & -k_2 & k_2 + n^2 k_4 & -n^2 k_4 \\ & & -n^2 k_4 & n^2 k_4 \end{bmatrix} \begin{Bmatrix} \theta_1 \\ \theta_2 \\ \theta_3 \\ \hat{\theta}_5 \end{Bmatrix} = \begin{Bmatrix} T_1 \\ T_2 \\ T_3 + nT_4 \\ nT_5 \end{Bmatrix} \qquad (2\text{-}13)
$$

In Xlrotor, when entering inputs for a model like this one, the input table could look similar to

Station #	Length (m)	OD (m)	ID (m)	Weight Density (kg/m³)	Shear Modulus (N/m²)	Added I_P (kg-m²)	Added K_T (N-m/rad)	Gear Ratio
1	1	0.01				0.24811	1.57E + 05	1
2	1	0.01				0.11965	1.57E + 05	1
3						0.94203		1
4	1	0.01				0.01938	1.18E + 05	−3
5						0.08974		−3

and on the Cplg's worksheet

Station #	Station #	Output loads	Coupling stiffness	Coupling damping
3	4		1.0E10	

In this example, the stiffness between stations 3 and 4 has been made high enough to behave as a rigid connection. The gear ratio has been entered as a negative value, so the two shafts rotate in opposite directions (see Fig. 2-16). From these inputs, Xlrotor will actually assemble a 5-dof system matrix, but the high stiffness of k_3 will cause the 5-dof system to behave like a 4-dof system.

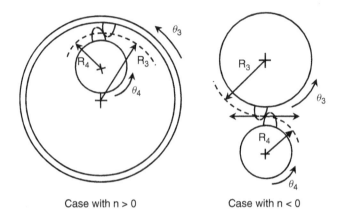

Case with n > 0   Case with n < 0

Figure 2-16 Modeling of gear ratios.

An alternative way to input the coupling data is

Station #	Station #	Output loads	Coupling stiffness	Coupling damping
3	4		rigid	

The only difference is the high stiffness for k_3 has been replaced with a rigid connection. Now Xlrotor will assemble a 4-dof system matrix. Numerical results with this model will match those obtained with a sufficiently stiff input for k_3.

Now consider the inclusion of gear tooth flexibility. We will refer to the stiffness of the teeth contact as k_g, which is a linear translational stiffness in units of force/length. The dof θ_4 will be retained in the model. Dof θ_4 and θ_5 will be replaced with equivalent dof that spin at the same speed as dof 1, 2, 3. The tangential contact force at the two mating gears is $P = k_g(R_4\theta_4 - R_3\theta_3)$ for $n > 0$. Note the strict right-handed dof definition.

Now let

$$\theta_4 = n\hat{\theta}_4 \qquad \theta_5 = n\hat{\theta}_5 \qquad (2\text{-}14)$$

where again

$$n = \frac{R_3}{R_4} \qquad (2\text{-}15)$$

is the gear ratio, which can be positive or negative. These will convert the high-speed shaft to an equivalent shaft spinning at the same speed as stations 1, 2, 3:

$$\text{KE} = \tfrac{1}{2}\left[I_1\dot{\theta}_1^2 + I_2\dot{\theta}_2^2 + I_3\dot{\theta}_3^2 + I_4 n^2\dot{\hat{\theta}}_4^2 + I_5 n^2\dot{\hat{\theta}}_5^2 \right] \tag{2-16}$$

$$\text{PE} = \tfrac{1}{2}\left[k_1\,(\theta_2 - \theta_1)^2 + k_2\,(\theta_3 - \theta_2)^2 + k_g\left(R_4 n\hat{\theta}_4 - R_3\theta_3\right)^2 \right.$$
$$\left. + n^2 k_4\left(\hat{\theta}_5 - \hat{\theta}_4\right)^2 \right] \tag{2-17}$$

$$\delta W = T_1\delta\theta_1 + T_2\delta\theta_2 + T_3\delta\theta_3 + nT_4\delta\hat{\theta}_4 + nT_5\delta\hat{\theta}_5 \tag{2-18}$$

The system equations can now be written:

$$\begin{bmatrix} I_1 & & & & \\ & I_2 & & & \\ & & I_3 & & \\ & & & n^2 I_4 & \\ & & & & n^2 I_5 \end{bmatrix} \begin{Bmatrix} \ddot{\theta}_1 \\ \ddot{\theta}_2 \\ \ddot{\theta}_3 \\ \ddot{\hat{\theta}}_4 \\ \ddot{\hat{\theta}}_5 \end{Bmatrix}$$
$$+ \begin{bmatrix} k_1 & -k_1 & & & \\ -k_1 & (k_1 + k_2) & -k_2 & & \\ & -k_2 & (k_2 + R_3^2 k_g) & -R_3^2 k_g & \\ & & -R_3^2 k_g & (R_3^2 k_g + n^2 k_4) & -n^2 k_4 \\ & & & -n^2 k_4 & (n^2 k_4) \end{bmatrix} \begin{Bmatrix} \theta_1 \\ \theta_2 \\ \theta_3 \\ \hat{\theta}_4 \\ \hat{\theta}_5 \end{Bmatrix} = \begin{Bmatrix} T_1 \\ T_2 \\ T_3 \\ nT_4 \\ nT_5 \end{Bmatrix}$$
$$\tag{2-19}$$

When entering the inputs for this model, as for every model, you define the stiffness as torque per radian of deflection of the flexible element, and it does not matter what the gear ratio is. Most of the time each stiffness will connect two stations that have the same gear ratio. When the gear ratios are not the same (as for our compliant gear teeth example), the input stiffness value should be torque/radian of deflection of the *primary* dof for which it is being input. As an example, in the following table the value 2E6 entered for k_3 is the torque of the gear teeth exerted on I_3 for a one radian deflection of θ_3. In our gear tooth example this value would have been calculated from $(R_3^2\,k_g)$.

Station #	Length (m)	OD (m)	ID (m)	Weight Density (kg/m³)	Shear Modulus (N/m²)	Added I_P (kg-m²)	Added K_T (N-m/rad)	Gear Ratio
1	1	0.01				0.24811	1.57E + 05	1
2	1	0.01				0.11965	1.57E + 05	1
3						0.94203		1
4	1	0.01				0.01938	1.18E + 05	−3
5						0.08974		−3

And on the Cplg's worksheet

Station #	Station #	Output loads	Coupling stiffness	Coupling damping
3	4		2.00E+06	

But, if you instead input the gear stiffness on the Cplg's worksheet as a coupling connecting station *4 to 3*, θ_4 is now the primary dof, so you would enter the torque exerted on I_4 due to a one-radian deflection at θ_4, which works out to be $R_4^2 k_g = 2E6/3^2 = 2.22E5$. Each of these alternatives for specifying the stiffness of the gear teeth will produce exactly the same set of eigenvalues and mode shapes.

What happens if you change the gear ratio from −3 to +3? Looking at the matrix equations above shows that they will not change if there are no external torques. So the natural frequencies will not change, and the mode shapes also will not change. However, when Xlrotor outputs mode shapes, it undoes the dof conversions used when assembling the matrix equations (i.e., $\theta_4 = n\hat{\theta}_4$, etc.), so the mode shapes displayed by the program will reflect the different direction implied by the sign of the gear ratio.

In either a harmonic or transient response calculation, the effect of external torques *does* depend on the sign of the gear ratio. In the above example, consider a positive driving torque of +50,000 N-m at I_1 being countered by a load torque at I_5. If the gear ratio is +3, the load torque needs to be negative so that the two torques counteract one another, and should be entered as −50,000/3 N-m. If the gear ratio is −3, then the load torque needs to be entered as +50,000/3 N-m.

TORSIONAL VIBRATION MEASUREMENT

Torsional vibration currently is not a common measurement made for turbomachinery diagnostics or for a general maintenance program. One reason is that torsional vibration is a difficult measurement to make. However, torsional vibration is probably the most common cause of gear tooth breakage, and it has also been identified as a cause of blade fatigue in steam turbines. In turbomachinery trains driven by electric motors and in turbogenerator sets, torsional vibration is often produced by variations in electromagnetic torque.

In contrast to the large number of complete instrumentation systems that can be purchased off the shelf to measure lateral (translational) vibrations in rotating machinery, only a few systems are available commercially for torsional vibration measurement. Most of them either require significant modifications to the rotating machine (which may be unacceptable for reasons of safety, reliability, or expense) or have significant limitations of application or performance. Consequently, engineers requiring measurements of torsional vibration often have no recourse but to design their own systems, using commercially available subcomponents wherever possible and tailoring the performance specifications to the particular application. Examples can be found in [6] and [7].

The oldest and probably the most widely known method of measuring torsional vibration utilizes strain gages bonded to the surface of the rotating shaft, oriented along the directions of principal strain. The method is accurate when properly installed and calibrated, but has some severe practical disadvantages. First, the signal must be transferred from the rotating shaft to a nonrotating frame, either by the use of slip rings or by radio telemetry. In the field, dirty brushes or radio interference can produce low signal/noise ratios. Improper applications of the strain gages can produce signals related to shaft bending rather than torsion. The gage bonding agent may be deteriorated by environmental conditions, such as temperature or process chemicals. If all of these problems are to be avoided, the requisite installation time for an accurate and reliable system may become prohibitive, especially in a troubleshooting situation.

Verhoef [8] describes several other types of instruments for measuring torsional vibration on existing shafts. One type, which was commercially available in past years, was known as a *torsiograph*. This instrument produced a voltage proportional to the oscillatory angular velocity of the shaft. It had the disadvantage of requiring an exposed end of the shaft for its attachment, and it did not respond accurately at frequencies below 10 Hz.

Prefabricated and precalibrated shaft sections are commercially available with preinstalled strain gages and slip ring assemblies, or with radio telemetry antennae installed. These devices are meant to replace some existing section of shaft, or shaft coupling, in the machine. However, retrofitting such devices into a machine often raises questions of a possible loss of machine reliability or safety. In many cases, space is not available for the instrumented shaft section with its couplings, slip rings, or antennae.

It is instructive to note that two of the measurement systems just described (strain gage and angular velocity measurement) will produce radically different measurements when applied at the same shaft location. The strain gages measure the twist in the shaft and consequently will produce a maximum signal in regions where the variation of angular velocity is minimum. Therefore, the selection of transducer types and locations should be guided by knowledge of the torsional mode shapes, or else a sufficient number of measurement locations should be used to determine the mode shape experimentally.

In machinery drive trains, sometimes the first evidence of torsional vibration can be damage or breakage of gear teeth. In fact, gear damage is probably the most common first incentive for measuring torsional vibration in rotating machinery. If gears are present, they offer an ideal source of a *carrier signal* for most of the measurement systems to be described here. Figure 2-17 shows the type of carrier signal that can be produced, for example, by a magnetic transducer excited by the passing gear teeth. Frequency or amplitude modulation of the signal can yield the torsional vibration characteristics, provided the frequencies of interest are much lower than the carrier wave frequency, which is typically the case.

A distinction must be made between steady-state and transient measurements. Transient measurements (when shaft speed and/or torsional vibration are changing with time) are much more difficult to make and to analyze. An application where transient measurements are required is the start-up of drive trains using synchronous electric motors as the driver [9–11]. These motors can produce torque pulsations of large amplitude and variable frequency during start-up that excite torsional vibration superposed onto the acceleration schedule.

The ideal torsional vibration measurement system would yield accurate measurements for both steady-state and transient conditions, would be applicable to any drive train with or without gears, and would be quickly and easily installed in the field even if only a short length of exposed rotating shaft or coupling spacer was accessible. The prototype of such a system, called TIMS for *time interval measurement system*, is described below after several more conventional systems have been described as tested by

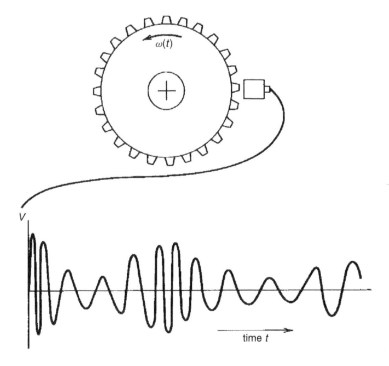

Figure 2-17   Carrier signal for torsional vibration measurements.

French. Then a later derivative of the TIMS system is described along with its application to measuring torsional vibration in a tumble mill test rig.

The first author supervised a study of torsional vibration measurement techniques with R. S. French and P. M. Barrios, and later with R. Kar and K. Toram, at Texas A&M University. The experiments with French were first published in [4] and later in [5].

FRENCH'S COMPARISON EXPERIMENTS

In all, five different types of instrumentation systems (including strain gages, used as a benchmark) were assembled by French [4,12] and used to measure torsional vibration in two experimental test rigs designed and constructed in the Turbomachinery Laboratories at Texas A&M University. The results were compared in terms of accuracy and ease of use. The most useful methods and results from that study are described here.

Strain Gages

The electrical resistance strain gage operates on the basic principle that certain metals exhibit a change in electrical resistance proportional to a

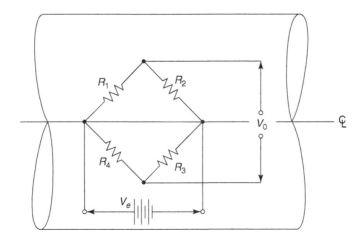

Figure 2-18 Wheatstone bridge on a shaft.

change in mechanical strain. Within the thermal and mechanical limits of
the particular strain gage, this relationship is linear according to Hooke's
law. For precision measurement of the small resistance changes associated
with the elastic strain of steel, one or more gages are bonded to the strained
element and incorporated into a Wheatstone bridge. With a typical input
of about 10 V, the output voltage is linearly proportional to the strains. For
torsional vibration measurements, a full Wheatstone bridge is bonded to
the surface of the shaft for measurement of the cyclic strain. The gages are
oriented along the directions of principle strains (zero shear), which are
at 45° to the shaft axis as shown in Fig. 2-18. This particular arrangement
eliminates the effects of bending strains and temperature changes at the
gage location.

In the study by French, the gages used were Micro-Measurements MA-
06-125TD-120, assembled in a 45° rosette. The power supply and signal
conditioner were connected to the bridge through a commercially available
slip ring assembly, and a voltage amplifier was used to boost the signal
to a more convenient level. This signal was used as the "true" value for
comparison.

Carrier Signal Transducers

If torsional vibration is to be used in a drive train containing gears, either a
magnetic transducer or a proximity probe can be installed close to the teeth
of a selected gear to produce a carrier signal, as illustrated in Fig. 2-17.
The predominant frequency in the carrier signal is the gear tooth passing
frequency.

Since a proximity probe produces output voltage proportional to the instantaneous probe–tooth gap, the amplitude of its carrier signal will not be modulated by torsional vibration; only the tooth passing frequency will be modulated. Conversely, the output signal from a magnetic transducer is both amplitude and frequency modulated, since the transducer produces a voltage proportional to the instantaneous velocity of the gear tooth in its close field, in addition to the probe-tooth gap. Instrumentation systems based on amplitude modulation of the carrier signal therefore require a magnetic transducer. Advantages of the magnetic pickup are that it requires no external power source, its output signal is generally clean with low noise levels, and it produces a strong signal that often requires no amplification.

In French's study, two B&K MM0002 magnetic transducers were used, one for each gear so as to determine the relative amplitudes, for comparison with the strain gage measurements of twist in the shaft. For the AM system described below, the magnetic transducers were velocity calibrated over the speed range 1000–2500 rpm to 0.48 MV (peak)/(degree/sec). In general, this calibration constant will be affected by the transducer gap to the gear tooth, so this setting must not be changed during a series of tests.

An FM carrier signal can also be optically transduced from alternating light and dark lines around the shaft, using light reflected to a photocell with fiber optics. A special reflective tape, which was crafted to meet this requirement, is described below. It was used with the TIMS and with one of the FM systems.

Frequency-modulated Systems

Since torsional vibration is simply a cyclic variation of shaft speed, it produces a variation of carrier signal frequency. One class of torsional vibration instrumentation systems utilizes the frequency modulation of this signal to produce an analog signal with dc amplitude proportional to instantaneous shaft speed. The ac component of the resulting signal is therefore torsional vibration velocity, which can be integrated to obtain torsional vibration displacement. The central and most critical component of this type of system is the electronic "box" that performs the frequency demodulation and signal conditioning. The electronic process is called frequency-to-voltage conversion, with the output voltage in this case proportional to the instantaneous rotational velocity of the shaft.

There are, or course, differences from one FM-type instrument to another in the filtering and calibration techniques, but the general design concept is similar for all. In French's study, two commercially available FM-type instruments were used. Both worked well, using a carrier wave produced by the magnetic transducer. Only one is still available at the time of this writing a general—purpose frequency-to-voltage converter, Model FC-62 manufactured by Validyne Engineering. The FC-62 is calibrated with a range adjustment screw, using a variable-frequency signal generator. For example, with the 120-tooth gear used in French's tests, the gear tooth passing frequency was 4000 Hz at a shaft speed of 2000 rpm. Using a 4000-Hz sine wave from a signal generator, the dc output of the FC-62 can be set to any desired value up to 10 V. The higher the dc set level, the greater the sensitivity to speed variations. This setting determines the calibration constant in millivolts per degree per second, calculated by dividing the dc output by the speed in degrees per second. If torsional vibration measurements are to be made at constant shaft speed using the FC-62, the dc level corresponding to the mean carrier frequency (shaft speed) can be suppressed to zero with a potentiometer on the front panel, leaving only the variations representing torsional vibration. In the tests by French, the mean shaft speed varied, so an external high-pass filter (ac coupling) was used instead that passes only the variations in voltage above a preset (low) frequency. This allows high-sensitivity settings to be used without exceeding the range of readout instruments.

Amplitude-modulated Systems

It has already been pointed out that a magnetic transducer installed close to passing gear teeth produces a carrier signal with amplitude (as well as frequency) modulated by the instantaneous gear velocity. An electronic process known as envelope detection can be used to produce a voltage analogous to the torsional vibration velocity.

Figure 2-19 illustrates the basic idea of envelope detection. In a typical circuit, the carrier signal is full-wave rectified with a bipolar precision diode detector and a differential amplifier. The rectified waveform then passes through a low-pass filter, where its level is averaged. The result is a dc level proportional to amplitude, which is analogous to the dc proportional to frequency described in the previous subsection. The output can be finally routed through another low-pass filter to reduce the running speed component of the signal induced by the inevitable minute deviations of gear teeth dimensions or magnetic properties.

With a moderate capability for wiring and packaging electronic circuits, it is entirely practical to custom-build an envelope detection system. Schematics are readily available, and the cost of components is low compared to the cost of a commercial instrument. In French's study, a commercial instrument was obtained and used since it had additional capabilities for other applications as well. It is no longer available at the time of this writing.

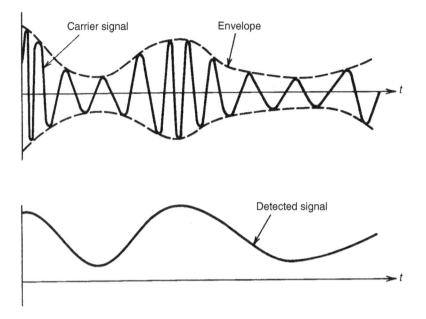

Figure 2-19   Envelope detection of a carrier signal.

The AM system as tested by French was found to be inferior to all the FM systems. The measurement errors from the AM system were larger and the signal was polluted to a much larger extent by lateral vibration of the shaft.

Frequency Analysis and the Sideband System

A frequency spectrum analyzer was used with each of the instrumentation systems tested. Figure 2-20 shows an example of a frequency-transformed signal, copied from the display of the dual-channel analyzer used in this study. The frequency spectrum of Fig. 2-20b showing the RMS amplitude of each frequency component in the signal is usually easier to interpret and more useful than the time trace of Fig. 2-20a. In measurements made on a typical drive train, the dominant frequency components are expected to be torsional excitation frequencies or natural frequencies; such a component can become destructively large when it is both of these.

If one accepts the spectrum analyzer (with zoom capability) as a necessary component of any of the analog-based instrumentation systems, then it can be stated that the sideband system is the simplest and least expensive. It consists of nothing more than a transducer to produce an FM carrier signal and the spectrum analyzer.

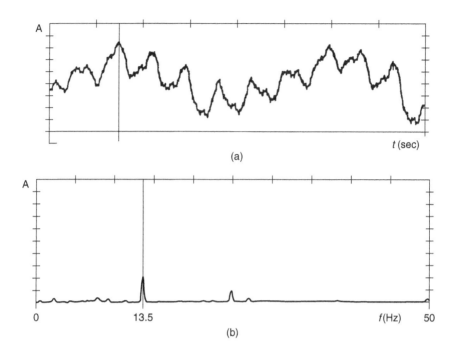

Figure 2-20 Time trace (a) and spectrum (b) of a torsional vibration signal.

The sideband system has been used for years to analyze the unwanted speed fluctuations ("wow") of tape recorders [13] and other precision constant-speed devices. It was suggested to the first author for measurement of torsional vibration in rotating machinery by Henry Bickel. The principle of operation is based on a Bessel series expansion of the simplest mathematical form of an FM carrier signal. The mathematical details are given by French in [5] and [12]. The method requires that the time-averaged running speed of the machine be constant during the measurement.

The spectrum analyzer must have the capability to expand about any center frequency with a narrow observation window. This is commonly referred to as zoom. About a 12:1 improvement in resolution is ample for most circumstances. By connecting the transducer signal directly to the FFT analyzer and setting the center frequency of the expansion window equal to the carrier frequency, a spike will appear in the center of the screen, flanked by a symmetrical array of spikes of decreasing magnitude. Figure 2-21 is an example of such a spectrum. The bordering components of the spectrum are called sidebands. The frequency of torsional vibration can be readily determined as the difference in frequency from the carrier to either of the first-order sidebands, both being equidistant from the carrier frequency.

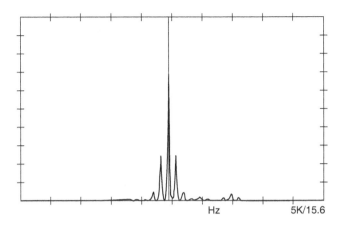

Hz   5K/15.6

Figure 2-21 Spectrum with sidebands from torsional vibration.

The amplitude, which is often of secondary interest, is determined from the ratio of the first-order (closest) sideband to the center or zero-order spike and requires a table of Bessel functions as described and provided in [5] and in Appendix B of [12].

French's Test Procedure and Results

French's test apparatus had two natural frequencies of torsional vibration: 8 Hz (lightly damped) and 13.5 Hz (strongly damped). He conducted steady-state tests for six different excitation amplitudes, with an excitation frequency of 8 Hz and a constant rotational speed of 2000 rpm. Tests were also conducted with other excitation frequencies, but in all cases the predominant test rig response was at 8 Hz. (The excitation torque was not purely harmonic.) Best results were therefore obtained at the resonance where the responding mode shape was well defined so that the strain gage signal could be accurately converted into an amplitude in degrees at the outboard gear of the test apparatus.

To make a typical steady-state measurement, the rig was brought up to a constant speed, a constant-amplitude excitation voltage was applied to the field of the motor, and the vibration allowed to stabilize. The strain gage signal was then read into one channel of the spectrum analyzer, and the signal from one of the test instruments was read into the other channel. Figure 2-22 shows an example, with the spectrum from the FC-62 FM system shown above the spectrum from the strain gages. The rotational speed is 2000 rpm and the excitation frequency is 8 Hz (resonant). The spikes on the spectral plots at 8 and 33.3 Hz are torsional vibration and lateral runout (or shaft whirling), respectively. Note that the 33.3-Hz lateral vibration component is synchronous with shaft speed and is

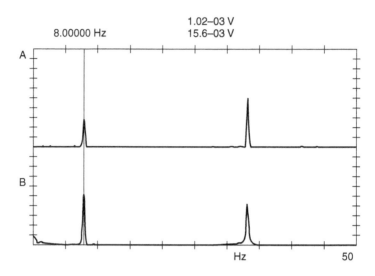

Figure 2-22 Steady-state spectrum, FC-62 (A) and strain gage (B).

produced by both the FM system and the strain gages. Since this is typical for measurements made on any rotating machine, it is fortunate that large-amplitude torsional excitations are usually not synchronous. Whenever they are, a method to separate the torsional vibration from the lateral vibration at the same frequency must be devised. One way to eliminate the lateral signal when using gear-excited transducers is to install two transducers on opposite sides (180° apart) of the same gear and sum the two signals electronically.

Table 2-3 is a summary of results from steady-state tests of the two carrier wave systems with best performance and accuracy, compared with the "true" values from the strain gages. French also conducted transient tests for a start-up condition in which the test rig was accelerated from 0 to 2500 rpm in 15–20 sec. Only the FC-62 FM system produced acceptable accuracy in these tests, with an average error of 5.3 percent. The sidebands method cannot measure a transient signal.

Table 2-3 Steady-state test results, 2000 rpm, 8 Hz excitation

FC-62		Sidebands	
True amplitude degrees peak	Fraction of true amplitude	True amplitude degrees peak	Fraction of true amplitude
0.035	1.03	0.034	1.00
0.079	1.00	0.078	1.00
0.131	1.00	0.133	1.00
0.164	1.00	0.169	1.00
0.319	1.01	0.315	1.00
0.683	1.00	0.674	1.02

A SPECIAL TAPE FOR OPTICAL TRANSDUCERS

When a gear or bladed wheel is not available for transducing a carrier signal, a reflective tape with alternating light and dark lines can be used to excite a fiberoptic photocell transducer. The tape can be wrapped around any section of accessible shafting in a drive train where torsional vibration is to be measured. A tape with equally spaced reflective and nonreflective lines was not commercially available at the time of this study. P. M. Barrios, from Zulia University in Maracaibo, Venezuela, made the tape shown in Figure 2-23 while visiting Texas A&M University. Utilizing a computer with graphics capability and a digital plotter, black lines were drawn on a white nonstretching fiberglass adhesive tape. The computer was programmed to produce optimum line widths for each application, depending on the shaft speed and diameter. Since the black ink tends to come off, it was coated with a clear varnish. This degraded the

Figure 2-23   Test apparatus with reflective tape for a fiber-optic transducer.

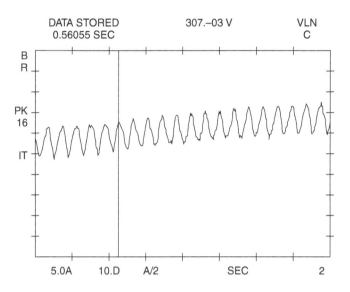

Figure 2-24 Torsional resonance measured from reflective tape with the FM system.

performance of the fiber-optic tachometer due to its high reflectivity. Nevertheless, good results were obtained using the laboratory test apparatus shown in Fig. 2-23. Its natural frequency was 10.2 Hz and torsional excitation was produced by gating a power transistor in the motor armature circuit with a signal generator.

One question that was resolved by tests on this apparatus is the possible effects of the tape overlap at the ends, as it is not practical to achieve perfect light/dark line spacing for every possible shaft circumference. Using the FC-62 FM system to demodulate the carrier signal, it was found that overlap (one unequally spaced line) produces a signal composed only of synchronous (shaft speed) frequency and its harmonics. This is the same effect that shaft runout produces from a magnetic transducer. By varying the overlap, it was found that the amplitude of the unwanted synchronous signal can be predicted from the amount of overlap. This allows subtraction of the overlap signal from the total synchronous component to give the torsional vibration in cases where the latter is synchronous. Figure 2-24 is a time trace of torsional resonance measured with this system while the test rig is accelerating at about 2000 rpm/min.

TIME-INTERVAL MEASUREMENT SYSTEMS

The development of the special reflective tape led to the design and testing of a promising prototype instrumentation system. This system, designated as TIMS, (time interval measurement system), is especially well suited to measurement of rapidly changing transients. The concept originated from a digital circuit for a tachometer designed by Mark Darlow at Rensselaer Polytechnic Institute. Figure 2-25 illustrates the system conceptually.

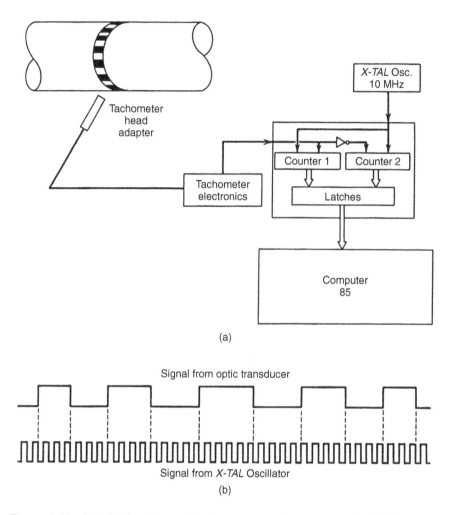

Figure 2-25 The TIMS concept: (a) schematic of TIMS components; (b) time trace of counting scheme. From [5].

The fiber-optic transducer produces a square wave that turns the counters on and off. While on, each counter counts the number of oscillations from a 10-MHz oscillator (see Fig. 2-25b). The latches store these numbers for input to the computer interface. The digits stored are equivalent to the passing times for each line on the tape. They are inverted by a computer code to produce angular velocity versus time.

Even with the computer-spaced lines on the tape, the extremely high resolution of this system produced a digital "noise" due to the uneven line spacing. This problem was magnified by the reflective varnish, which reduced the contrast and definition of the interface between white and black lines. It was overcome by recording the noise at constant speed with no excitation and subtracting it later from each digital record, leaving only the torsional vibration as "corrected data."

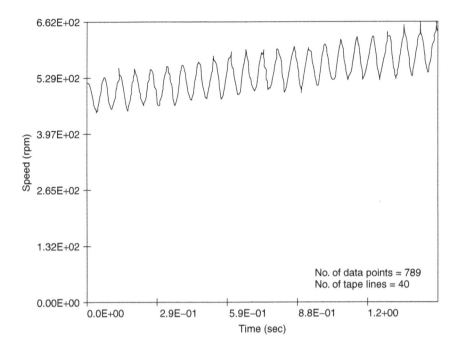

Figure 2-26   Resonant torsional vibration with transient shaft speed, recorded by TIMS. From [5].

Figure 2-26 is an example of transient torsional vibration recorded from the test rig by feeding the corrected data to a digital plotter. The resonant (10.2-Hz) torsional vibration is shown superposed onto the accelerating speed of the test rig. In a comparison discussion below, this method is referred to as the Barrios/Darlow method.

A number of similar TIMS concepts have been developed since this one was first published in 1986. See [14] for examples. The method gets more attractive with the advancement of the capability of computers and data acquisition hardware. The development of an improved TIMS was revived at the Texas A&M Turbomachinery Laboratory in 2005 and was used by Toram [15] to measure torsional vibration of a scaled-down ball mill. A magnetic transducer was used, with a metal chain encoder on the ball mill drum (see Fig. 2-27). Toram used a variation of the TIMS method in which the number of carrier wave pulses in a specific time period is counted, rather than accurately measuring the time from pulse to pulse (as in the description above). Toram's method requires the shaft speed to be constant and produced a signal rich in harmonics and noise.

Later, comparison measurements on the test rig of Fig. 2-23, using two different encoders, showed the original method to be superior, as follows. Referring to Fig. 2-23, the optical tape was removed and a series of metal strips were attached using aluminum tape to the top disk.

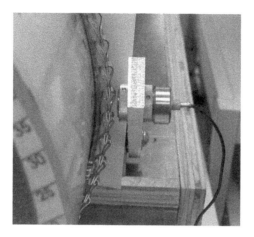

Figure 2-27   Toram's encoder and transducer.

Twenty-eight strips were spaced as close to equally apart as possible. These strips act as the carrier wave generator (encoder). The tachometer signal was generated using a B&K MM0002 magnetic probe. This encoder was used to represent an unequally spaced encoder that will introduce noise into the measurement system.

On the bottom disk, 40 quarter-inch segments were machined, creating another encoder. This encoder is more accurate than the one on the top disk due to the high-precision procedure used in machining the segments. This tachometer signal was created using an eddy–current proximity probe.

Torsional vibration was excited in the test rig by superimposing an ac signal on the dc power supplied to the motor. The result is a variation in the voltage supplied to the motor, thus causing a fluctuation of driving torque. The superimposing of the ac signal was accomplished by connecting the dc power supply and the motor in series with a function generator. To achieve appreciable amplitudes for the ac signal, the signal from the function generator was fed into a power amplifier before connecting to the circuit.

Schomerus used Labview by National Instruments for the data acquisition and analysis of the signals described below. The software application, LVTorsion, is available from the Turbomachinery Laboratory at Texas A&M University.

Results from Toram's Method

Toram's method for measuring the instantaneous speed of the shaft counts the number of pulses in a given time period and divides it by the number of encoder segments. The main issue with this method is risking the chance

Time-Series of Instantaneous Speed

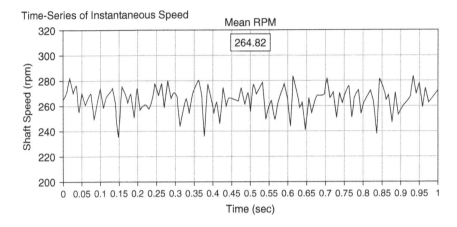

Figure 2-28   Instantaneous speed using Toram's method.

FFT of the Instantaneous Speed

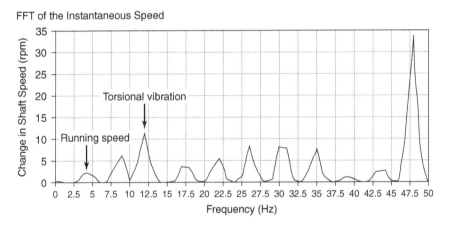

Figure 2-29   Frequency spectrum of shaft speed using Toram's method.

of being between pulses at the end of the time period. This method for calculating speed assumes that the encoder segments are equally spaced apart. To check the accuracy in measuring torsional vibration using this method, the test rig was run at constant speed with a constant torsional excitation of 12 Hz. The tachometer signal was generated using the magnetic probe and the encoder at the top disk of the test rig. The computed time-series signal of the instantaneous shaft speed is displayed in Fig. 2-28.

The generated signal showed speed fluctuations at multiple frequencies; this is evident in the frequency spectrum displayed in Fig. 2-29. The only fluctuating frequency should have been the 12-Hz frequency from the signal generator. Much of the noise is seen to be at harmonics of the shaft speed (4.4 Hz).

Results from the Barrios/Darlow Method

In the original Barrios/Darlow method as implemented by Schomerus [17], the oscillator was not used. Instead (in the time domain), the instantaneous shaft speed is measured by first computing the pulse position of each encoder segment. Next, the pulse width is found by taking the time difference between each pulse position. The pulse width of each encoder segment is inverted and then divided by the number of encoder segments in one revolution. This calculation method assumes that the encoder segments are equally spaced.

To check the improvement in using this speed calculation method, the test rig was run at the same constant speed as before with a constant

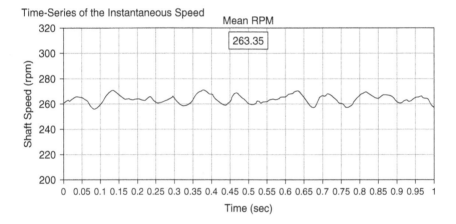

Figure 2-30   Instantaneous speed using the Barrios/Darlow method.

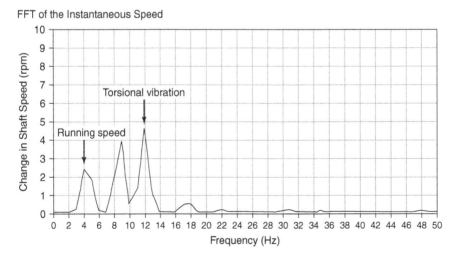

Figure 2-31   FFT of shaft speed using the Barrios/Darlow method.

12-Hz excitation. The results yielded the time-series signal of the instanta-neous shaft speed displayed in Fig. 2-30. This graph showed fewer speed fluctuating frequencies than the one shown in Fig. 2-28. The result was a cleaner frequency spectrum as shown in Fig. 2-31. A few harmonics of the nominal shaft speed still appear in the spectrum but most of the high-order components were eliminated. Because this calculation method uses the pulse width of each encoder segment to calculate shaft speed, the chance of missing pulses is dramatically reduced. As a result, the noise generated in the time-series signal of shaft speed is reduced. The false speed fluctua-tions occur due to the encoder spacing errors. These errors manifest them-selves as harmonic components of shaft speed in the frequency spectrum of the time-series signal of shaft speed. Schomerus also developed addi-tional algorithms to account for the unequally spaced encoder segments.

REFERENCES

[1] Den Hartog, J. P. *Mechanical Vibrations*, 4th ed., New York: McGraw-Hill, 1956, pp. 184–224.

[2] Rao, J. S. *Rotor Dynamics*. New York: Wiley, 1983, Chapter 3.

[3] Eshleman, R. Torsional vibration of machine systems. Proceedings of the 6th Turbomachinery Symposium, Texas A&M University, 1977, p.13.

[4] Vance, J. M. and French, R. S. Measurement of torsional vibration in rotating machinery. *Journal of Mechanisms, Transmissions, and Automation in Design* 108:565–577 (1986).

[5] Vance, J. M. *Rotordynamics of Turbomachinery*. New York: Wiley, 1988, Chapter 8.

[6] Mruk, G. K. Halloran, L. D., and Kolodziej, R. M. New method predicts startup torque. *Hydrocarbon Processing*, 229–234 (May 1978).

[7] Ramey, D. G., and Harold, P. F. Measurements of torsional dynamic characteristics of the San Juan No.2 turbine–generator. *Journal of Engineering for Power*, 378–384 (July 1977).

[8] Verhoef, W. H. Measuring torsional vibration. *Instrumentation Technology*, 61–66 (November 1977).

[9] Mruk, G. K., Halloran, L. D., and Kolodziej, R. M. New method predicts startup torque. *Hydrocarbon Processing*, 229–234 (May 1978).

[10] Ramey, D. G., and Harold, P. F. Measurements of torsional dynamic characteristics of the San Juan No.2 turbine–generator. *Journal of Engineering for Power*, 378–384 (July 1977).

[11] Sohre, J. S. Transient torsional criticals of synchronous motor-driven, high-speed compressor units. ASME Paper 65-FE-22, presented at the Applied Mechanics and Fluids Engineering Conference, Washington, DC, June 7–9, 1965.

[12] French, R. S. *An Experimental Study of Torsional Vibration Measurement*. Thesis, Texas A&M University, 1981.

[13] Savage, D. Effects of flutter on tape recorded data. *Sound and Vibration*, 18–24 (November 1969).

[14] Resor, B. R., Trethewey, M. W., and Maynard, K. P. Compensation for encoder geometry and shaft speed variation in time interval torsional vibration measurement. *Journal of Sound and Vibration* 286: 897–920 (2005).

[15] Vance, J. M., and Toram, K. K. Grinding media oscillation: effect on torsional vibrations in tumble mills. 30th Annual Meeting of the Vibration Institute, 2007, and Abstracts from the Current Literature, *The Shock and Vibration Digest* 39: 21–70. (2007).

[16] Vance, J. M. *Rotordynamics of Turbomachinery*. New York: Wiley, 1988, p. 59.

[17] Schomerus, A. Investigation of LabView as a tool for rotordynamic measurements and diagnostics. Turbomachinery Research Consortium Report TRC-RD-1-07, May 2007, Texas A&M University.

EXERCISES

2-1. This exercise shows the mathematical analogy of torsional vibration to lateral vibration and demonstrates that the natural frequencies of torsional vibration are independent of shaft speed. Consider the single degree of freedom system in Fig. 2-7 with every component (including the wall) rotating at constant speed O. Assume that the speed of the wheel with inertia I_{p2} is perturbed by a small amount $\dot{\theta}_2$.

a. Show that the kinetic energy T and strain energy V of the system are

$$T = \frac{1}{2}I_{P1}\Omega^2 + \frac{1}{2}I_{P2}(\Omega + \dot{\theta}_2)^2 \qquad (2\text{-}20)$$

$$V = \frac{1}{2}K\theta_2^2 \qquad (2\text{-}21)$$

b. Use Lagrange's equation (see [16]) to derive the differential equation of free and undamped torsional vibration. Note that it is the same equation as (1-2) in (with different symbols and $c = 0$, $F = 0$) and note that it does not contain the speed Ω.

c. Show that the natural frequency is given by Eq. 2-3.

2-2. Show that the differential equations for the two degrees of freedom system in Fig. 2-4 can be written as

$$\ddot{\theta}_1 + \frac{K}{I_{P1}}(\theta_1 - \theta_2) = 0 \qquad (2\text{-}22)$$

and

$$\ddot{\theta}_2 + \frac{K}{I_{P2}}(\theta_2 - \theta_1) = 0 \qquad (2\text{-}23)$$

where θ_1 and θ_2 are the twist angles of I_{p1} and I_{p2}, respectively. Note that the twist in the shaft is $\theta_1 - \theta_2 = \theta$. Subtract Eq. 2-23 from Eq. 2-22 and show that the resulting differential equation in θ produces the eigenvalue of Eq. 2-1. This model is analogous to the one in Exercise 1–9.

2-3. Use expressions for kinetic energy and strain energy to show that Fig. 2-6 is an equivalent dynamic model for Fig. 2-5.

2-4. For the torsional system of Fig. 2-11, the output shaft of the motor shaft and input shaft of the pump have the properties shown in the table. Include the stiffnesses of these shafts in the calculation of the torsional natural frequencies.

Motor output shaft	Pump input shaft
$L = 180\,mm$	$L = 300\,mm$
OD $= 80\,mm$	OD $= 60\,mm$
Shear modulus $= 8E10\,N/m^2$	Shear modulus $= 8E10\,N/m^2$

Answer: 68.0 Hz and 222.5 Hz.

Flexible Coupling Model

Element	Inertia (lb-in-s^2)	Stiffness (lb-in/rad)
Armature	$I_{P1} = 42$	
		$K_1 = 3.6E6$
Coup. Hub	$I_{P2} = 0.39$	
		$K_2 = 0.14E6$
Coup. Hub	$I_{P3} = 0.24$	
		$K_3 = 1.09E6$
Blower	$I_{P4} = 300$	

Reduced Flexible Coupling Model

Element	Inertia (lb-in-s^2)	Stiffness (lb-in/rad)
Armature + Coup. Hub	$I_{P1} + I_{P2} =$ $42 + 0.39$	
		$K_2 = 0.14E6$
Blower + Coup. Hub	$I_{P3} + I_{P4} =$ $300 + 0.24$	

Rigid Coupling Model

Element	Inertia (lb-in-s^2)	Stiffness (lb-in/rad)
Armature	$I_{P1} = 42$	
		$K_1 = 3.6E6$
Coup. Hub	$I_{P2} = 0.67$	
		$K_2 = 38E6$
Coup. Hub	$I_{P3} = 0.52$	
		$K_3 = 1.09E6$
Blower	$I_{P4} = 300$	

Reduced Rigid Coupling Model

Element	Inertia (lb-in-s^2)	Stiffness (lb-in/rad)
Armature	$I_{P1} = 42$	
		$K_1 = 3.6E6$
Rigid Coupling	$I_{P2} + I_{P3} =$ $0.67 + 0.52$	
		$K_2 = 1.09E6$
Blower	$I_{P4} = 300$	

3

INTRODUCTION TO ROTORDYNAMICS ANALYSIS

Rotordynamics is the dynamics of rotating machinery. Rotordynamics is different from structural vibrations analysis because of gyroscopic moments, cross-coupled forces, and the possibility of whirling instability. These difference makers are all due to rotation of the rotor assembly. Some of these difference makers have been recently incorporated into some of the popular FEM computer programs for structural vibration analysis. Meanwhile, computer codes have been developed that are specifically designed for rotordynamics analysis, with all of the difference makers included.

The power density in a rotating machine strongly affects its rotordynamics. Modern turbo machines and energy storage flywheels produce or absorb an amazing amount of power in a relatively small package. Undoubtedly, the most impressive turbomachine example is NASA's Space Shuttle main engine turbopumps, which produce 70,000 hp in two turbine stages about the size of a Frisbee. An energy storage flywheel developed by the Center for Electromechanics has discharged energy at a rate of 1 GW (1,341,000 hp). It weighs about the same as a modern automobile. In more common applications, turbojet engines provide propulsion for supersonic airplanes, turbine–compressor trains accomplish astounding rates of process in petrochemical industries, and steam turbines produce megawatts of electrical power for utilities.

A property of these machines that allows these high-power densities and flow rates to be accomplished is high shaft speed, relative to other types of machines of the same physical size. Along with high speeds come high inertial loads, and potential problems with shaft whirl, vibration, and rotordynamic instability, the subjects to which this book is addressed.

The engineering design challenge presented by aerodynamic and hydro-dynamic flows, design of blading, and other performance issues, some-times causes the rotordynamic requirements of a design to be overlooked. Too often it has happened that expensive turbomachines have been built and found to be incapable of producing their rated performance, or even of running at all, because of an assumption that making the rotor run smoothly and reliably at the design speed is a trivial problem. Some other potential problems in the engineering phase include the use of scaling when designing a new machine, or not fully modeling all the relevant parameters that influence the rotordynamic characteristics. Some of the more complex machinery trains involve the integration of equipment and components made by different manufacturers. While each machine or component may function properly alone, the integration with other equip-ment or re-rating it for different use will also require proper rotordynamic modeling and evaluation.

Even when the rotordynamics problems are considered, empirical and intuitive methods will get the rotordynamics engineer into trouble. Two characteristics of rotordynamics analysis are that its predictions are quite accurate when compared against experimental measurements (as long as accurate values for machine parameters are used in the mathematical model) and that its predictions are also quite often contrary to human intuition. One example of the latter characteristic is the prediction (veri-fied by experiment) that the unbalanced mass of a rotor will not "fly to the outside" of the shaft whirl orbit at high speed. (It will come around to the inside and stay there.) Another example is the verified prediction that *damping* in the rotor[1] of a turbomachine can produce a violently unstable whirling motion at high speed.

Given the potential accuracy of rotordynamics analysis and the fal-libility of human intuition to replace it, the reader should recognize the usefulness of learning some of its mathematical predictions and familiariz-ing him/herself with the principal results of the experimental investigations that have been made.

OBJECTIVES OF ROTORDYNAMICS ANALYSIS

In designing, operating, and troubleshooting rotating machinery, rotordy-namics analysis can help accomplish the following objectives:

---

[1]In this book the word "rotor" will be used to designate the assembly of rotating parts in a turbomachine, including the shaft, turbine wheels, compressor disks, and pump impellers.

1. *Predict critical speeds.* Speeds at which vibration due to rotor imbalance[2] is a maximum can be calculated from design data, so as to avoid them in normal operation of the machine.

2. *Determine design modifications to change critical speeds.* Whenever design engineers fail to accurately accomplish objective 1 above, or it becomes necessary to change the operating speed range of a machine, design modifications may be required to change the critical speeds.

3. *Predict natural frequencies of torsional vibration.* This objective usually applies to the entire drive train *system* in which the machine is employed. For example, a centrifugal compressor rotor driven by a synchronous electric motor through a gearbox may participate in a mode of torsional vibration excited by pulsations of the motor during start-up. In such a case, it might be desirable to change the natural frequency to a value which has the least possible excitation (in magnitude and/or time duration).

4. *Calculate balance correction masses and locations from measured vibration data.* This capability allows "in-place" rotor balancing to be accomplished, thereby reducing the amplitude of synchronous vibration.

5. *Predict amplitudes of synchronous vibration caused by rotor imbalance.* This is one of the most difficult objectives to accomplish accurately since the amplitude of rotor whirling depends on two factors that are both very difficult to measure: (a) the distribution of imbalance along the rotor, and (b) the rotor–bearing system damping. What can be done, however, is to predict the relative effects of rotor imbalance and system damping at specific locations.

6. *Predict threshold speeds and vibration frequencies for dynamic instability.* This objective is another challenging one at present, since a number of the destabilizing forces are still not understood well enough for accurate mathematical modeling. However, the instability caused by journal bearings, known as oil whip, can be predicted quite accurately.

7. *Determine design modifications to suppress dynamic instabilities.* This objective can be met more readily than objective 6, since computer of simulations can predict the relative stabilizing effect of various hardware modifications, even if the models for destabilizing force are only approximations.

---

[2]Imbalance is the proper English noun, but "unbalance" is used as a noun by a large number of machinery engineers in the literature.

THE SPRING–MASS MODEL

The simplest possible model for vibration analysis is a rigid mass mounted on a linear spring, with only one degree of freedom (see Fig. 3-1). The first critical speed of some rotor–bearing systems can be approximated by the natural frequency of this model converted to revolutions per minute:

$$N_1 = \frac{60}{2\pi} \sqrt{\frac{k}{m}} \tag{3-1}$$

where k is the effective stiffness for the first mode of whirling and m is the effective mass.

If the bearing support stiffness K_B in Fig. 3-1a is very soft relative to the shaft bending stiffness, the shaft does not bend and the rotor becomes a *rigid rotor*. In this case the effective mass m in Eq. 3-1 is the total mass of the rotor, and the effective stiffness k is the stiffness of the two bearing supports taken in parallel $(2 K_B)$.

For a rotor that has a relatively flexible shaft, compared to the bearing support stiffness, Fig. 3-1b shows that the effective stiffness is determined by the bending stiffness of the shaft. In this case, only a portion of the shaft mass contributes to the effective mass of the single-degree-of-freedom model, since the shaft mass near the bearing supports does not fully participate in the vibratory motion. These models are helpful in understanding natural frequencies of the first mode, but they have a number of serious limitations for more advanced rotordynamics analysis. First, the single-degree-of-freedom spring–mass model can execute a translational motion in only one direction, whereas the rotor–bearing system can execute whirl orbits, which may have complex shapes and patterns. This shortcoming can be partially removed by considering a spring–mass system with two degrees of freedom, allowing it to vibrate simultaneously in two directions, say X and Y.

The combination of vibrations in two orthogonal directions can produce several different types of motion of the mass. The type of motion produced depends on the relative amplitudes and phase relationship of the X and Y motions (see Fig. 3-2a). If the vibration has a single frequency, the motions produced are circular orbits (Fig. 3-2b), elliptical orbits (Fig. 3-2c), and straight line motions at any angle to the X axis (Fig. 3-2d).

The two-degrees-of-freedom model in Fig. 3-3 was used by Rankine in 1869 for the first published analysis of machinery rotordynamics [1] in an attempt to explain the *critical speed* behavior of rotor–bearing systems. The system model consisted of a rigid mass whirling in a circular orbit, with an elastic spring acting in the radial direction (see Fig. 3-3). Rankine used Newton's second law incorrectly in a rotating coordinate system,

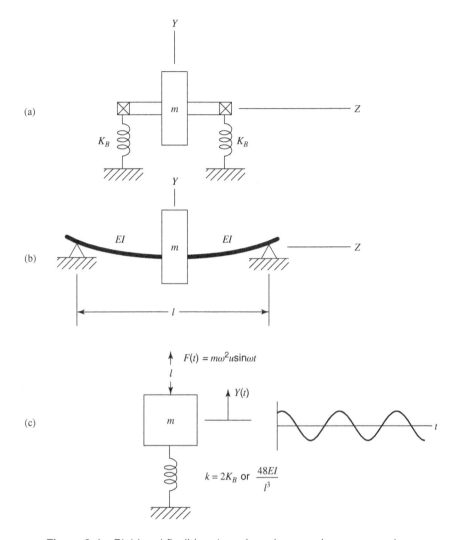

Figure 3-1   Rigid and flexible rotors viewed as a spring–mass system.

and predicted that rotating machines would never be able to exceed their first critical speed.

Although the two-degrees-of-freedom spring–mass model can execute the orbital motions of a rotor–bearing system, it does not contain a realistic representation for the rotating imbalance in the rotor. Since a perfectly balanced rotor never occurs in real machines, and since it is the rotating imbalance that excites the most commonly observed type of vibration (synchronous), it follows that the rotating imbalance is an essential ingredient of one of the most useful models for rotordynamic analysis. This model is called the Jeffcott rotor (see Fig. 3-4), named after the English

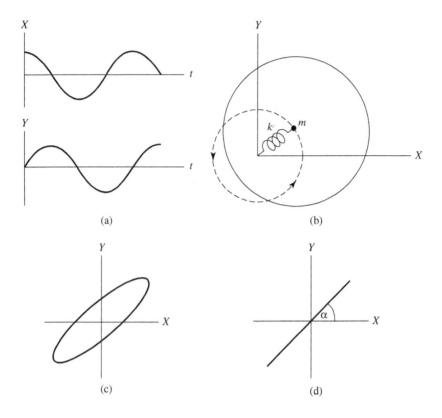

Figure 3-2 How X and Y vibrations combine to produce orbits. (a) time traces, (b) circular orbit, (c) elliptical orbit, (d) translational vibration.

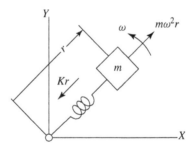

Figure 3-3 Rankine's model.

dynamicist who first used the model in 1919 to analyze the synchronous response of high-speed rotating machines to rotor imbalance [2]. It consists of a massive unbalanced disk mounted midway between the bearing supports on a flexible shaft of negligible mass. The bearings are rigidly

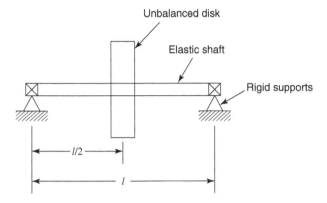

Figure 3-4 The Jeffcott rotor.

supported, and viscous damping acts to oppose absolute motions of the disk.[3]

Jeffcott's analysis explained how the rotor whirl amplitude becomes a maximum value at the critical speed but diminishes as the critical speed is exceeded due to the critical speed inversion of the imbalance.

SYNCHRONOUS AND NONSYNCHRONOUS WHIRL

The frequencies present in the measured vibration signal constitute some of the most useful information obtainable for diagnosing rotordynamics problems. For example, a common source of shaft whirling is rotor imbalance, and imbalance always produces whirling which is synchronous with shaft speed. Hence, large amplitudes of synchronous vibration usually indicate a rotor imbalance problem. Synchronous whirl excited by imbalance was the problem addressed by Jeffcott's analysis, described above, and presented in detail in the next section. But not all shaft whirling is synchronous; in fact, the more destructive rotordynamic problems involve nonsynchronous whirl.

Figure 3-5 shows an end view of a whirling rotor and describes the essential difference between the two types of motion. The shaded element represents an unbalanced mass. In Fig. 3-5a the time rate of change of the angle ϕ (i.e., $\dot{\phi}$) is the whirl speed. The angle β remains constant, so the whirl speed and the shaft speed are the same (synchronous whirl). Thus, the rotor imbalance U leads the rotor whirl vector V by a constant angle β. In Fig. 3-5b the time rate of change of the angle β (i.e., $\dot{\beta}$) is the spin

[3]In the Jeffcott model, the only source of this type of damping is air drag on the disk. The viscous representation is a useful approximation.

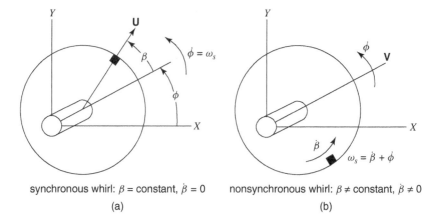

synchronous whirl: β = constant, $\dot{\beta}$ = 0   nonsynchronous whirl: $\beta \neq$ constant, $\dot{\beta} \neq 0$

(a)   (b)

Figure 3-5 Synchronous (a) versus nonsynchronous (b) whirl.

velocity of the rotor, relative to the rotating whirl vector V, so the shaft speed is the sum of $\dot{\beta}$ and $\dot{\phi}$. In this case, the whirl speed and shaft speed are not the same (nonsynchronous whirl).

The distinction between these two types of rotor motions provides the most basic classifying factor for frequency spectrum analysis. Critical speed and imbalance response analysis deals only with synchronous whirl. Rotordynamic instability deals almost exclusively with nonsynchronous whirl.

ANALYSIS OF THE JEFFCOTT ROTOR

Figure 3-6 shows an end view of the whirling Jeffcott rotor, with coordinates that describe its motion. The center of mass of the unbalanced disk is at M. The point C locates the geometric center of the disk. Thus, the amount of static imbalance is denoted $u = \bar{C}\bar{M}$ (inches), and the shaft bending deflection due to dynamic loads is $\bar{O}\bar{C}$. Gravity loads are neglected in this analysis. They are insignificant compared to the dynamic (inertial) loads in many high-speed machines.[4]

The shaft has a bending stiffness of k (lb/in), the disk has a mass of m (lb − sec^2/in), and air drag on the whirling disk and shaft is approximated by a viscous damping coefficient of c (lb-sec/in). The dynamic system has three degrees of freedom; an assumption of constant speed reduces them to two. The polar coordinates r, ϕ, β have an advantage of

---

[4]Large multistage steam turbines in electric utility plants constitute the most notable exception, in which gravity effects can cause a peak response to occur at a speed of about one-half the first critical speed.

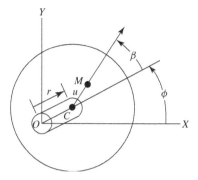

Figure 3-6 End view of the Jeffcott rotor, with coordinates.

giving the synchronous whirl solution in terms of constants that are readily interpreted, but the equations of motion are nonlinear and hence are not well suited to an analysis of rotordynamic instability (introduced in a later section of this chapter). The Cartesian coordinates X, Y (of the shaft center) along with the angle β produce linear differential equations. Furthermore, the solution in terms of X and Y as functions of time correlates better with what vibration displacement probes (eddy–current proximity probes) usually measure in a typical rotor installation.

The differential equations and their solutions in both sets of coordinates are presented here, for the case of negligible gravity and constant shaft speed ω.

Polar Coordinates

The differential equations of motion are [3]

$$\ddot{r} + \frac{c}{m}\dot{r} + \left(\frac{k}{m} - \dot{\phi}^2\right) r = \omega^2 u \, \cos(\omega t - \phi) \tag{3-2}$$

$$r\ddot{\phi} + \left(2\dot{r} + \frac{c}{m}r\right)\dot{\phi} = \omega^2 u \, \sin(\omega t - \phi) \tag{3-3}$$

The solution for synchronous whirling is

$$r_s = \frac{\omega^2 u}{\sqrt{\left(k/m - \omega^2\right)^2 + (c\omega/m)^2}} \tag{3-4}$$

$$\omega_s t - \phi_s = \beta_s = \tan^{-1}\left(\frac{c\omega}{m\left(k/m - \omega^2\right)}\right) \tag{3-5}$$

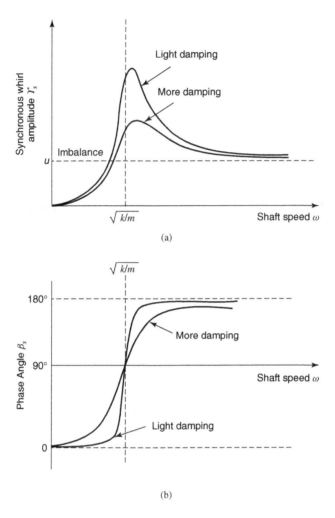

Figure 3-7 Imbalance response of a Jeffcott rotor: (a) amplitude, and (b) phase.

The constant whirling amplitude r_s and phase angle β_s satisfy the differential equations (3-2) and (3-3) for any constant shaft speed ω. Typical plots of r_s and β_s versus ω are shown in Fig. 3-7 for two different values of damping.

Cartesian Coordinates

The differential equations of motion are [4]

$$m\ddot{X} + c\dot{X} + kX = m\omega^2 u \cos \omega t \tag{3-6}$$

$$m\ddot{Y} + c\dot{Y} + kY = m\omega^2 u \sin \omega t \tag{3-7}$$

The solution for synchronous whirling is

$$X = \frac{\omega^2 u}{\sqrt{\left(k/m - \omega^2\right)^2 + \left(c\omega/m\right)^2}} \cos\left(\omega t - \beta_s\right) \tag{3-8}$$

$$Y = \frac{\omega^2 u}{\sqrt{\left(k/m - \omega^2\right)^2 + \left(c\omega/m\right)^2}} \sin\left(\omega t - \beta_s\right) \tag{3-9}$$

$$\beta_s = \tan^{-1}\left(\frac{c\omega}{m\left(k/m - \omega^2\right)}\right) \tag{3-10}$$

From the geometry of Fig. 3-6 it can be seen that the shaft deflection is

$$r = \sqrt{X^2 + Y^2} \tag{3-11}$$

Inspection of Eq. 3-4, 3-8, and 3-9 show that the solutions satisfy Eq. 3-11. Hence, Fig. 3-7a is also a typical plot of the amplitudes of horizontal vibration x (Eq. 3-8) or of vertical vibration y (Eq. 3-9).

Physical Significance of the Solutions

Figure 3-7a shows how the amplitude of synchronous whirl increases with speed as the critical speed is approached, and then decreases after traversing the critical speed and approaches the value of static imbalance at supercritical speeds. Thus, at high speed, the synchronous whirl amplitude can be made arbitrarily small by precision balancing of the rotor. At speeds near the critical speed, it can be seen that the most important parameter for reducing the whirl amplitude is damping. Figure 3-7a also provides the most useful definition of a critical speed: the "speed at which synchronous response to imbalance is maximum." Note that increased damping raises the critical speed slightly (but lowers the free vibration eigenvalue).

Inspection of Fig. 3-6 in conjunction with the solution for the synchronous phase angle β_s (Fig. 3-7b) yields an explanation for the asymptotic approach of the whirl amplitude toward u. As the critical speed is traversed, the angle β_s passes through $90°$ and approaches $180°$ at highly supercritical speeds. Figure 3-6 shows that β is the angle by which the imbalance leads the whirl vector. Thus, at high speed the center of mass M comes around to the inside of the whirl orbit, stands still, and the shaft center C whirls around the center of mass M. This phenomenon is called the *critical speed inversion* or *mass center inversion*.

Note that the center of mass stays to the outside of the whirl orbit only at low speed, and that the imbalance leads the whirl vector by exactly $90°$ when the shaft speed equals the undamped critical speed ($\sqrt{k/m}$). This latter fact is the basis of a method for accurate measurement of the undamped critical speed (which would have an unbounded whirl amplitude, difficult to measure).

Three Ways to Reduce Synchronous Whirl Amplitudes

A review of the Jeffcott rotor analysis yields three approaches to the problem of minimizing amplitudes of synchronous whirl: (1) balance the rotor, (2) change the operating speed (away from the critical speed), and (3) add damping to the rotor–bearing system. Although the Jeffcott rotor is a simple model, the same three approaches are effective for controlling synchronous whirl in more complex machines. The proper approach to use depends on the practical constraints of the problem at hand.

Balancing the rotor is the most direct approach, since it attacks the problem at its source. It should be pointed out here, however, that in practice a rotor cannot be balanced perfectly, no matter what method is used, and that the best achievable state of balance tends to degrade during extended operation. The second approach, moving machine operation farther away from the critical speed, can be achieved either by changing the operational shaft speed or by changing the critical speed itself. In practice the latter is usually accomplished by modifying rotor support stiffness. This parameter is not included in the Jeffcott model, but it has the same effect on critical speeds as the shaft stiffness k. A detailed analysis of the effect of flexible bearing supports is presented in a section to follow. In general, changing the critical speed is most useful for constant speed machines or for machines with a narrow range of operational speeds.

If a critical speed must be traversed slowly or repeatedly, or if machine operation near a critical speed cannot be avoided, then the most effective way to reduce the amplitude of synchronous whirl is to add damping. This would be difficult in the Jeffcott rotor, since the only source of damping is aerodynamic drag, but fortunately most turbomachines have flexible bearing supports in which damping can be added or oil-film bearings in which damping is inherent and can be changed by design modifications. Note that internal damping, or hysteresis, in the rotor shaft does not provide the type of damping modeled by Eqs. 3-2, 3-3, 3-6, and 3-7, since it acts only on motions relative to the whirl vector OC. In fact, internal friction in rotating parts is a source of self-excited subsynchronous whirling (rotordynamic instability).

SOME DAMPING DEFINITIONS

Since damping forces are difficult to measure directly and since a rotor–bearing system can have several different sources of damping, it has become common practice to quantify the total damping in terms of a percentage of *critical damping*. The critical damping coefficient c_{cr} is the value required to completely suppress any free vibration of the system, but no more. Thus, the damping ratio ξ is c/c_{cr} and the percent damping is 100ξ. For the Jeffcott rotor, the critical damping coefficient c_{cr} has the value $2\sqrt{km}$ and is assumed to be concentrated at the central disk. Using these definitions, Eq. 3-4 gives the whirl amplitude at $\omega = \sqrt{k/m}$ as $u/2\xi$ and the critical speed as $\sqrt{k/m(1 - 2\xi^2)}$. The imbalance multiplier $1/2\xi$ is sometimes referred to as the *magnification factor* or *Q factor* of the rotor–bearing system. For small (<10 percent) damping, it gives a good approximation to the maximum whirl amplitude at the critical speed. The damping ratio ξ can be calculated from a measured logarithmic decrement δ, $\xi = \delta/[(2\pi)^2 + \delta^2]^{1/2}$. See the discussion associated with Fig. 1-4.

Equation 3-4 can be put into dimensionless form by using the damping ratio ξ and the natural frequency $\omega_n = \sqrt{k/m}$ as parameters. The result is

$$
\frac{r_s}{u} = \frac{(\omega/\omega_n)^2}{\sqrt{\left[1 - (\omega/\omega_n)^2\right]^2 + (2\xi\omega/\omega_n)^2}}
\tag{3-12}
$$

THE "GRAVITY CRITICAL"

Large massive rotors with a long bearing span, such as utility turbines, and electric generators and motors, sometimes exhibit an apparent critical speed at a shaft speed of one-half the true first critical speed. The vibration is characterized by both synchronous and supersynchronous (twice synchronous) frequencies. There are two potential causes: 1) if the rotor has different stiffness in orthogonal directions (x and y rotating with the rotor, such as two-pole motors and generators), then gravity will bend the rotor a different amount twice per revolution; and (2) the fluctuating moment caused by gravity and rotor imbalance, a moment that was ignored in the Jeffcott rotor analysis where gravity was ignored and constant speed was assumed in order to reduce the number of degrees of freedom to two. See [5] for details of this analysis.

CRITICAL SPEED DEFINITIONS

The definition of a critical speed depends on who you are talking to. Communication is enhanced if you understand the definition of the term being used by others. One misconception held by some is that a critical speed is an instability (which is true only in the imaginary case of no damping). A list of sometimes used definitions of the critical speed follows:

1. Speeds at which response to unbalance (synchronous whirl) is a maximum (used here).
2. If the frequency of any harmonic component of a periodic forcing phenomenon is equal to or approximates the natural frequency of any mode of rotor vibration, a condition of resonance may exist; if resonance exists at a finite speed, that speed is called a critical speed (API 613, 1995, 4th Edition).
3. "A speed at which vibrations occur of such character and magnitude as to be of real significance in relation to the life or function of the machine" (Wilson, *Vibration*, Griffin (London) 1959, pp. 155–160).
4. When the synchronous rotor frequency equals the frequency of a rotor natural frequency, the system operates in a state of resonance, and the rotor's response is amplified if the resonance is not critically damped (API 684, August 20, 2005). See Fig. 3-8 to evaluate the validity of this statement.

EFFECT OF FLEXIBLE (SOFT) SUPPORTS

The bearing supports (the bearing and the structure that supports it) in any real machine are necessarily flexible, since every engineering material has elasticity. Furthermore, it is desirable from a rotordynamics standpoint to have the supports more flexible than the rotor. The two major reasons are as follows:

1. Low support stiffness reduces the dynamic loads transmitted through the bearings to the nonrotating structure, thus prolonging bearing life and minimizing structural vibration.
2. Low support stiffness allows the damping in hydrodynamic bearings or dampers to operate more effectively, thus attenuating rotor whirl amplitude at the critical speed. (This is assuming that the stiffness is in parallel with the damping.)

The first reason can be illustrated by an analysis of a short rigid rotor[5] on symmetric flexible supports with damping in the supports (see Fig. 3-9).

---

[5]From the standpoint of transmitted bearing loads, the rigid rotor is the worst case.

EIGENVALUES AND CRITICAL SPEEDS

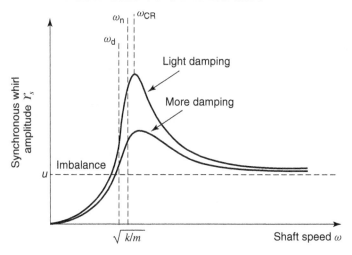

ω_d is idenified in rap tests

ω_n is undamped

ω_{CR} is identified from imbalance response

In lightly damped machines the difference is insignificant

In heavily damped machines (e.g. pumps) the difference can be very large

Figure 3-8   Natural frequencies and critical speeds on a Bode plot.

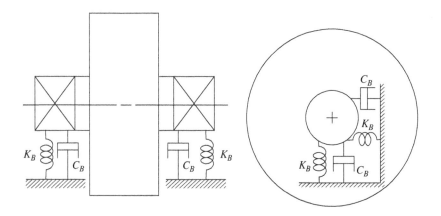

Figure 3-9   Short rigid rotor on damped flexible bearing supports.

The differential equations of motion and their solution for this model are exactly the same as for the Jeffcott rotor (Eqs. 3-2 through 3-11). Only the definitions of stiffness k and damping C are changed: $k = 2K_B$ and $c = 2C_B$ (the sum of the stiffness and damping of both bearing supports).

Figure 3-6 still defines the coordinates that describe the motion, but now $r = \bar{OC}$ is the deflection of the bearing supports rather than of the rotor shaft. The damping in bearing supports is produced by the oil film in hydrodynamic bearings or squeeze film dampers, by specially designed wire mesh or elastomeric dampers, and/or by internal friction in the bearing and its housing structure assembly.

Consider the force F_B transmitted through each bearing to the machine structure during a synchronous whirl motion in which point C (Fig. 3-6) traces out a circular orbit of radius r_s. Force F_B is the vector sum of the radial stiffness force F_k and the tangential damping force F_c. Figure 3-10 shows these two force components

$$F_k = K_B r_s \quad \text{and} \quad F_C = C_B \omega r_s \tag{3-13}$$

which have a resultant of

$$F_B = \sqrt{F_k^2 + F_c^2} = r_s \sqrt{K_B^2 + (C_B \omega)^2} \tag{3-14}$$

where r_s is given by Eq. 3-4. Thus,

$$F_B = \frac{1}{2}m\omega^2 u \sqrt{\frac{(2K_B)^2 + (2C_B\omega)^2}{\left(2K_B - m\omega^2\right)^2 + (2C_B\omega)^2}} \tag{3-15}$$

is an equation that gives the dynamic bearing load as a function of shaft speed and rotor–bearing system parameters.

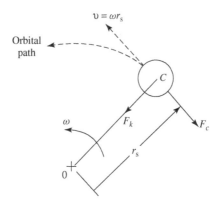

Figure 3-10   Forces on a bearing journal executing synchronous whirl.

Consider that the bearing load F_B would be given by

$$F_\infty = \tfrac{1}{2}m\omega^2 u \qquad (3\text{-}16)$$

if the supports were rigid. A few sample numerical calculations of this rigid support–bearing force F_∞ for typical high-speed turbomachines will convince the reader that it is intolerably high. With the proper selection of rotor–bearing parameters, the flexible support–bearing force F_B can be made considerably smaller than the rigid support force F_∞. The ratio of these forces can be obtained by dividing Eq. 3-15 by Eq. 3-16. The result, expressed in terms of dimensionless ratios, is

$$\frac{F_B}{F_\infty} = \sqrt{\frac{1 + (2\zeta\omega/\omega_n)^2}{\left[1 - (\omega/\omega_n)^2\right]^2 + (2\xi\omega/\omega_n)^2}} \qquad (3\text{-}17)$$

where $\omega_n = \sqrt{2K_B/m}$ is the undamped critical speed and $\xi = C_B/m\omega_n$ is the damping ratio.

Equation 3-17 gives the transmissibility of the imbalance force to the bearing support structure. Figure 3-11 is a plot of Eq. 3-17 as a function of speed ratio for two values of damping. Note that the two curves intersect at a speed ratio of $\sqrt{2}$. This is because Eq. 3-17 gives the same

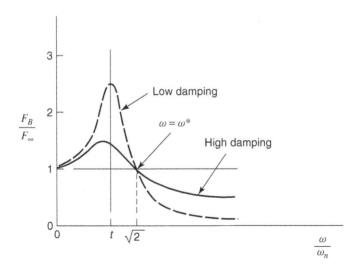

Figure 3-11 Bearing force transmissibility versus shaft speed ratio for two values of damping.

transmissibility value (1.0) for any value of damping at the shaft speed $\omega^* = \sqrt{2}\omega_n$. The following are some observations of practical interest from Fig. 3-11:

1. Bearing support flexibility can greatly reduce the dynamic load transmitted through the bearings, provided that the supports are made soft enough to keep the undamped critical speed considerably less than 70 percent of the operating speed. In Fig. 3-11, this corresponds to an operating speed range of $\omega/\omega_n \gg \sqrt{2}$ to keep the transmissibility low.
2. Bearing support damping increases the dynamic load transmitted through the bearings at high speeds ($\omega > \omega^*$) where the effect of support flexibility is favorable.
3. Bearing support damping may be necessary to keep the transmitted load within acceptable limits while traversing the critical speed.

Low support stiffness is not an unconditional panacea; improperly chosen support parameters can produce dynamic bearing loads in excess of the rigid support values, or can produce unacceptable static deflections.

Note also that the effects of damping on transmitted force is different from its effect on rotor whirl amplitude. Remember that Fig. 3-7 gives the whirl amplitude for the rotor–bearing model of Fig. 3-9, and shows that the effect of damping on whirl amplitude is favorable over the entire speed range.

If shaft flexibility is incorporated into the model (e.g., a Jeffcott rotor modified to include damped flexible supports), as shown in Fig. 3-12 the analysis becomes more complicated, even when internal damping in the rotor shaft is neglected.[6]

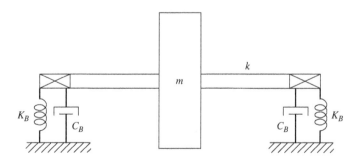

Figure 3-12 Jeffcott rotor modified to include damped flexible bearing supports.

[6]Internal friction in a shaft is usually insignificant compared to bearing support damping. If not, its major effect is to produce a subsynchronous whirl instability.

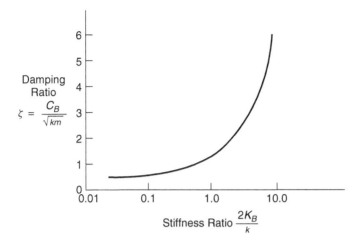

Figure 3-13 Optimum support damping for the rotor–bearing system of Fig. 3-12. From [6].

Reference [6] gives a solution for optimum support damping from the standpoint of minimizing whirl amplitude, as a function of the ratio of support stiffness to rotor shaft stiffness. This solution is represented here by Figs. 3-13 and 3-14. Figure 3-13 shows how the optimum amount of support damping varies with stiffness ratio. Note that the optimum support damping coefficient is less than the critical value[7] when the support stiffness is less than the shaft stiffness. Too much damping (more than the optimum value) "locks up" the supports, allowing the shaft, which has no damping, to act as the dominant spring in the system.

Figure 3-14 shows how the synchronous whirl amplitude r_s increases with support stiffness, even with the optimum amount of damping. This is because increasing support stiffness inhibits the motion of the support and therefore does not allow the damping to operate effectively. Many turbomachines have been designed with stiffness ratios of 5 or larger. Figure 3-14 shows that these machines cannot have synchronous whirl amplitudes less than ten times the unbalance at the critical speed, even with optimum damping at the bearings. (Reference [6] shows that Fig. 3-12 can accurately represent a multi-stage compressor at its first critical speed.)

It is useful to learn and memorize how the various force coefficients (for bearings, seals, impellers, etc.) affect rotordynamic performance

---

[7]The critical value of damping is defined here as $2\sqrt{km}$, the amount required to completely suppress free vibration of the spring–mass system of Fig. 3-4 with rigid supports (i.e., the Jeffcott rotor with damping on the central disk). The total bearing damping is $2C_B$ in Fig. 3-12.

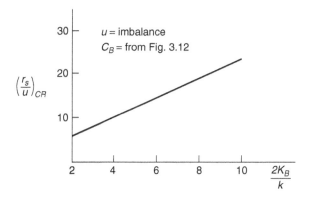

Figure 3-14 Whirl amplitude at the critical speed, with optimum damping, versus stiffness ratio. From [6].

as follows, since they are required inputs to computer codes for rotordynamics.

ROTORDYNAMIC EFFECTS OF THE FORCE COEFFICIENTS — A SUMMARY

The X and Y (orthogonal Newtonian) components of any force on the rotor can be expressed in terms of linearized force coefficients for small perturbations about a stationary equilibrium at a given shaft speed. These coefficients are the partial derivatives of the forces with respect to the displacements and velocities. Thus, the forces are expanded as follows:

$$F_X = -K_{XX}X - K_{XX}Y - C_{XX}\dot{X} - C_{XX}\dot{Y} \qquad (3\text{-}18)$$

$$F_Y = -K_{YX}X - K_{YY}Y - C_{YX}\dot{X} - C_{YY}\dot{Y} \qquad (3\text{-}19)$$

where $\dot{X}'$, $\dot{Y}'$ are velocities. In the following discussion, a circular whirl orbit is assumed for simplicity.

The Direct Coefficients

Direct stiffness coefficients $K_{XX} = K_{YY}$ produce a radial force directed inward and collinear with the rotor deflection vector. If the coefficients are negative the direction of the force reverses (outward). *Direct damping coefficients* $C_{XX} = C_{YY}$ produce a tangential force normal to the rotor deflection vector and opposing the whirl velocity. If the coefficients are negative the force is in the direction of whirl velocity.

The Cross-coupled Coefficients

Cross-coupled stiffness coefficients $K_{XY} = -K_{YX}$ produce a tangential force normal to the deflection vector and with direction dependent on the algebraic signs. The combination $K_{XY} > 0$, $K_{YX} < 0$ pushes forward whirl and can be destabilizing. Cross-coupled stiffness coefficients with the same sign are slightly stabilizing, as they make the orbit elliptical. *Cross-coupled damping coefficients* $C_{XY} = -C_{YX}$ produce a radial force collinear with the deflection vector and with direction dependent on the algebraic signs, i.e., either stiffening or destiffening. The combination $C_{XY} > 0$, $C_{YX} < 0$ is stiffening.

The coefficients that produce radial forces (direct K and cross-coupled C) have little effect on stability. They mainly affect the natural frequencies and location of critical speeds. The coefficients that produce forces tangential to the whirl orbit (cross-coupled K and direct C) have little effect on natural frequencies. They mainly affect stability and amplification factors at critical speeds.

Two different dimensionless ratios have been used to evaluate the effect on stability of a bearing, seal, or process wheel, based on its force coefficients. The *effective damping* $C_{\text{eff}} = C_{XX} - K_{XY}/\Omega$ gives the magnitude and sign of the "follower force" when the whirl frequency is Ω (the natural frequency, not the running speed). A negative value shows that the component being evaluated is destabilizing. The *whirl frequency ratio* WFR $= K_{XY}/C_{XX}\omega$ gives the expected ratio Ω/ω at the onset of instability, where ω is the running speed. These ratios are meaningless unless $K_{XY} = -K_{YX}$.

ROTORDYNAMIC INSTABILITY

The great majority of rotordynamic problems encountered involve synchronous whirl, i.e., response to imbalance. The approaches to these problems are straightforward, as described earlier in this chapter, even though the typical machine has a rotor–bearing system that is more complex than the models that have been analyzed here.

The remaining minority of problems involving nonsynchronous whirl or vibration can be subdivided into three classifications:

1. Supersynchronous vibrations due to shaft misalignment (the dominant frequency is often twice shaft speed)
2. Subsynchronous and supersynchronous vibrations due to cyclic variations of parameters, mainly caused by loose bearing housings or shaft rubs, or by nonlinear force coefficients

3. Nonsynchronous (usually subsynchronous) rotor whirling that becomes unstable, or has the potential to become unstable, typically when a certain speed called the threshold speed is reached

Problems of the first and second classifications have solutions that are obvious: align the shafts, tighten the bearing housings, or eliminate the rub. Problems of the third classification, although relatively uncommon, have a history of causing some very expensive failures, with elusive causes and cures. This is the main classification of problems referred to as rotordynamic instability.

The reader may note that rotordynamic instability occurs mostly in high-speed turbomachinery, but not in reciprocating internal combustion engines, as used in automobiles. Reciprocating machines are characterized by (1) lower speeds, (2) multiple interior bearing supports, and (3) high natural frequencies of the rotor (crankshaft). "High speed" and "lower speeds" are relative terms here. High speed could be only a few hundred revolutions per minute, provided it is significantly higher than a natural whirling frequency (eigenvalue) of the rotor–bearing system. Rotordynamic instability is manifested by shaft whirling, and the shaft will tend to whirl at its natural frequency (as modified by gyroscopic moments). Since it has already been said that instability frequencies are subsynchronous, it follows that they almost always occur when shaft speeds are higher than the natural whirling frequency.

There is a well-documented history of severe rotordynamic instability problems in centrifugal compressors used by the process industries (see Chapter 7), so that it is now commonly addressed in their design specifications. Historically, the occurrence of serious instability problems in aircraft turbine engines has been sufficiently infrequent to preclude its consideration as a primary factor in their design. There have been cases involving Alford's force, however (see below), and the U.S. Army funded a research program to quantify destabilizing forces from shaft splines in turboshaft engines for helicopters after a few spline-induced instabilities occurred [8]. During the period of development of turbopumps for the cryogenic fuels used in rocket engines, approximately 50 percent of these machines suffered rotordynamic instability problems at some stage during their development [9].

Rotordynamic instability is a special case of the more general theory of dynamic instability, or instability of dynamic systems. Both the classical theory and experiments show that the amplitude of free vibration in a linear system grows exponentially with time if the damping is negative. In mathematical terms, the system is unstable if the real part of the eigenvalue is positive.

Rotordynamic instability is seldom produced by negative direct damping. Instead, it is usually produced by a *follower force* that is modeled by cross-coupled stiffness coefficients. Follower forces are tangential to the rotor whirl orbit, acting in the same direction as the instantaneous velocity and following the rotor around in its orbit. In the rare case where the magnitude of the follower force is proportional to the instantaneous whirl velocity, it is classified as a negative direct damping force, just as in classical vibration theory. More typically the force is proportional to the rotor *displacement* (instantaneous orbit radius), and therefore is classified as a cross-coupled stiffness force. The "cross-coupled" terminology comes from the form of the force expressions in a nonrotating $X-Y$ coordinate system. A rotor displacement in the X direction produces a force in the Y direction, and vice versa. Figure 3-15 shows how a tangential follower force F_Φ produced by cross-coupled stiffness coefficients $K_{XY} = -K_{YX}$ is resolved into F_x and F_y components.

Forces acting on the rotor are usually fluid forces from bearings, seals, impellers, and turbine stages. These forces have radial and tangential (or normal to radial) components relative to the whirl orbit. The force component F_ϕ as shown in Fig. 3-16 is destabilizing to forward whirl, because it is acting in the direction of instantaneous velocity. Direct damping produces F_ϕ opposite to the whirl direction.

Most rotordynamic computer codes require F_r and F_ϕ to be converted into X-Y (Newtonian coordinates). Thus, we have stiffness and damping force coefficients K_{XX}, K_{XY}, C_{XX}, C_{XY}, etc., as described in a previous section. The representation of forces on the rotor by these stiffness and

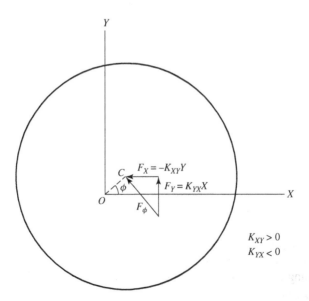

Figure 3-15 Cross-coupled stiffness representation of destabilizing force on a deflected rotor disk.

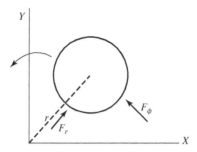

Figure 3-16   Components of a fluid force on a rotor.

damping coefficients implies that the forces are linearly proportional to the rotor displacement or velocity, which further implies that the rotor motions are small. The usefulness of this assumption for analysis is based on the fact that motions growing larger are generally unacceptable. If there is any interest in accepting or analyzing larger rotor motions they require nonlinear analysis.

Figure 3-17 illustrates *Alford's force* [10], a type of destabilizing force that can occur in axial-flow turbomachinery. It is the resultant of aerodynamic forces on the blades, produced by the variation of blade tip clearance around an unshrouded axial flow stage. A simple rotordynamics model to illustrate the destabilizing effect of Alford's force can be constructed by mounting the bladed disk (of mass m) on a flexible shaft midway between two hard-mounted bearings, i.e., a Jeffcott rotor with a bladed disk (see Figs. 3-4 and 3-17).

If the direct stiffness and damping properties are completely symmetrical (i.e., the same in X and Y directions), then the free motion of this rotor–bearing system is described by solutions to the following two coupled differential equations:

$$m\ddot{X} + c\dot{X} + kX + K_{XY}Y = 0 \tag{3-20}$$

$$m\ddot{Y} + c\dot{Y} + kY + K_{YX}X = 0 \tag{3-21}$$

where c is the external aerodynamic damping coefficient, k is the shaft stiffness, and the coefficients $K_{XY} = -K_{YX}$ produce Alford's force. Alford hypothesized [10] that it would be proportional to the eccentricity (X or Y), the stage torque (T), and the efficiency factor β, and inversely proportional to the pitch diameter (D) and vane height (H) of the blades. Thus, the magnitude κ of the cross-coupled stiffness coefficients K_{XY} and K_{YX} in this case can be expressed as

$$\kappa = \beta T / DH \tag{3-22}$$

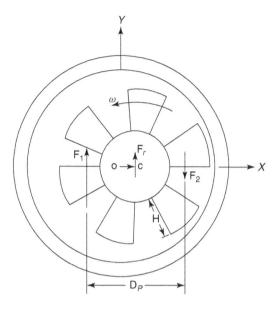

Figure 3-17   An axial flow stage eccentric in its housing.

Equations 3-20 and 3-21 are linear, homogeneous, and coupled, and have constant coefficients. Their general solution is [11]

$$X = A_1 e^{st} \tag{3-23}$$

$$Y = A_2 e^{st} \tag{3-24}$$

where s is the eigenvalue and A_1, A_2 are determined by the initial amplitude of a perturbation.

Substitution of Eqs. 3-23 and 3-24 into Eqs. 3-20 and 3-21 transforms the differential equations into algebraic equations with unknowns A_1 and A_2. Expressed in matrix form, these equations are

$$\begin{bmatrix} (ms^2 + cs + k) & K_{XY} \\ K_{YX} & (ms^2 + cs + k) \end{bmatrix} \begin{Bmatrix} A_1 \\ A_2 \end{Bmatrix} = \begin{Bmatrix} 0 \\ 0 \end{Bmatrix} \tag{3-25}$$

Since the equations are homogeneous, nonzero solutions for the ratio A_1/A_2 can exist only if the determinant of the matrix is zero. Equating the determinant to zero gives the *characteristic polynomial* in the eigenvalue s:

$$(ms^2 + cs + k)^2 - K_{XY} K_{YX} = 0 \tag{3-26}$$

The eigenvalues of the rotordynamic system are the roots of the characteristic polynomial. They are generally complex numbers. That is, each root will have the form

$$s = \lambda + i\omega_d \tag{3-27}$$

where λ is the damping exponent, and ω_d is the damped natural frequency (i.e., the whirling frequency) due to the form of the solutions (3-23) and (3-24). For example, Eq. 3-23 becomes

$$X(t) = A_1 e^{\lambda t}(\cos \omega_d t + i \sin \omega_d t) \tag{3-28}$$

If $\lambda > 0$, the perturbed motion grows exponentially with time and is therefore said to be unstable. The algebraic sign of λ depends on the relative magnitude of the cross-coupled stiffness (i.e., the destabilizing force) in Eq. 3-26. This follower force is modeled with the cross-coupled stiffness coefficients

$$K_{XY} = -K_{YX} \tag{3-29}$$

Representing the magnitude of K_{XY} and K_{YX} by κ, the real and imaginary parts of the roots (3-27) are found to be

$$\lambda = -\frac{c}{2m} \pm \sqrt{\left(\frac{c}{2m}\right)^2 + \left(\omega_d^2 - \frac{k}{m}\right)} \tag{3-30}$$

$$\omega_d^2 = \frac{k}{2m} - \frac{c^2 \pm \sqrt{(c^2 - 4km)^2 + 16\kappa^2 m^2}}{8m^2} \tag{3-31}$$

Figures 3-18 and 3-19 show how the whirl frequency ω_d and the damping exponent λ vary with the strength of the cross-coupled stiffness κ for three different values of damping ratio ξ. The latter is defined by

$$\xi = \frac{c}{2\sqrt{km}} \tag{3-32}$$

where c is the direct damping coefficient. Figure 3-18 helps to explain why the measured whirling frequency in violently unstable machines is usually higher than the associated critical speed, since the whirling frequency increases with the magnitude of the destabilizing force. Figure 3-19 shows that a rotor-bearing system with 5 percent damping (a typical value) can become unstable with a cross-coupled stiffness κ of only 10 percent of the effective shaft stiffness k. (Both stiffness values must be measured at the same point on the rotor).

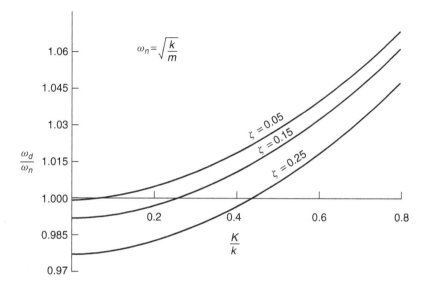

Figure 3-18 Effect of cross-coupled stiffness κ on whirl frequency.

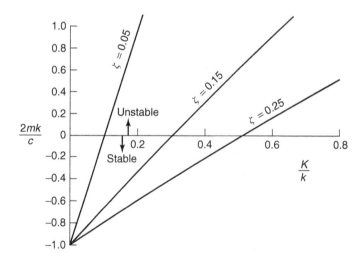

Figure 3-19 Effect of cross-coupled stiffness κ on stability

Using Eqs. 3-30 and 3-31 or Routh's method [12] it can be shown that a necessary condition to prevent the real part λ of the complex eigenvalues from becoming positive (i.e., a necessary condition for stability) is

$$\kappa < c\sqrt{k/m} \qquad (3\text{-}33)$$

which can also be written as

$$\frac{\beta T}{DH} < c\omega_n \qquad (3\text{-}34)$$

where ω_n is the undamped critical speed.

If the aerodynamic load torque T of the wheel increases with speed, then Eq., 3-34 shows that there could be a *threshold speed of instability*, above which the inequality is no longer satisfied. Note that the threshold speed can be raised by stiffening the shaft (raising ω_n). Increasing the effective damping coefficient c would also raise the threshold speed, but this would require a damper seal near midspan since this model has rigid bearings. Reducing the bearing support stiffness could increase the effective damping coefficient by allowing any bearing damping to become effective.

Most destabilizing forces in rotating machines can be represented by cross-coupled stiffness, as in the example just presented. Many engineers and the API specifications for rotating machinery now refer to *all* of them as Alford forces and many use Alford's equation to model them, even though Alford applied the equation only to axial-flow compressor and turbine stages.

A number of destabilizing mechanisms have been identified or hypothesized to explain incidents of rotordynamic instability. Some of the known or hypothesized sources of destabilizing forces are listed below, along with a pertinent reference for each case (not included are "false instabilities," subsynchronous vibration caused by the time variation of parameters):

1. Hydrodynamic bearings (oil whip) [13] and Chapter 5
2. Fluid ring seals (similar to oil whip) [14]
3. Internal friction in rotating parts (cross-coupled internal moment stiffness) [15]
4. Aerodynamic forces due to blade-tip clearance eccentricity (Alford's force) [10]
5. Trapped liquids inside a hollow shaft or rotor [16]
6. Dry friction whip (backward whirl driven by rubbing friction between rotor and stator) [17]
7. Labyrinth seals [18] and Chapter 6
8. Torquewhirl (the direct effect of very high-stage torque when misaligned by the mode shape) [19]

The destabilizing effect of internal friction is especially surprising and difficult to model correctly. Some researchers (including the first author) have modeled it as a follower force represented by cross-coupled stiffness K_{XY}, $-K_{YX}$. Lund [20] and Jafri and Vance [21] have shown that this model is physically incorrect. If the friction is from the rotor material hysteresis, a correct model is to express the elastic modulus as a complex number and use a loss coefficient [22]. However, the internal friction developed in most rotors is not a material hysteresis effect. Rather, it is due to micro-slipping that occurs between two shrink-fitted components of the rotor assembly. Jafri's analysis shows clearly that the internal friction developed through shrink fits is actually a system of internal *moments* developed at the interface due to friction forces that drive the rotor in forward whirl when the rotor–bearing system becomes unstable. This system of friction forces and equivalent couple moments are internal to the system and cannot be modeled correctly as a bearing to the ground (as an external force). This system of internal friction moments, however, can be modeled correctly using moment coefficients available in XLRotor software [23]. A significant challenge is to determine the correct magnitude of these coefficients. The interface friction is subject to a large number of variables that are difficult to know, e.g., surface roughness, oxidation, temperature, lubricant, and cleanliness. Each disassembly may change these variables.

EFFECT OF CROSS-COUPLED STIFFNESS ON UNBALANCE RESPONSE

The effect of cross-coupled stiffness on synchronous response to unbalance is surprising since it can either act like direct damping or cancel the existing damping, depending on its magnitude. The simplest possible model that can show this is the Jeffcott rotor used in the Alford's force analysis above, with unbalance added to the model (Fig. 3-4 with cross-coupled stiffness at the disk). With the unbalance term added on the right hand side, the equations become

$$m\ddot{X} + c\dot{X} + kX + K_{XY}Y = m\omega^2 u \cos(\omega t) \qquad (3\text{-}35)$$

$$m\ddot{Y} + c\dot{Y} + kY + K_{YX}X = m\omega^2 u \sin(\omega t) \qquad (3\text{-}36)$$

The particular solution for the whirl amplitude $r = (X^2 + Y^2)^{1/2}$ and phase β (with $K_{XY} = -K_{YX} = K$) is

$$r = \frac{m\omega^2 u}{\sqrt{\left(k - m\omega^2\right)^2 + (\omega c - K)^2}} \qquad (3\text{-}37)$$

$$\beta = \arctan\left[\frac{\omega c - K}{k - m\omega^2}\right] \qquad (3\text{-}38)$$

Comparison with the solution (3-4) and (3-5) for the Jeffcott rotor without cross-coupling shows that the effect of cross-coupled stiffness on response to imbalance is to modify the effective damping in the system. The new effective damping coefficient is

$$C_e = c - K/\omega \tag{3-39}$$

Note that there is a value of cross-coupling $K = \omega c$ that completely removes damping from the system, but if $K > 2\omega c$ or if $K < 0$ (backward driving) the effective damping is increased and the synchronous whirl amplitude is reduced. This analysis shows that the identification of system damping from measured vibration response plots (Bode plots) is really identification of direct damping and cross-coupled stiffness combined, whenever the latter is present in a turbomachine.

ADDED COMPLEXITIES

Although analyses of symmetric single-disk rotor models, as heretofore presented, provide valuable insight into many of the fundamental questions of rotordynamics, there are additional characteristics of real machines that are not included in these models. Some of them are the following:

1. Gyroscopic effects, which modify the eigenvalues and critical speeds.
2. Asymmetric stiffness properties of the bearing supports, which can produce backward whirl at the synchronous frequency and can stabilize an unstable system.
3. More than one disk and a significant distributed mass of the shaft, which produce a multiplicity of critical speeds and which may require multiplane balancing.
4. Asymmetric mass and stiffness properties of the rotor, which produce parametrically excited whirl.
5. Speed-dependent rotor support stiffness and damping, produced by oil film bearings and fluid seals, which modify the critical speeds and the response to imbalance.

Although all of these characteristics (and more) can be put into a computer simulation of rotordynamics, it is difficult to gain an understanding of their individual effects from a model that is too complex. A more enlightening approach is to analyze the simplest possible model that includes the factor or characteristic of special interest. The understanding thus obtained can be used to interpret the output from a computer simulation of the machine with all factors included, or vibration measurements from the machine itself. Some simplified examples follow here.

GYROSCOPIC EFFECTS

It is often instructive to analyze a model in which the factor of interest has a magnified effect. For example, gyroscopic effects are most pronounced in models with an overhung (cantilevered) disk. It is significant that one of the most common simple models analyzed in the rotordynamics literature is the single overhung disk on a cantilevered shaft. Such a model is illustrated in Fig. 3-20. It actually simulates a certain class of real machines fairly well, such as overhung compressors and pumps with no bearing outboard of the wheels.

The shaft is supported at the left end by two closely spaced bearings, each with axisymmetric stiffness. To simplify the analysis let us assume that the stiffness K_1 is very stiff so that the point O at the left end is fixed, while the stiffness K_2 is fairly soft. This combination will produce an axisymmetric moment stiffness $K_\omega = K_2 L^2$ (in-lb/rad or N-m/rad). Denote the two angular degrees of freedom for the conical motion of the fundamental mode as α and β (practically valid only for small angles of three-dimensional rotations). The rotor has a polar mass moment of inertia I_P and a transverse moment of inertia I_T about the left end at point O. If shaft bending is neglected, the differential equations of motion for the free angular motion without unbalance excitation are

$$I_T\ddot{\beta} + I_P\omega\dot{\alpha} + K_2L^2\beta = 0 \tag{3-40}$$

$$I_T\ddot{\alpha} - I_P\omega\dot{\beta} + K_2L^2\alpha = 0 \tag{3-41}$$

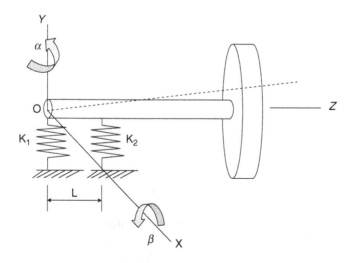

Figure 3-20 A cantilevered rotor model.

The middle term in the equations is the gyroscopic moment. $I_P \omega$ is a cross-coupled damping coefficient since it appears in the damping matrix (see Chapter 4). It acts like stiffness. The purely imaginary eigenvalues of this system are

$$\omega_1 = \left[\frac{I_P}{2I_T}\omega + \sqrt{\frac{K_2 L^2}{I_T} + \left(\frac{I_P}{2I_T}\omega\right)^2} \right] \tag{3-42}$$

$$\omega_2 = \left[\frac{I_P}{2I_T}\omega - \sqrt{\frac{K_2 L^2}{I_T} + \left(\frac{I_P}{2I_T}\omega\right)^2} \right] \tag{3-43}$$

These are the undamped natural frequencies of the rotor–bearing system. The following notes regarding them are instructive, even for more complex systems:

1. If the rotor angular speed ω is zero, the natural frequency becomes $\omega_T = \sqrt{K_2 L^2 / I_T}$. In this case the vibration modes are planar: pitching and yawing, respectively. Nonzero shaft speed ω changes the ω_1 and ω_2 frequencies to produce forward and backward whirl in conical mode shapes.

2. Shaft speed ω raises the ω_1 frequency of forward whirl above the planar pitching vibration value ω_T and lowers the ω_2 frequency of backward whirl.

3. Inspection of Eqs. 3-40 and 3-41 shows that the gyroscopic terms prevent either α or β from remaining zero whenever the other is nonzero. Thus, the rotor whirls when perturbed and cannot execute a planar pitching vibration when gyroscopic moments are acting (unless additional forces are acting or the bearing supports are asymmetric).

4. The strength of the gyroscopic moment is determined by the ratio $P = I_p / I_T$, which becomes larger as the rotor is shortened, until $P = 2$.

Figure 3-21 shows how the dimensionless natural frequencies ω_1 / ω_T and ω_2 / ω_T vary with dimensionless rotor speed ω / ω_T, for four different values of $P = I_p / I_T$. The value $P = 0$ is a fictitious case with no gyroscopic moment. The forward synchronous excitation frequency of rotating imbalance is shown by the dashed line, and its intersection with the $P = 0$ and $P = 0.5$ curves are the conical critical speeds for these two cases. Note that if P is large enough (e.g., short rotors) there is no intersection

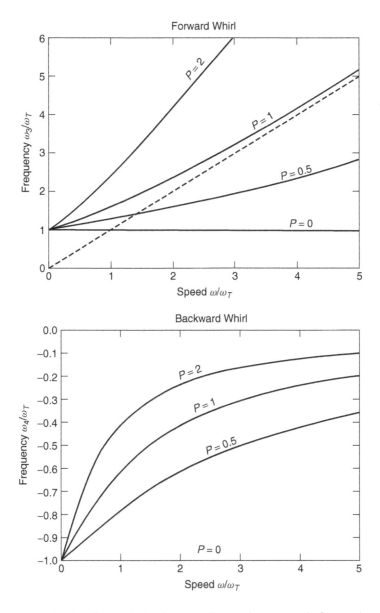

Figure 3-21 Eigenvalues of the cantilevered rotor vs. shaft speed.

and therefore no conical critical speed. In this case the eigenvalue exists but is never excited by rotating imbalance.

The foregoing eigenvalue analysis has determined the undamped critical speeds of the system without including imbalance forces in the equations, even though a critical speed is defined above as the speed at which synchronous response to imbalance is maximum. This was done by noting the speeds at which the eigenvalues are excited by the synchronous frequency. It is more efficient computationally to preconstrain the frequency in the analysis to be synchronous with shaft speed, as follows.

Forward synchronous whirl in the conical mode results from synchronous vibration of the angular coordinates α and β in Eqs. 3-40 and 3-41, described by

$$\alpha = -A \cos \omega t \quad \beta = B \sin \omega t \tag{3-44}$$

where ω is the rotor speed and $A = B$ with X-Y symmetric bearing supports. Substitution of (3-44) into (3-40) or (3-41) gives

$$(I_T - I_P)\,\omega^2 A = \tfrac{1}{2} K L^2 A \tag{3-45}$$

or

$$\omega_{con}^2 = \frac{\omega_T^2}{1 - P} \tag{3-46}$$

where

$$\omega_{con}^2 = \text{the conical critical speed}$$

$$\omega_T^2 = K L^2 / I_T, \text{ as above}$$

$$P = I_P / I_T, \text{ as above}$$

Equation 3-46 gives the loci of the intersections of synchronous excitation with the eigenvalue curves in Fig. 3-21. It is thus referred to as a critical speed analysis, as opposed to an eigenvalue analysis (which can also give the critical speeds as special cases when there is no damping).

None of the analyses presented to this point for the cantilevered rotor are capable of predicting amplitudes of synchronous whirl. Damping, which has been omitted from the model so far, has a strong effect on the amplitudes near critical speeds but not on the critical speeds. Likewise, the rotor imbalance, also omitted so far, affects the amplitude but not the critical speed. An analysis of the response to imbalance requires both imbalance and damping to be included in the equations of motion. When this is done, Eqs. 3-40 and 3-41 can be decoupled (only for the present case of symmetric bearing supports) and expressed as

$$(I_T - I_P)\,\ddot{\alpha} + \frac{C L^2}{2}\dot{\alpha} + \frac{K L^2}{2}\alpha = (I_T - I_P)\,\omega^2 \theta \cos \omega t \tag{3-47}$$

$$(I_T - I_P)\,\ddot{\beta} + \frac{C L^2}{2}\dot{\beta} + \frac{K L^2}{2}\beta = (I_T - I_P)\,\omega^2 \theta \sin \omega t \tag{3-48}$$

where C is the symmetric bearing damping and θ is the couple imbalance angle, defined as the angular misalignment of the principal z axis of inertia (one way to express couple imbalance). Equations 3-47 and 3-48 for the conical mode are analogous to Eqs. 3-6 and 3-7.

The particular solution to the nonhomogeneous Eqs. 3-47 and 3-48 is

$$\alpha = A\cos(\omega t - \gamma) \quad \beta = B\sin(\omega t - \gamma) \tag{3-49}$$

where

$$A = B = \frac{(I_T - I_P)\theta\omega^2}{\sqrt{\left[KL^2 - (I_T - I_P)\omega^2\right]^2 + \left(CL^2\right)^2\omega^2}} \tag{3-50}$$

$$\gamma = \arctan\left[\frac{\left(CL^2/2\right)\omega}{KL^2/2 - (I_T - I_P)\omega^2}\right] \tag{3-51}$$

The response plots of conical whirl amplitude A versus rotor speed ω, and phase angle γ versus ω, look similar to Fig. 3-7 for any given inertia ratio. Figure 3-21 shows that increasing P moves the amplitude peak to a higher critical speed, and this will be seen on the Bode plots as well.

Figure 3-22 shows an overhung turbocharger rotor. The two ball bearings are both located on the left end of the 0.3''-diameter shaft. The outboard bearing is O-ring supported to provide a small amount of damping and a soft support. The simplifying assumptions of the previous analysis are not valid so we resort to a finite element or transfer matrix method to solve for the eigenvalues and forced response in XLRotor.

The two wheels have a combined weight of 0.31 lb and the lowest rap test frequency is 15,751 cpm, so the modal stiffness is $k = m\omega_n^2 = 2183$ lb/in. With ball bearings and no gyroscopic moments the critical speed would be independent of shaft speed at about 15,751 rpm (very slightly higher due to the damping). But the wheels have significant polar

Figure 3-22   An overhung turbocharger rotor.

moments of inertia, which produce gyroscopic moments increasing with speed. They are stiffening for forward whirl and destiffening for backward whirl. A finite element model in XLRotor is used here to calculate the eigenvalues and critical speeds (see Chapter 4). Figure 3-23 shows how the eigenvalues (natural frequencies) of forward (◇) and backward (□) whirl vary with shaft speed. The synchronous (dashed) line crosses the forward eigenvalue at about 16,500 rpm. The critical speed on the Bode plot (Fig. 3-24) is slightly higher at 16,800 rpm due to the damping. The backward eigenvalue is seldom excited by unbalance.

Gyroscopic effects are also significant in a long rigid rotor on flexible supports, as illustrated in Fig. 3-25. In reference [24], a long rigid rotor

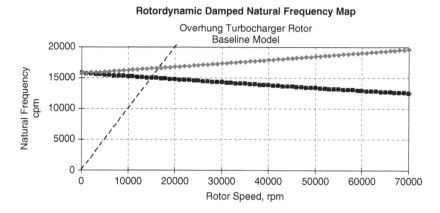

Figure 3-23 Forward and backward eigenvalues of the turbocharger rotor.

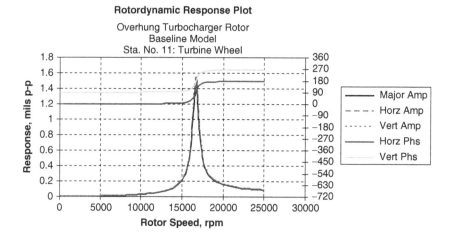

Figure 3-24 Synchronous unbalance response of the turbocharger rotor.

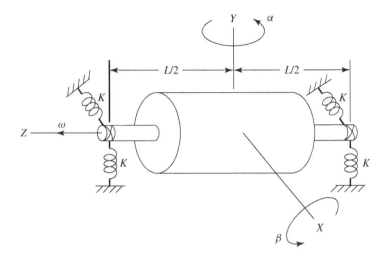

Figure 3-25   The long rigid rotor model.

model is modeled with the differential equations of motion, including the gyroscopic terms. The solution is used to illustrate the gyroscopic effects, conical whirl modes, backward whirl, and the distinction between critical speeds and eigenvalues. Figure 3-26 shows the cylindrical and conical mode shapes of the long rigid rotor on symmetric and identical flexible supports. If the rotor is symmetric, the cylindrical mode is modeled like the short rigid rotor. Notice in Fig. 3-26 that the shaft appears flexible (small diameter) but is not bending. This can occur if the bearing supports are soft enough. Figure 3-27 shows the bending modes of the same rotor that are produced by hard supports. This is a dangerous design, difficult to balance and prone to instability.

EFFECT OF SUPPORT ASYMMETRY
ON SYNCHRONOUS WHIRL

If the direct stiffness coefficients of the supports are different in orthogonal directions, they can be represented by K_{XX} and K_{YY}. With low damping this will produce a *split critical*, i.e., two closely spaced critical speeds. Backward whirl can then be excited by rotor imbalance when the rotor speed is between the two natural frequencies split by bearing support stiffness asymmetry, for instance, $(\omega_1 < \omega < \omega_2)$, where ω_1 is the eigenvalue associated with the lower stiffness and ω_2 is associated with the higher. The speed range $\omega_1 - \omega_2$ is narrowed by damping. Sufficient damping will make the backward whirl disappear. The simplest model capable of illustrating this is that of Fig. 3-9, with K_B replaced by $K_{XX} < K_{YY}$. The

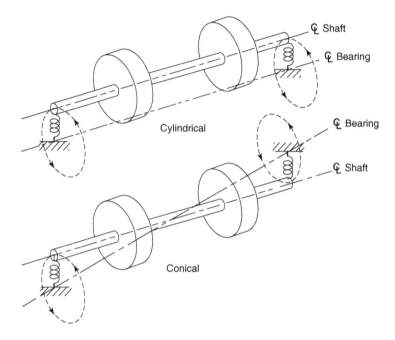

Figure 3-26 Symmetric whirl modes of the long rigid rotor.

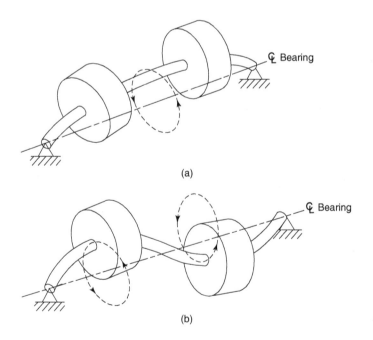

Figure 3-27 The rotor of Fig. 3-26 on hard supports.

equations of motion are (dropping the subscripts on $C_B = C_{XX} = C_{YY}$)

$$m\ddot{X} + 2C\dot{X} + 2K_{XX}X = m\omega^2 u \cos(\omega t) \tag{3-52}$$

$$m\ddot{Y} + 2C\dot{Y} + 2K_{YY}Y = m\omega^2 u \sin(\omega t) \tag{3-53}$$

with particular solutions

$$X(t) = X_C \cos(\omega t) + X_S \sin(\omega t) \tag{3-54}$$

$$Y(t) = Y_C \cos(\omega t) + Y_s \sin(\omega t) \tag{3-55}$$

Expressions for the amplitudes X_C, X_S, Y_C, and Y_S can be obtained by substituting Eqs. 3-54 and 3-55 into Eqs. 3-52 and 3-53, and solving the resulting set of linear algebraic equations.

Referring to Fig. 3-6, $X(t)$ and $Y(t)$ are the coordinates of the point C and the precession or whirl angle is

$$\phi = \arctan\left(\frac{Y}{X}\right) \tag{3-56}$$

Backward whirl is described by $d\phi/dt < 0$. Taking the time derivative in (3-56) and using (3-54) and (3-55) gives the condition for backward whirl as

$$X_C Y_S - Y_C X_S < 0 \tag{3-57}$$

A test for this condition can be programmed into any computer code for imbalance response to determine whether the whirl is forward or backward. This should be done at each inertia station in a multidisk model, since the rotor can whirl forward and backward simultaneously at different locations!

For the simple example here X_C, Y_S, Y_C, and X_S are available as functions from the solution described above. Substitution of these functions into (3-23) yields the condition for backward whirl as ($\omega_1 < \omega < \omega_2$), where

$$\omega_1 = \frac{2K_{XX}}{m} - \frac{1}{2}\left(\frac{2C}{m}\right)^2 + \frac{1}{2}\left(\frac{K_{YY} + K_{XX}}{K_{YY} - K_{XX}}\right)\left(\frac{2C}{m}\right)^2$$

$$\omega_2 = \frac{2K_{YY}}{m} - \frac{1}{2}\left(\frac{2C}{m}\right)^2 - \frac{1}{2}\left(\frac{K_{YY} + K_{XX}}{K_{YY} - K_{XX}}\right)\left(\frac{2C}{m}\right)^2 \tag{3-58}$$

The actual occurrence of backward whirling was doubted by some in the past [7], but modern instrumentation confirms that it does occur. The first author has observed it often in his laboratory at Texas A&M University. Support stiffness asymmetry has a different effect on rotordynamic instability. It acts as a stabilizing influence by making the whirl orbits more elliptical. Backward whirl can also be nonsynchronous at a natural frequency. This was seen in the section above on gyroscopic effects.

FALSE INSTABILITIES

As described in a section above, most "instabilities" are due to destabilizing cross-coupled forces from variable fluid dynamic pressure around a rotor component, acting in the direction of forward whirl and causing subsynchronous orbiting of the rotor. However, not all subsynchronous whirling frequencies are unstable. The onset of a true whirl instability usually occurs at some threshold speed beyond the critical speed, and the frequency of the whirling is the first natural frequency. It is an eigenvalue of free vibration that does not die out, unlike passing through a resonance where the amplitude of vibration increases and then decreases at higher speeds. According to linear theory for the vibration of an unstable system, its amplitude grows exponentially and becomes unbounded. However, often in the field and in lab experiments, the eigenvalue vibration becomes bounded in *limit cycles*. If it is large in amplitude, this vibration can cause damage to seals, bearings, or process wheels. If it is small, the question arises as to whether it has the potential to grow larger (a true "instability") or whether it is benign and harmless. Subsynchronous vibration is often observed in rotor–bearing systems, but in many cases it is a benign and harmless phenomenon. References [25] and [26] describe diagnostic tools, or indicators, that were developed in the Turbomachinery Laboratory at Texas A&M University. A summary of some of these tools follows here.

One of the most basic diagnostic indicators is whether the subsynchronous vibration "tracks" the running speed. If the vibration tracks it is not an instability. Tracking means that the subsynchronous vibration is always a fixed fraction (like 0.5 or 0.33) of the shaft speed. Of course, if the machine is constant speed it may not be possible to determine if tracking exists. Some of the other diagnostic indicators of stability for subsynchronous frequencies are as follows.

1. *Agreement of the subsynchronous frequency with known eigenvalues of the system.* Rotordynamic instability occurs at a damped natural frequency of the rotor–bearing system. The frequency can often be determined from an accurate rotordynamic model or from bump tests if the bearings are rolling-element.

2. *Presence of higher harmonics or multiple frequencies.* Subsynchronous frequencies caused by loose bearing clearances, loose bearing caps, or loose foundations may be excited by intermittent impact as the separated surfaces come together repeatedly as the machine runs. These impacts will excite a number of natural frequencies, none of which are unstable. Spectral analysis of the complex signal may expose a rich spectrum.

3. *Orbit Shape—Ellipticity.* Destabilizing follower forces, modeled by cross-coupled stiffness, are always normal to the instantaneous rotor deflection vector (orbit radius). As Fig. 3-28 shows, the follower force F_∞ will be always collinear with the velocity V only if the orbit is circular. The force normal to the orbit radius becomes more oblique to the orbital velocity as the orbit becomes more elliptical. The energy fed into the whirl motion is the dot product of the vectors F_∞ and V, so a very elliptical orbit is an indicator of stability.

4. *Subsynchronous frequency exactly one-half of shaft speed.* Although it is possible for a whirl frequency of order 0.5 to be unstable, this is often an indication of benign vibration caused by an intermittent rotor rub or nonlinear bearing support stiffness. See [25] for details. Note that the whirl frequency of the true instability due to fixed-geometry oil film bearings is of order 0.47, not 0.5. Precise instrumentation and signal analysis is required.

5. *Subsynchronous frequency equal to a torsional natural frequency.* Interaction between lateral and torsional vibration can occur, especially in machines with gears. A torsional natural frequency can appear in lateral measurements as a false instability. See an example in [27].

6. *A change in the measured synchronous phase angle due to cross-coupled stiffness.* Reference [26] shows how the synchronous phase angle changes as cross-coupled stiffness is added to a rotor–bearing system. Figure 3-29 is an example from a computer simulation of a

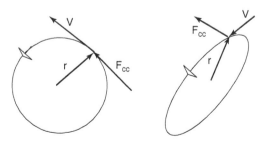

Figure 3-28   Orbit shapes and the follower force. From [25].

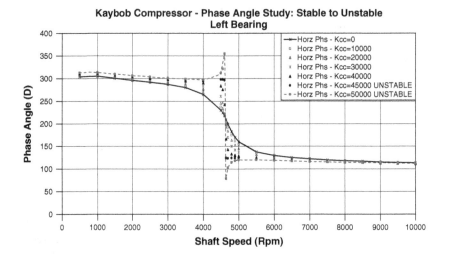

Figure 3-29 Phase angle as affected by cross-coupled stiffness. From [26].

multistage compressor. The various graphs are for added values of cross-coupled stiffness. If phase angle measurements are available and accurate enough, they may be used as an indicator that a follower force is present.

REFERENCES

[1] Rankine, W. A. On the centrifugal force of rotating shafts. *Engineer (London)* 27:249 (1869).

[2] Jeffcott, H. H. The lateral vibration of loaded shafts in the neighborhood of a whirling speed: the effect of want of balance. *Philosophical Magazine*, Ser. 6, 37:304 (1919).

[3] Gunter, E. J., Jr. *Dynamic Stability of Rotor–Bearing Systems*. NASA SP-113, 29 (1966).

[4] Rieger, N. F. *Vibrations of Rotating Machinery*, Pt. I: *Rotor–Bearing Dynamics*. Clarendon Hills, IL: The Vibration Institute, 1982, Sec. 1.6.

[5] Vance. J. M. *Rotordynamics of Turbomachinery*. New York: Wiley, 1988, pp. 27–32.

[6] Barrett, L. E., Gunter, E. J., and Allaire, P. E. Optimum bearing and support damping for unbalance response and stability of turbomachinery. *Journal of Engineering for Power*, 89–94 (January 1978).

[7] Den Hartog, J. P. *Mechanical Vibrations*, 4th ed. New York: McGraw-Hill, 1956, p. 265.

[8] Marmol, R. A. *Engine Rotor Dynamics: Synchronous and Nonsynchronous Whirl Control*, USARTL- TR- 79-2, U.S. Army Research and Technology Laboratories, Ft. Eustis, VA, pp. 69–125 (February 1979).

[9] Ek, M. C. Dynamics problems in high performance rocket engine turbomachinery. Workshop on Rotordynamics Technology for Advanced Turbopumps, NASA- Lewis Research Center, Cleveland (February 23, 1981).

[10] Alford, J. S. Protecting turbomachinery from self-excited whirl. *Journal of Engineering for Power*, 333–344 (October 1965).

[11] Beachley, N. H., and Harrison, H. L. *Introduction to Dynamic System Analysis*. New York: Harper & Row, 1978, pp. 43–48.

[12] Pipes, L. A. *Applied Mathematics for Engineers and Physicists*, 2nd ed. McGraw-Hill, 1958, pp. 239–242.

[13] Pinkus, O., and Sternlicht, B. *Theory of Hydrodynamic Lubrication*. New York: McGraw-Hill, 1961.

[14] Kirk, R. G., and Miller, W. H. The influence of high pressure oil seals on turbo–rotor Stability. *ASLE Transactions*, 22: 14–24 (1979).

[15] Jafri, S., *Shrink Fit Effects on Rotordynamic Stability: Experimental and Theoretical Study*. Thesis, Texas A&M University, August 2007.

[16] Ehrich, F. F. The influence of trapped fluids on high speed rotor vibration. *Journal of Engineering for Industry*, 806–812 (November 1967).

[17] Ehrich, F. F. The dynamic stability of rotor/stator rubs in rotating machinery. *Journal of Engineering for Industry*, 1025–1028 (November 1969).

[18] Childs, D. W. *Turbomachinery Rotordynamics*. New York: Wiley, 1993, pp. 297–306.

[19] Vance. J. M., *Rotordynamics of Turbomachinery*. New York: Wiley, 1988, pp. 292–310.

[20] Lund, J. W. Destabilization of Rotors from Friction in Internal Joints with Micro-slip. International Conference in Rotordynamics, JSME, 1986, pp. 487–491.

[21] Jafri, S. M., and Vance, J. M., Shrink fit effects on rotordynamic stability: theoretical study. GT2008-50412, Proceedings of the ASME-IGTI-Turbo Expo, Berlin, Germany, June 9–13, 2008.

[22] Lund, J. W. Stability and damped critical speeds of a flexible rotor in fluid-film bearings. *Journal of Engineering for Industry* 96:509–517 (1974).

[23] www.XLRotor.com

[24] Vance. J. M., *Rotordynamics of Turbomachinery*. New York: Wiley, 1988, pp. 125–130.

[25] Vance, J. M. and Kar, R. Sub-synchronous Vibrations in Rotating Machinery—methodologies to identify potential instability. GT2007-27048, Proceedings of ASME Turbo Expo 2007: Power for Land, Sea and Air, May 14–17, 2007, Montreal, Canada.

[26] Rajagopalan, V. N., and Vance, J. M., Diagnosing sub-synchronous vibration: unstable or benign. Proceedings of the ASME 2007 International Design Engineering Technical Conferences & Computers and

Information in Engineering Conference, IDETC/CIE 2007, September 4–7, 2007, Las Vegas, Nevada, USA.

[27] Rajagopalan, V. N., and Vance, J. M., Diagnosing coupled lateral–torsional vibrations in turbomachinery. GT2008-50125, Proceedings of ASME Turbo Expo 2008: Power for Land, Sea and Air, June 9–13, 2008, Berlin, Germany.

EXERCISES

3-1.   Refer to Fig. 3-2. Let $x(t) = A\cos(\omega t - \theta)$ and $y(t) = B\cos(\omega t - \phi)$. Program Excel spreadsheets to construct and view the orbits created when A, B, θ, and ϕ are varied:

$$m\ddot{X} + c\dot{X} + kX = m\omega^2 u \cos(\omega t)$$

$$m\ddot{Y} + c\dot{Y} + kY = m\omega^2 u \sin(\omega t)$$

3-2.   The differential equations of motion for the Jeffcott rotor in Cartesian coordinates are shown above. Derive these equations from Newton's second law, $F = ma$. Derive the accelerations in the two equations by writing the expressions for X and Y from Fig. 3-6 and taking two derivatives. Note that X and Y are the coordinates of the geometric center of the disk (not the center of mass). Do not use vectors. Notice that the first term in each differential equation is not the complete expression for the acceleration of the mass.

3-3.   See the differential equations of motion for the Jeffcott rotor in polar coordinates as given by Eqs. 3-2 and 3-3.

    a. Show that they cannot be derived from $F = ma$, where $a = d^2r/dt^2$ and $d^2\omega/dt^2$. Be sure that you know the reason why. Identify the Coriolis term in the equations.

    b. Derive the functional solution for response to unbalance (amplitude r and phase β) under the assumption of constant speed. Notice that the equations are nonlinear, so "intelligent guessing" will be the only feasible approach. Be sure that you can clearly show all the steps of the derivation to get the solution. *Hint*: Start out by guessing a circular orbit.

    c. Use the coordinate transformation $X = r\cos(\varphi)$, $Y = r\sin(\varphi)$ to show the equivalence of Eqs. 3-2 and 3-3 to Eqs. 3-6 and 3-7.

    d. Modify the equations to include the effect of gravity. Attempt to find a solution equivalent to what you got above. What is the main difficulty?

    Question to ponder: *Is the Jeffcott rotor a linear system or a nonlinear system?*

3-4. Refer to Eq. 3-12. Show that the critical speed, i.e., the speed at which response to imbalance is maximum, is given by the expression $\omega_{cr} = \omega_n/\sqrt{1 - 2\zeta^2}$.

3-5. See Chapter 1, page 5, for the definition of the damped natural frequency ω_d in terms of ω_n and ζ. Show that the critical speed ω_{cr} is 1.63 times higher than the damped natural frequency ω_d when the system has 50 percent of critical damping ($\zeta = 0.5$).

3-6. A short rigid turbine rotor weighs 30 lb and operates at 30,000 rpm on two ball bearings. The rotor unbalance amounts to an offset of the C.G. from the bearing centerline by .001''. The two bearing supports are to be identical, so the whirl mode is symmetric. The machine must be capable of operating inverted. The maximum allowable blade/housing clearance is .030'' (radial).

   a. Determine the bearing stiffness and damping to minimize the dynamic bearing forces without producing blade rub under any steady-state vibration condition or under the condition of a 1g vertical shock (step function acceleration). Specify the stiffness in lb/in for each bearing, the required percentage of critical damping, and the damping as lb-sec/in for each bearing.

   b. Determine the resulting critical speed.

   c. Determine the resulting bearing forces at 30,000 rpm.

   d. How much smaller than the rigid support forces are they? What is the difference in bearing life for the two cases?

Hint: Determine the bearing stiffness first, by looking at the transmissibility curves and then applying the constraint on static deflection from gravity and shock.

Answers: (these numbers can have some variance, depending on the allowed static deflection):

   a. 1000 lb/in, 3.3 percent damping, 0.416 lb-sec/in (each bearing)

   b. 1535 rpm

   c. 1.63 lb

   d. 1.63 lb/383 lb, bearing life is 13 million times longer with the soft bearings. (The life ratio is force ratio cubed for ball bearings; the life with rigid bearing supports is only a few minutes.)

3-7. Assuming circular centered orbits:

   a. Prove that positive direct stiffness coefficients $K_{XX} = K_{YY}$ produce a radial force directed inward and collinear with the rotor deflection vector, i.e., stiffening.

  b. Prove that positive direct damping coefficients $C_{XX} = C_{YY}$ produce a tangential force normal to the rotor deflection vector and opposing the whirl velocity.

  c. Prove that cross-coupled stiffness coefficients $K_{XY} = -K_{YX}$ produce a tangential force normal to the deflection vector and with direction dependent on the algebraic signs. Show the force direction for each sign case and note that its vector is collinear with the whirl velocity vector.

  d. Prove that cross-coupled damping coefficients $C_{XY} = -C_{YX}$ produce a radial force collinear with the deflection vector and with direction dependent on the algebraic signs, i.e., either stiffening or destiffening. Show the direction for each sign case.

  Hint: One way to do the above proofs is by writing vector or dot products of the rotor deflection vector with the force vector, or the velocity vector with the force vector.

3-8. Consider a short rigid rotor running on two identical squeeze film dampers with no mechanical centering springs. Thus, the rotor has no direct support stiffness, that is $K_{XX} = K_{YY} = 0$. The two combined squeeze film damper coefficients are $C_{XY} = -C_{YX}$ (with $C_{XY} > 0$), and $C_{XX} = C_{YY}$. The static unbalance (C.G. offset) = u. Write the differential equations of motion in X-Y coordinates and use the solution to show that the resonant speed (with $90°$ phase lag) is C_{XY}/m (rad/sec) and the synchronous amplification factor is C_{XY}/C_{XX} instead of $1/2\xi$.

  Hint: The algebraic equations are much shorter if you use the complex amplitude $R = X + iY$ and reduce the two equations to one. See Chapter 5 for a description of squeeze film dampers.
  Note: The purpose of this exercise is to show that the cross-coupled damping of a squeeze film damper acts like stiffness. Some authors have classified it as stiffness, but an oil film with no journal spin cannot produce pressure without journal velocity.

3-9. Derive the differential equations of motion in X-Y coordinates for an undamped Jeffcott rotor with cross-coupled stiffness $K_{XY} = -K_{YX}$ (to produce a follower force) at the disk. Use the solution for the homogeneous equations to derive the characteristic matrix (2 $\times$ 2) in terms of the eigenvalue. Derive the characteristic equation (polynomial) and solve it for the eigenvalues. Show that the system is unstable since the eigenvalue has a positive real part. Solve for the eigenvector and use it to describe the unstable motion.

3-10. Consider a Jeffcott rotor with a seal installed on the disk. The seal has a damping coefficient $C = 11.4$ lb $=$ sec/in constant at all rotor speeds. This is the only damping in the system. The seal also has cross-coupled stiffness $K_{XY} = -K_{YX}$ that increases linearly with speed. Its value is given by 0.57 lb/in per rpm. The modal mass of the rotor (disk + half the shaft) weighs 190 lb. The shaft stiffness at the disk location is 11,000 lb/in. It is desired to run the rotor at 3000 rpm. This exercise refers to the effect of the seal on rotordynamic stability.

a. Use any method to calculate the damped eigenvalue at 3000 rpm. Give the value and units of the whirl frequency. Give the value of the logarithmic decrement. Is the rotor stable or unstable?

b. Calculate the effective damping at 3000 rpm and use the result to evaluate the stability of the system.

c. Calculate the value of the whirl frequency ratio WFR for the subject rotor at 3000 rpm. What onset speed of instability does it predict?

Solution:

a. The equations of motion for the rotor/seal system are

$$m\ddot{X} + c\dot{X} + kX + K_{xy}Y = 0$$
$$m\ddot{Y} + c\dot{Y} + kY + K_{yx}X = 0$$

Let s represent the eigenvalues of the system

$$s = \lambda + i\omega_d$$

As given in Eqs. 3-30 and 3-31, the real and imaginary parts of the eigenvalues are found by

$$\lambda = \frac{-c}{2m} \pm \sqrt{\left(\frac{c}{2m}\right)^2 + \left(\omega_d^2 - \frac{k}{m}\right)}$$

$$\omega_d^2 = \frac{k}{2m} - \frac{c^2 \pm \sqrt{(c^2 - 4km)^2 + 16\kappa^2 m^2}}{8m^2}$$

where κ is the value of K_{XY} (and K_{YX}).
Substituting the given numerical values for a rotor speed of 3000 rpm yields $\lambda = 0.037$, $\omega_d = 149.51$ rad/sec. The existence of a positive real part (the damping exponent) indicates rotordynamic instability. It can be converted to the logarithmic decrement using $\delta = -2\pi\lambda/\omega_d = -0.0016$. The negative value shows instability. The eigenvalues can also

be found by substituting Ae^{st} into the differential equations, using Cramer's rule to generate the characteristic polynomial, and rooting it for the complex values of s.

b. The effective damping at 3000 rpm is $C_{\text{eff}} = C_{XX} - K_{XY}/\omega_d$. The value of K_{XY} at 3000 rpm is $0.57 \times 3000 = 1710$ lb/in. So $C_{\text{eff}} = 11.4 - 1710/149.51$, which gives $C_{\text{eff}} = -0.037$ lb $-$ sec/in. The value is negative so the system is unstable.

c. Converting 3000 rpm to ω (rad/sec) gives $\omega = (3000)(2\pi/60) = 314.159$ rad/sec. The whirl frequency ratio at 3000 rpm is WFR $= K_{XY}/C_{XX}\omega = 1710/(11.4)(314.159) = 0.4775$. Note that $\omega_d/\omega = 149.51/314.159 = 0.4759$, so the onset speed of instability is correctly predicted to be just under 3000 rpm. The whirl frequency ratio is more useful when K_{XY} is constant.

Note that the *system* stability would not be correctly predicted by either C_{eff} or WFR if there were other components in the system with damping or cross-coupling.

4

COMPUTER SIMULATIONS OF ROTORDYNAMICS

This chapter covers the use of a computer to analyze the dynamics of rotating machinery. The focus will be on how rotors are modeled and analyzed. In the particular field of turbomachinery, many rotordynamic analyses address only the lateral vibration of the shaft supported by its bearings. There are a number of different types of lateral analyses that can be performed with just the rotor and its bearings, depending on the particular machine and its intended use, such as showing if a rotor is rigid, semi-rigid, or flexible; locating critical speeds and their balance sensitivities; selecting and designing fluid film bearings for good rotordynamic behavior; predicting or troubleshooting subsynchronous instabilities; analysis to show compliance with API purchase specifications; and design audits.

For turbomachinery, torsional vibration can also be important and that topic is covered in Chapter 2.

DIFFERENT TYPES OF MODELS

Analysis with a computer starts with creation of a model. The natural starting point in rotordynamics is a model of the rotating assembly, or shaft. Figure 4-1 shows a multistage compressor rotor along with an example graphical rendition of a corresponding computer model. There is much more to a model than just a picture, however. Material properties and bearing stiffness and damping parameters are also required. The model needs to be appropriate for the analysis task at hand. The model depicted in Fig. 4-1 could be entirely satisfactory for computing the compressor's

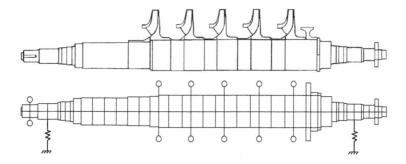

Figure 4-1 Example of a rotor and corresponding shaft model.

first lateral critical speed. On the other hand, this model would probably have much more detail than needed for a torsional or axial analysis, and not enough detail to accurately compute higher-order lateral modes. The model as shown has several dozen elements, allowing it to very closely follow the contour of the underlying shaft. The impellers are represented by simply adding their inertia properties directly to the shaft. This rotor would be driven by another machine, to which it is coupled at its left end. In the figure you can see a keyway at the left end of the shaft where a coupling hub would be mounted, but the coupling hub itself is not shown. The weight and rotational inertia of a coupling hub plus half of a spacer shaft (of moderate length) have been added (i.e., lumped) at this end. For a torsional or axial analysis this rotor could probably be adequately modeled with just a few elements (perhaps as few as one!), but the model would need to include the entire coupling and driving machine.

If an increased level of detail were desired for a lateral analysis, the next logical step would be to model the impellers so as to include contributions the hubs may provide to the bending stiffness of the shaft, particularly if they are mounted with interference fits which remain tight at all conditions. Figure 4-2 shows how the model might appear with additional elements added for the impeller hubs. These elements for the hubs are treated as extra layers superimposed on the elements of the underlying shaft. Internally, the computer program will merely sum the properties of coincident elements to produce the total properties for the station. In

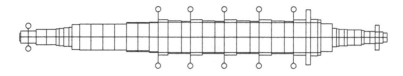

Figure 4-2 Rotor model with impeller stiffness contributions.

both Figs. 4-1 and 4-2 the impeller inertia is added directly to the shaft (in which case the newly added *impeller hub* elements in Fig. 4-2 would be assigned a zero value for density and modulus values for steel). The five impeller inertias are treated as rigid disks attached rigidly to the shaft at the five positions indicated.

An additional step to make the model more detailed would be to model each impeller as a rigid disk attached flexibly to the shaft [1]. This ordinarily would have little if any affect on the calculation of the first one or two modes of a shaft like this one, but could be significant for higher-order bending modes.

In addition to modeling the shaft, the bearings also need to be modeled. For nearly all machines the bearings are modeled as stiffness and damping elements that connect the shaft either to ground or to a model of the bearing pedestals or machine housing. The modeling of the bearings is generally an analysis done separately that provides stiffness and damping inputs to the overall rotordynamic system model. This may entail designing the bearings (selection of type, clearance, diameter, length, etc.). The influence of the bearing model on the final rotordynamic results is case dependent, and can range from very little to very great.

High-pressure seals are known to affect the rotordynamics of pumps, compressors, and turbines. An example is the balance piston seal near the right most impeller in the compressor of Fig. 4-1. Such seals are modeled, analyzed, and incorporated into a rotor system model in the same fashion as bearings. By computing their stiffness and damping properties, and placing these at the corresponding location on the rotor. Other fluid handling elements, such as pump impellers, compressor impellers, and turbine stages, can also be sources of significant stiffness and damping to be included in the rotor system model [2].

In Fig. 4-1 the bearings are depicted as connecting the shaft to ground. Figure 4-3 illustrates the modeling of bearing pedestals, as with some machines the bearings are housed in such pedestals separate from the main machine housing. If pedestals have sufficient compliance, this can affect the calculated critical speeds of the machine. The two pedestals shown are in turn connected to ground by stiffness and damping determined for the pedestal.

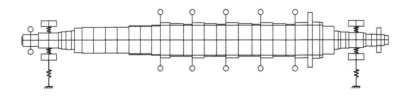

Figure 4-3 Rotor model with bearings connecting the shaft to bearing pedestals.

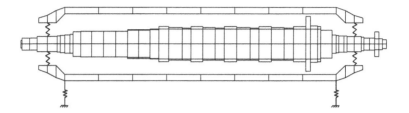

Figure 4-4 Rotor model with flexible housing model.

Figure 4-4 depicts how the model might appear if the machine housing were to be modeled as a separate entity. For many machines the housing is massive enough compared to the rotor such that it is essentially immobile and can be considered as ground, in which case the model of Fig. 4-1 is appropriate. If the housing is not rigid, or if it is rigid but not immobile, then it might affect the critical speeds of the rotor, or there might be critical speeds dominated by housing motion. Machine housings are typically complicated structures. The housing model in Fig. 4-4 is obviously a simplified version of the actual housing. Here it has been modeled like the shaft as a series of beam elements since many computer programs for rotordynamics are limited to a concentric assembly of beam elements (i.e., all elements share a common centerline). It is often adequate for the beam element housing model to be effectively rigid (for instance, with a high elastic modulus) and just match the total inertia properties of the actual housing, and to have the housing ground springs and dampers be set to appropriate values. This would then account for the housing's effect on the rotor's critical speeds, as well as allow for rigid housing critical speeds to appear in the results. Having a housing included in the model also then makes it possible to calculate housing vibration amplitudes in units of g's, ips, or mm/sec for direct comparison to field measured values (this could also be done for the pedestals in Fig. 4-3).

Situations in which the housing cannot be considered rigid can often be addressed by adjusting the properties of the simplified housing model to match some basic known properties of the actual housing—for instance, the first natural frequency of the housing or its stiffness at certain locations. These values may come from experience, tests conducted on the housing, or a three-dimensional finite-element structural dynamics model of the housing.

Let us now direct our attention to the basic mathematics of modeling and analyzing rotors. Decades of experience have confirmed that relatively simple beam element models are entirely adequate for analyzing a large majority of rotors. For lateral analysis, beam elements are used to model bending type flexure of the shaft. Some would refer to this as a 1D model because only an axial dimension is defined for each station (or node). Others might refer to this as a 2D model because the rotor can flex in two planes. Beam models are accurate as long as the basic assumptions inherent in beam flexure formulations are satisfied—that planar cross sections remain plane, and round sections remain round. There are, of course, situations where beam flexure will not accurately model what the shaft is doing. Common examples are trunnions and thin-walled cones. In these cases, plane sections do not remain plane. When plane sections do not remain plane, one can still use a beam element rotordynamic program if the actual behavior of the rotor is approximated some other way. Quite often the effect which is not adequately modeled with beam elements results in the rotor being less stiff. So by estimating the decrease in stiffness, and incorporating this decrease into a beam element model, the rotordynamic analysis can still be done with a beam element code. Otherwise, one can resort to a three-dimensional solid model. There are two main benefits of using a model constructed with 3D solid elements or axisymmetric elements: (1) it entirely eliminates the beam flexure assumptions, and (2) it enables analysis of more than just lateral vibration of the shaft, like turbine and impeller blade vibration or modes of a flexible disk. Commercially available general purpose finite element codes like NASTRAN and ANSYS provide this capability. However, there is a substantial cost of increased model complexity and time to create it, and increased computer calculation time.

Returning now to the compressor shaft beam element model of Fig. 4-1, Fig. 4-5 illustrates how 2D beam elements are defined. Beam finite elements in widespread use today have two or three nodes per finite element, with two nodes being most common since they make for easy modeling of typical industrial rotor geometries. Having four degrees of freedom (dof) at each node enables the simultaneous modeling of beam deflection in two planes, for example horizontal and vertical planes. Figure 4-6 shows a common style of defining the degrees of freedom at each node in the model. The basic mechanical properties of a beam finite element are

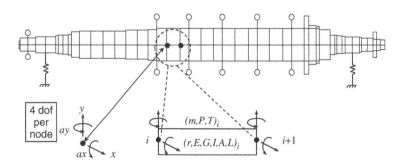

Figure 4-5 Two-node beam element used to model rotors.

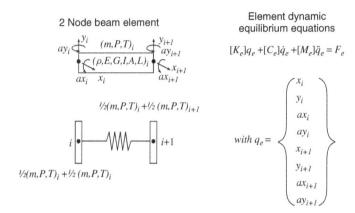

Figure 4-6 Definition of degrees of freedom in a 2-node lumped mass beam element.

used to construct a mathematical description of how it behaves. Through rigorous application of the formal theory of finite element formulations, this results in stiffness, damping and inertia matrices on the element level. Each node (or station) in the model has 4 dof, which may often be referred to as x, y, about-x, and about-y (i.e., right-handed rotations). The following notation is used:

L = element length

A = cross section area

I = cross section second moment of area

ρ = mass density

E = elastic modulus

G = shear modulus

m = beam mass

P = beam polar inertia (about centerline)

T = beam transverse inertia (about diameter through centroid)

x = lateral displacement along x axis of right-handed xyz system

y = lateral displacement along y axis of right-handed xyz system

ax = right-handed rotation about x axis

ay = right-handed rotation about y axis

Note that the two dof x and ay specify deflection of the rotor in the x plane. Deflection in the y plane is specified by dof y and ax.

A question often asked is, how many shaft modeling stations are needed for accurate results? There is no single answer to this question. It depends on the rotor, and also on the computer program used to the model the rotor. One general statement that can be made is that the more flexing taking place in the rotor, the more stations it will take to model it accurately. Another statement that ought to be true in general is that simple element formulations will require more elements than a more sophisticated element formulation (for example, a simple 2-node lumped mass beam element versus a 3-node isoparametric beam element with consistent mass). When relatively simple 2-node lumped mass elements are employed, at least about 8 shaft modeling stations should be used for every half sine wave in a mode shape. For example, if the first mode has the shape of the jump rope (i.e., one half sine wave), then about 8 or more stations should be sufficient for good accuracy in predicting the frequency of that mode. If the second mode looks like a full sine wave (i.e., two half sine waves), then about 16 or more stations should be sufficient. Shaft modeling programs that employ a more sophisticated element than a 2-node lumped mass element should be able to deliver comparable accuracy with fewer stations. Experience has shown that for many industrial rotors, creating a model that endeavors to reproduce the rotor's actual cross-section geometry will likely end up with more than enough stations to accurately predict the lowest two or three flexible modes, regardless of the program being used. Another rule of thumb often followed is to have the length to diameter ratio (L/D) for all shaft modeling elements not exceed 1. In most situations station L/D does not need to be less than about 0.5 strictly for the purpose of modeling accuracy. Employing a large number of very short elements could be done in order to follow a very intricate rotor cross section, but this probably will not significantly affect the accuracy of predicting the first few rotor modes.

Another question sometimes asked is, how many modes should be calculated? Some API purchase specifications require computing all critical speeds up to 125 percent of maximum continuous operating speed. Calculating critical speeds at speeds higher than this is generally not necessary because they will not respond significantly to a synchronous excitation like imbalance.

There are situations when modeling higher modes accurately is required. In this context, higher modes means modes whose natural frequencies are as much as 1.5 to 2 times (sometimes even 3 times) the maximum frequency of rotation. Examples are when doing design of actively controlled magnetic bearings [3], and when analyzing a rotor for shaft differential heating instabilities (aka Morton effect [4]).

BEARING AND SEAL MATRICES

In lateral rotordynamic models there are 4 dof per station (or node): two displacements x and y, and two rotations ax and ay. The bearing K matrix shown in Eq. 4-1 is specifically for a dof ordering scheme of (x, y, ax, ay), where ax and ay are right-handed rotations about the x and y coordinate axes, respectively. This is very nearly a universally accepted form for defining the dof and their order in the matrix, but be aware that there is no standard.

Sometimes the word "interconnection" might be used instead of the word "bearing." This is because bearings, seals, impellers, and all types of physical connections between modeling stations, as well as from modeling stations to ground, are specified with the same type of stiffness matrix as Eq. 4-1. Along with a stiffness matrix there is a corresponding damping matrix, which takes the same form as the stiffness matrix, but contains damping constants. Some rotordynamic codes may also allow input of an inertia matrix for a bearing or other type of interconnection. An inertia matrix would be used to input values which are sometimes called *effective mass*. High-pressure annular pump seals, for example, may have significant effective mass values, which should be input to a rotordynamic model:

$$[K_{\text{brg}}] = \begin{bmatrix} K_{xx} & K_{xy} & K_{xax} & K_{xay} \\ K_{yx} & K_{yy} & K_{yax} & K_{yay} \\ K_{axx} & K_{axy} & K_{axax} & K_{axay} \\ K_{ayx} & K_{ayy} & K_{ayax} & K_{ayay} \end{bmatrix} \qquad (4\text{-}1)$$

In a vast majority of cases, the only values in this matrix that will not be zero are the four in the upper left corner. Indeed, some rotordynamic computer codes only allow input of these four. In the simplest of bearings only K_{xx} and K_{yy} will be nonzero, and often they are equal. Ball bearings often are modeled this way. Having $K_{xx} \neq K_{yy}$ means the horizontal and vertical stiffnesses are different, a common case for fluid film bearings.

For fluid film elements like oil film bearings and annular pump seals, the four "primary" stiffnesses $(K_{xx}, K_{xy}, K_{yx}, K_{yy})$ are all distinct nonzero values. This is how fluid-generated cross-coupling effects enter into a rotordynamic model. The bearing stiffness matrix can attach a shaft model station to ground, or to a station in a housing or bearing pedestal. When computing response amplitudes due to externally applied forcing functions (e.g., imbalance), bearing connections to ground are necessary to prevent the system stiffness matrix from being singular (i.e., unsupported in one or more coordinate axes).

The following stiffness and damping matrices are for a two-lobe sleeve bearing with oil supply grooves on the sides:

$$K = \begin{bmatrix} 2.3 & 1.5 \\ -5.5 & 7.2 \end{bmatrix} \times 10^5 \frac{\text{lbf}}{\text{in}} \qquad C = \begin{bmatrix} 668 & -296 \\ -293 & 2288 \end{bmatrix} \frac{\text{lbf}-\text{s}}{\text{in}} \qquad (4\text{-}2)$$

The stiffness and damping properties for any type of fluid film bearing are highly dependent on shaft speed and static load. For these example values, the direction of rotation is counterclockwise and the static load on the bearing is in the negative Y axis. If either of these conditions is changed, the matrices will change. For example, in the *Journal Bearing Data Book* by Someya [5], bearing matrices are given with the static load applied in positive X axis and rotation is counterclockwise. Regardless of whether the above matrices are used or Someya type matrices are used, eigenvalues will be identical. The difference between using the above and Someya will be the mode shapes. The orientation of whirl orbits will simply be 90 degrees different. The whirl directions will be the same because both conventions assumed counterclockwise shaft rotation.

The input of bearing matrices should always use the same convention for direction of rotation as that used to define rotor gyroscopics, and this nearly always is counterclockwise (but again, there is no accepted standard). The most important point of this discussion is that the two conventions should not be mixed in the same model. Conversion from the Someya convention to the convention of Eq. 4-2 is done by applying a coordinate transformation matrix. This amounts to swapping K_{xx} and K_{yy}, and swapping K_{xy} and K_{yx} and changing the signs on both K_{xy} and K_{yx}. These same changes would also be used to convert going the other way.

TORSIONAL AND AXIAL MODELS

The chapter up to this point has focused on lateral dynamics of the rotor system, whose models are constructed with two-dimensional beam elements having 4 dof per node. For torsional analysis, the twisting action of the shaft is adequately modeled with 1D beam elements having 1 dof per node. The torsional stiffness of a shaft beam element can be simply the familiar JG/L (assuming J is constant). The inertia of a beam element can be divided equally between the two nodes at the ends of the beam element (assuming uniform cross section). Torsional analysis of turbomachinery does often require dealing with rotors running at different speeds through use of a gearbox, and sometimes accounting for finite stiffness of gear teeth. See Chapter 2 for more information on torsional vibration analysis. Once a lateral analysis model has been made for an individual rotor, that model can often be used directly in constructing a torsional model.

Axial analysis of rotor systems is similar to torsional analysis. One-dimensional 2-node beam elements having 1 dof per node will generally suffice. The beam element stiffness would now simply be the familiar AE/L, and the beam mass can be split evenly between the two nodes of the beam. Unlike torsional analysis, the axial stiffness of a thrust bearing(s) will generally be required. Also unlike torsional analysis, the machine housing may form an important part of the axial model.

Lateral, torsional, and axial models, and the analyses done with them, are generally done independently and are thereby assumed to have no affect on each other. This would be termed *uncoupled* analysis. In actual machines there will always be some degree of coupling, or cross-talk, between the different modes of vibration. A motor direct coupled to a pump should exhibit negligible lateral–torsional coupling. However, a motor–gearbox–pump rotor system most definitely can exhibit lateral–torsional coupling. The presence of torsional vibration in a meshing pair of gears will produce some measure of lateral motion of the gears by way of the oscillating gear mesh force. The reverse is also true. Coupled analysis will not be discussed in this book. See a reference such as Schwibinger et al. [6] for information on this type of modeling and analysis.

DIFFERENT TYPES OF ANALYSES

Once a model has been constructed, there are several different basic types of analyses that can be done. Rotordynamic analyses can be broadly categorized in the following ways:

- Linear or nonlinear
- Static or dynamic
- If dynamic, then steady-state or transient

This chapter focuses primarily on linear, steady-state dynamic analysis. Nearly all, perhaps 99%, of lateral rotordynamic analyses are of this type. Within this classification, there are two basic types of analysis: *eigenanalysis* and *forced-response analysis*. Both of these analysis types start with the same set of system dynamical equations, which result from the same system model of the types discussed earlier in this chapter. The individual element matrices described in Fig. 4-6 are assembled to produce global system matrices, which in turn are used to form the following system level dynamic equilibrium equation. The column vector q is simply a list of all the dof in the model, which, for lateral models constructed with 2-node beam elements, will amount to 4 dof for each node. The total number of degrees of freedom, NDOF, is then 4 times the number of nodes. Each of the three system matrices in Eq. 4-3 will be a square matrix with NDOF rows and columns:

$$[K]q + [C]\dot{q} + [M]\ddot{q} = F(t) \tag{4-3}$$

This equation represents a set of NDOF second-order differential equations with constant coefficients. We will solve this equation for two different important cases: first for eigenvalues and eigenvectors (natural frequencies and mode shapes) and second for response amplitudes (and phases) due to external forces.

EIGENANALYSIS

In eigenanalysis all externally applied forces, static and/or dynamic, which are contained in the $F(t)$ term of Eq. 4-3, are identically zero. This results in the homogenous form of the dynamic equilibrium equation:

$$[K]q + [C]\dot{q} + [M]\ddot{q} = 0 \tag{4-4}$$

This equation leads to the standard eigenproblem, where the solutions we seek are the unforced free vibration modes of the system, also called natural modes or normal modes (unforced because $F(t) = 0$). As discussed in Chapter 3, normal modes are characterized by three things: a natural frequency, a damping ratio (or log dec), and a mode shape.

To determine the solutions to Eq. 4-4 we assume a harmonic motion. This means every point (every dof) on the rotor vibrates in a pure sine wave at exactly the same frequency, but each with its own amplitude and phase. In a vibrating undamped system the sine wave will persist indefinitely with constant amplitude. In a damped system the sine wave will decay exponentially because there is no external force to keep it going (i.e., $F(t) = 0$). In an unstable system the sine wave will exponentially grow unbounded. The general expression for this harmonic motion, which solves Eq. 4-4, has the same form for every dof in the model:

$$x(t) = Xe^{\lambda t} \cos(\omega_d t - \phi) \tag{4-5}$$

where

$x(t) =$ the displacement of the dof as a function of time (inches or radians)

$X =$ the constant real valued amplitude (inches or radians)

$\phi =$ the phase angle of the motion (radians, denoted "lagging" phase)

$\omega_d =$ the frequency of the harmonic motion (radians/sec), same for all dof

$\lambda =$ reciprocal of the time constant of decay (1/sec), same for all dof

Equation 4-5 is plotted in Fig. 4-6 and contains only real valued constants, and produces a real value for the motion of the dof, as it should. In the course of performing a computer solution it is advantageous, although not necessary, to use complex variables. The following expressions show how the real solution can be expressed using complex variables:

$$x(t) = e^{\lambda t} \frac{\hat{X} e^{i\omega_d t} + (\hat{X} e^{i\omega_d t})^*}{2} \tag{4-6}$$

$$\hat{X} = x_r + ix_i \qquad * = \text{conjugate}$$

$$x(t) = e^{\lambda t} \frac{(x_r + ix_i)(\cos \omega_d t + i \sin \omega_d t) + (x_r - ix_i)(\cos \omega_d t - i \sin \omega_d t)}{2}$$

$$x(t) = e^{\lambda t}(x_r \cos \omega_d t - x_i \sin \omega_d t) = e^{\lambda t} X \cos(\omega_d t - \phi)$$

$$X = |\hat{X}| = \sqrt{x_r^2 + x_i^2} \qquad \tan \phi = \frac{-x_i}{x_r}$$

An alternative way to make use of complex variables is as follows (where Re means take the real part):

$$x(t) = \text{Re}(\hat{X} e^{st}) \qquad s = \lambda + i\omega_d$$

$$\hat{X} = x_c - ix_s$$

$$x(t) = e^{\lambda t}(x_c \cos \omega t + x_s \sin \omega t) \tag{4-7}$$

$$x(t) = e^{\lambda t} X \cos(\omega t - \phi)$$

$$X = |\hat{X}| = \sqrt{x_c^2 + x_s^2} \qquad \tan \phi = \frac{x_s}{x_c}$$

The two approaches defined by Eqs. 4-6 and 4-7 are seen to be equivalent. Both will convert a complex solution value to the same real valued amplitude and phase.

With every dof having a solution of this form, the solution for all the dof in the model can be written in vector form as (dropping the Re( ) notation for brevity if utilizing the notation of Eq. 4-7):

$$\{q(t)\} = \{\hat{q}\}e^{st}$$

$$\{\dot{q}(t)\} = s\{\hat{q}\}e^{st} \tag{4-8}$$

$$\{\ddot{q}(t)\} = s^2\{\hat{q}\}e^{st}$$

This solution form is substituted into Eq. 4-4, resulting in

$$([M]s^2 + [C]s + [K])\{\hat{q}\}e^{st} = 0 \tag{4-9}$$

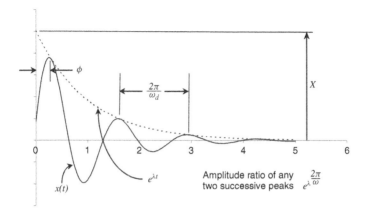

Figure 4-7 Graphical form of general harmonic solution for any single dof.

This still represents NDOF second-order differential equations. These will be replaced with an equivalent set of $2 * \text{NDOF}$ first-order differential equations as follows:

$$\begin{bmatrix} M & 0 \\ 0 & 1 \end{bmatrix} \begin{Bmatrix} \ddot{q} \\ \dot{q} \end{Bmatrix} = \begin{bmatrix} -C & -K \\ 1 & 0 \end{bmatrix} \begin{Bmatrix} \dot{q} \\ q \end{Bmatrix} \qquad (4\text{-}10)$$

Equation 4-10 is ready for computer solution of the eigenanalysis problem. General-purpose finite-element programs like NASTRAN and ANSYS solve this equation using matrix solution algorithms suitable for arbitrarily complex system models. For rotordynamics, a very specialized and highly optimized method for solving Eq. 4-10 numerically for its eigenvalues and eigenvectors is the *polynomial transfer matrix method* [7]. This is not a different formulation of the problem, but a numerical method for computing the solution. If the rotor system model satisfies certain criteria, this method is the fastest known way to compute the eigensolution.

Regardless of which numerical method is used, solving the matrix eigenproblem defined by Eq. 4-10 yields the three things mentioned earlier: the damped natural frequency ω_d, the damping constant λ, and the mode shape q. These are found by finding matched pairs of s and $\hat{q}$ that solve Eq. 4-10. These can be determined for each "mode" of the system. The number of modes that the model has is equal to NDOF. The number of eigenvalues that can be computed is equal to the total order of the system, which is $2 * \text{NDOF}$ (i.e., NDOF second-order differential equations in Eq. 4-4, or the size of each matrix in Eq. 4-10). Each mode then has associated with it a *pair* of eigenvalues. For any mode that is not critically damped (critical damping is discussed below), its two eigenvalues are complex conjugates of each other. For modes that are critically damped, each of its two eigenvalues is a distinct real value. In system models having many dof, our interest will be limited to frequencies that are usually no higher than about two or three times the maximum rotational frequency of the rotor.

Before finishing this introduction to eigenanalysis, we'll summarize a few things and present a few other useful relationships. Recall that the eigenvalue, s, is a complex number:

$$s = \lambda \pm i\omega_d \quad i = \sqrt{-1}$$

where

$\lambda =$ damping exponent, real part of s (rads/sec)
$\omega_d =$ damped natural frequency, imaginary part of s (rads/sec)

- $\lambda < 0$ means the mode corresponding to this eigenvalue is stable
- $\lambda > 0$ means the mode is unstable
- $\omega_d = 0$ means the mode is critically damped

The damping of a mode can be expressed in several other ways:

$$\zeta = \frac{-\lambda}{\sqrt{\lambda^2 + \omega_d^2}} \quad \text{Damping ratio (note minus sign)}$$

$$\delta = \frac{-2\pi\lambda}{\omega_d} = \frac{2\pi\zeta}{\sqrt{1 - \zeta^2}} \approx 2\pi\zeta \quad \text{Logarthmic decrement (note minus sign)}$$

$$\text{A.F.} = Q = \frac{\omega_{cr}}{\Delta\omega_{0.707}} = \frac{1}{2\zeta\sqrt{1 - \zeta^2}} \quad \text{Amplification factor from half power points}$$

$$\text{A.F.} = \frac{\pi}{360} N_c \frac{\Delta\theta}{\Delta N} \quad \text{A.F. from slope of phase plot at a critical speed}$$

$$\omega_{cr} = \frac{\omega_d}{\sqrt{1 - \zeta^2}} \frac{1}{\sqrt{1 - 2\zeta^2}} \quad \text{Location of synchronous critical speed}$$

$$\# \text{ cycles} = \frac{1}{\delta} \quad \text{Number of cycles in 1 time constant}$$

Note that a natural frequency and its associated critical speed frequency are not equal whenever there is damping present in the model. A natural frequency is an aspect of force-free vibration. A critical speed frequency is an aspect of forced response vibration (i.e., resonance), discussed in the next section. The last relationship above can be used with ring down time waveforms by counting the number of cycles for the waveform amplitude to decay by about 2/3 (or more precisely, $1 - e^{-1} = 0.632$).

The expression given above for damping ratio restricts its value from -1 to $+1$. A common textbook definition of damping ratio for single dof systems is the ratio of the viscous damping constant to its value at the threshold of critical damping. The textbook definition obviously permits values greater than 1. For a single dof system, the two definitions are equivalent when damping is equal to or less than critical. The textbook definition is really applicable only to purely single dof systems. For example, multi-dof systems will generally have multiple damping constants and these can have different values, so the concept of a damping constant corresponding to critical damping does not really apply. As any mode becomes more heavily damped, its frequency will drop. When a mode becomes critically damped, by definition the frequency has dropped all the way to zero. This is because critically damped modes do not oscillate, and thus their natural frequency is therefore identically zero.

The relationships presented above are exact only for *single* dof systems. For multi-dof systems these relationships provide approximations. Single dof concepts and formulas are often used for individual modes of multimode systems, particularly when the level of damping is light ($\zeta < 0.1$), and the modes are not too close to one another in frequency (i.e., within about 5 or 10 percent of each other).

LINEAR FORCED RESPONSE (LFR)

In linear forced response (LFR) analysis the externally applied forces term $F(t)$ of Eq. 4-3 will consist of harmonic forces. The harmonic forces are applied to any or all of the dof in the model simultaneously. Each force can have a distinct magnitude and phase angle, but they must all have exactly the same frequency:

$$F(t) = f \cos(\omega t + \beta) \tag{4-11}$$

where

f = magnitude of the force (lbf or in-lbf)
ω = frequency of the applied force (rad/sec)
β = phase angle of the force (radians, denoted *leading* phase)

The term "linear" is prefixed to indicate that the rotor system model *must* be linear, which means the stiffness, damping, and inertia matrices in the dynamic equilibrium equation (4-3) are each constant. The solution to Eq. 4-3 due to these applied forces will again be a harmonic motion of the same form as for the eigenanalysis. Now, however, the frequency is known beforehand, and we are specifically seeking a steady-state solution. So the exponential decay factor $e^{\lambda t}$ is not needed because it has to be identically 1 (i.e., $\lambda = 0$):

$$x(t) = X \cos(\omega t - \alpha) \tag{4-12}$$

where

$x(t) =$ the displacement of the dof as a function of time (inches or radians)

$X =$ the constant real valued amplitude (inches or radians)

$\alpha =$ the phase angle of the motion (radians, denoted "lagging" phase)

$\omega =$ the frequency of the harmonic motion (rad/sec), same for all dof

Note that ω is known, and $x(t)$, X, and α are unknown.

The exact same approach of utilizing complex variables can be used to solve this problem numerically on a computer. Equations 4-7 and 4-8 still apply, except that now s is a known value. The dynamic equilibrium equation to be solved is then

$$\left([M]s^2 + [C]s + [K]\right)\{\hat{q}\}e^{st} = \hat{F}e^{st} \tag{4-13}$$

$$\{\hat{q}\} = \left([M]s^2 + [C]s + [K]\right)^{-1}\hat{F} \tag{4-14}$$

With $s = i\omega$ being a given value, all values on the right-hand side of Eq. 4-14 are known. The problem is then NDOF linear complex equations to be solved for NDOF complex response amplitudes. The complex response amplitudes are then equated to real valued amplitudes and phase angles via Eq. 4-6. The numerical solution can be computed by any of a large number of available linear equation solvers [8]. For rotordynamics analysis the fastest known method of computing the linear response solution is again the polynomial transfer matrix method [9].

TRANSIENT RESPONSE

There are many situations in which the desired response of the rotor system cannot be calculated using the LFR analysis method described above. Two prime examples are unsteady forcing functions and nonlinear system models. The former will arise whenever the forcing function is not purely harmonic (e.g., a blade loss, rapid speed ramp, aircraft maneuvers, earthquakes). The later will arise when any aspect of the rotor system model is not linear (e.g., seal rubs, ball bearing deadband, uncentralized squeeze film dampers). The most common approach to solving such problems is a transient response analysis, formally referred to as a *single-point boundary value* problem. In the case of rotordynamics the single point will be the starting point, making this an *initial value* problem. Starting with the rotor system in a given initial state at a given time (usually $t = 0$), arbitrarily defined forcing functions can be applied, and system response is computed forward in time in a stepwise fashion at suitably small incremental steps in time.

This is an extremely powerful method of analysis, and it is the most practical approach to many analysis problems. But it is more difficult and time-consuming to use than either eigenanalysis or LFR analysis. It can be time-consuming because sometimes (particularly with nonlinear models) many separate response cases may need to be analyzed. Extra difficulties come about because results can be significantly affected by additional calculation parameters that must be carefully specified, which are not required for eigenanalysis or LFR analysis.

Torsional vibration is an area in which transient analysis can be used very effectively. During start-up of turbomachinery driven by synchronous motors, motor torque pulsates with a variable frequency, which can have an extremely large amplitude. A transient analysis is used to compute shaft stresses during start-up [1]. If stresses exceed the endurance limit, then the number of allowable machine starts is limited. Another transient torsional condition sometimes analyzed is of electrical faults in generators. This is because the transient electromagnetic torques produced during the fault can be large enough to lead to equipment failures.

Additional information about transient response is presented in the books by Childs [2], Lalanne and Ferraris [10], and Genta [11].

SHAFT MODELING RECOMMENDATIONS

When creating shaft models of compressors like the one shown in Fig. 4-1, or of pumps, turbines, generators, and other types of machines, common sense is often the best guideline. This section presents some suggestions aimed at helping you create accurate shaft models.

How Many Elements

One of the most common questions about shaft modeling is, how many elements or stations are required to get good results? As was discussed earlier in this chapter no single answer is going to be correct for every machine, but an often cited value is that the number of stations should be at least 4 times the number of modes desired. But this depends on the machine, the modes, and the computer program being used. Figure 4-8 shows a uniform steel beam 20 in. long, 1 in. in diameter, and with an undamped bearing at each the end having $K = 1000$ lb/in. The mode shapes of the first four modes are also shown. The first two modes are fairly rigid. Modes like these don't require many elements for good accuracy. The second two modes look like a half sine wave and a full sine wave. Figure 4-9 shows that accuracy depends not only on the number of elements, but also on the formulation of the elements. Notice in Fig. 4-9 that the rigid modes have very small percentage errors. This is because there is little to no flexure of the elements. The flexible modes, however, illustrate the effect of the number of elements and also the difference between the four different element types.

For our uniform beam test rotor, the four-element case corresponds to an L/D of 5. For this case the first mode, which is a rigid shaft mode, has frequency errors less than 1 percent. Even the second mode, which is also a rigid shaft mode, has errors less than 1 percent for all but the uncorrected 2-node element. For the first flexible mode, however, four elements are not enough and the errors exceed 10 percent. Based on the errors computed for this example, a guideline of "use elements with $L/D < 1$" will result in acceptable accuracy for all cases considered. Another reason to maintain an element $L/D < 1$ is that when using a transfer matrix-based calculation procedure, the calculations can become numerically ill-conditioned if the L/D of very many elements is much

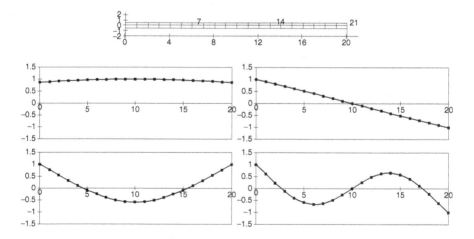

Figure 4-8   Model of uniform beam used for mode accuracy study, and mode shapes of the first four normal modes.

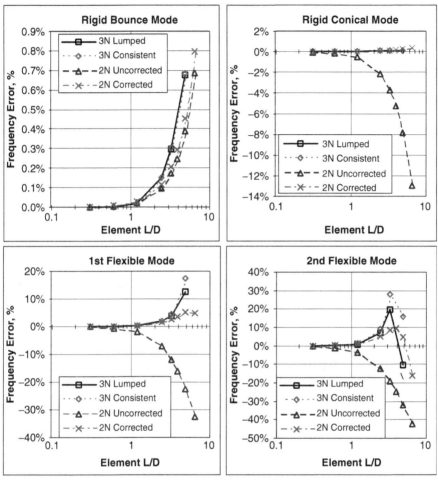

The 4 element types compared are:

    2 Node lumped mass element without inertia correction term [2]

    2 Node lumped mass element with inertia correction term [19]

    3 Node consistent mass element [20]

    3 Node lumped mass element with inertia correction (unpublished)

Figure 4-9   Mode accuracy versus number of elements for four different element definitions.

longer than one. This is an aspect specific to the algorithm, and not the rotor system model.

45-Degree Rule

A common situation encountered in rotor modeling is a large abrupt change in diameter. The bending stiffness of the shaft is not able to change

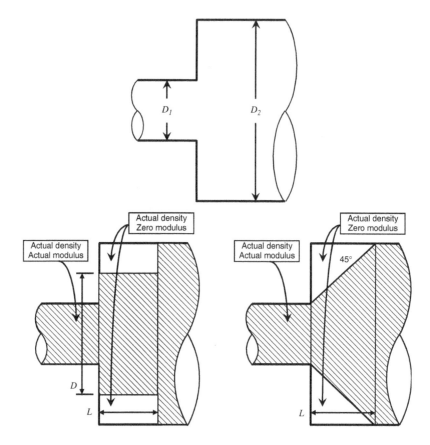

Figure 4-10 Two methods of the 45-degree rule for modeling abrupt diameter changes.

quite as sharply as the diameter. A step change in diameter of one-third or more is considered abrupt. The material outside the 45° line shown in Fig. 4-10 will not contribute significantly to the bending stiffness of the section. Two ways to model this section are shown, one using cylindrical beams and one using conical beams. Both approaches model the transition section with a layered pair of beams (inner and outer layers). The inner beam has the actual density and modulus of the shaft material. The outer beam has the actual density, but zero modulus so as to not contribute any stiffness.

Interference Fits

Many rotor assembly parts, such as impellers, spacers, and sleeves, are thermally shrunk or pressed on with sufficient interference to remain tight under all conditions. When this is the case, the part should be considered to contribute fully to the bending stiffness of the underlying shaft. If the parts have a loose fit on the shaft, the density should be retained but the modulus should be set to zero. The compressor impellers in Fig. 4-2 have tight interference fits. Some of the hub portion of each impeller has been included in the bending stiffness model of the rotor (with the extra layer of elements shown in the figure). The spacers between the impellers were also included since they, too, are tight.

Laminations

Lamination stacks are present in many generator rotors and rotating components of active magnetic bearings. Even though these are often mounted on the shaft with a tight interference fit, the stack usually contributes little stiffness to the shaft in a bending sense. If no stiffness is expected from the stack, then beam elements for the stack would be given its actual density value and a modulus value of zero. If some stiffness contribution is known to be present (perhaps indicated by rap test frequency measurements), then a modulus value that is an appropriately reduced percentage of the lamination material modulus could be input to the model.

Laminations can create difficulties when using 3D models in programs like NASTRAN or ANSYS. Even though the material properties of elements comprising the laminations can be given their own density and modulus values, the modulus cannot be set too small or else the laminations behave like jelly.

In some rotor designs it may be desired to increase the stiffness contribution of a lamination stack. This can be achieved to a modest degree by applying an axial compressive force to the stack and locking this compression into the rotor assembly. Figure 4-11 shows results of an axial

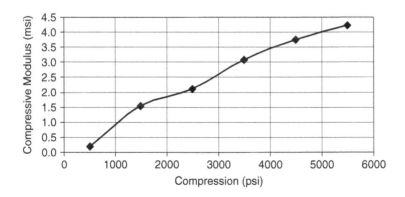

Figure 4-11   Sample test results for lamination stack stiffness as a function of axial compression [12].

compression test performed on a stack of 14-mil steel laminations [12]. The data illustrate that for the tested range of compressive stress, the effective modulus of the stack increases in a fairly linear sense. Even though the compressive force is substantial and would be difficult to build into an assembled rotor structure, the effective modulus is well below that of the parent material.

Trunnions

Figure 4-12 illustrates a rotor with trunnion sections that cannot be modeled as a beam in bending. To address this with a modeling program limited to beam elements requires determining the net rotational stiffness at the hub of the trunnion plate, and entering this directly into the rotor model. Figure 4-13 presents formulas for estimating the rotational stiffness of a trunnion [13, 14]. The reference to a pinned and clamped boundary condition refers to where the plate joins with the large diameter cylindrical section. The radial stiffness of the trunnion would be considered either very high or defined as a pin joint.

Impeller Inertias via CAD Software

In lateral rotordynamic analysis the gyroscopic influence of impellers and turbine stages can be significant. To properly account for this, it is necessary to have accurate values for their polar and transverse inertia, in addition to weight. In many cases accurate calculated values can be provided from a 3D CAD model of the impeller by using the CAD software to compute the mass properties of the part. At other times only a drawing may be available in either electronic or hardcopy form. In these situations

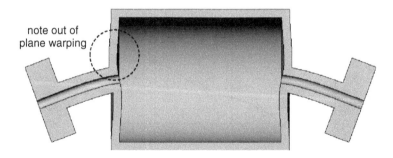

Figure 4-12 Example of plate like deformation in a trunnion.

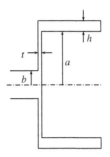

$$K_{\text{pinned}} = \frac{\pi E t^3}{3(1 - \mu^2)} \left(\frac{2(\beta^2 - 1) + \mu(\beta^2 - 1)^2}{(3 + \mu) + (1 - \mu)\beta^4} - \ln(\beta) \right)^{-1}$$

$$K_{\text{clamped}} = \frac{\pi E t^3}{3(1 - \mu^2)} \left(\frac{(\beta^2 - 1)}{(\beta^2 + 1)} - \ln(\beta) \right)^{-1}$$

$$\beta = \frac{b - \frac{1}{2}t}{a + \frac{1}{2}t} \quad \text{or if} \ \ h < t \ \ \beta = \frac{b - \frac{1}{2}t}{a + \frac{1}{2}h}$$

- Plate bending predominates if $(a - b) \gg t$ and $a \gg b$
- Use K_{pinned} to be conservative
- Otherwise use K_{clamped} if $h \geq t/2$

$$E = \text{elastic modulus,}$$
$$\mu = \text{Poisson's ratio}$$

Figure 4-13   Calculation of trunnion rotational stiffness [13, 14].

CAD software can still be used to get values of adequate accuracy by the following procedure (refer to Figs. 4-14 and 4-15):

1. If not already available, use a scanner or digital camera to get an electronic image of the part. The image should show a cross section of the part and its centerline. At least one reference dimension must be known, for example, its outer diameter.
2. Paste the image into CAD software, which can calculate mass properties for 3D parts, and scale the size of the image to match the known size of the part.
3. Using your computer's mouse, make a closed curve by tracing the outline of the part with the CAD software. This may need to be done separately for multiple portions of the part, for example, the hub and shroud. Also create a line over the centerline.

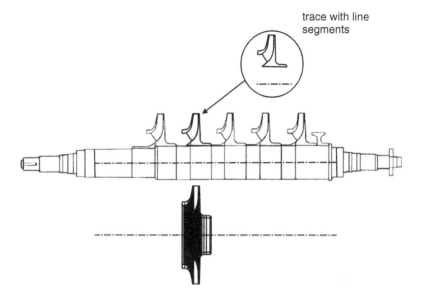

Figure 4-14 Determining impeller mass properties from sketch or drawing.

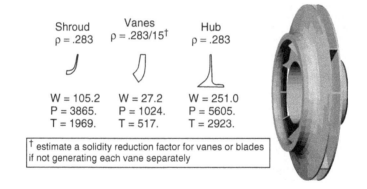

Figure 4-15 Dealing with multiple portions in a mass property calculation.

4. Direct the CAD software to create solids of revolution from each closed curve, about the centerline.

5. Finally, use the CAD software to compute the mass properties of the part(s).

Stations for Added Weights

The inertia properties of compressor and pump impellers are most often added to the rotor shaft model as added concentrated weights and inertias. This is the method employed to model the compressor shown in Fig 4-2. When constructing the model of the underlying shaft, be sure to provide stations at or close to the center of gravity of each impeller. These will be the locations where the concentrated masses will be placed.

It is also possible to model impellers in a more direct fashion that eliminates the need to add concentrated masses. Another approach to modeling impellers is illustrated in an integral turbine generator depicted in Fig. 4-16. Here the stiffness and mass contributions of various portions of the two impellers (these particular impellers have no shrouds) are dictated by the material properties specified for the elements as shown in the figure. Here both the stiffness and inertia of each impeller is being provided by the element definitions, and no concentrated weights are necessary.

Rap Test Verification of Models

When a rotor like the one in Fig. 4-16 is designed to operate above or close to any critical speeds that exhibit significant shaft flexing (i.e., flexible rotor critical speeds), then the accuracy of the shaft model could be verified by comparing predicted and measured free–free natural frequencies via a rap test (for examples of this see [1] or [15]).

Stations for Bearings and Seals

The stiffness and damping properties of items like bearings and seals should be located at their midplanes. So stations should be provided in the shaft model at these axial locations.

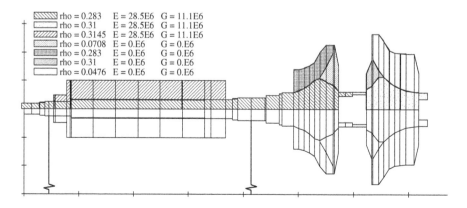

Figure 4-16 Example of another method of modeling impellers.

Flexible Couplings

Many turbomachines are driven by flexible couplings. Disk pack and diaphragm style couplings are commonly used because they do a good job of isolating two rotors from each other, while also accommodating modest amounts of misalignment between the centerlines of the two machines during operation. Ideally, the coupling will have only a minor influence on the critical speeds of a machine, and the critical speeds of the coupling itself will be very much higher than maximum operating speed. Coupling manufacturers often provide the total weight and total polar inertia (also called WR2) of the complete coupling assembly including the two coupling hubs. Sometimes the weight and inertia of each half of the coupling might be provided, especially if the two coupling hubs are of different styles (e.g., one standard and one reduced moment). Flexible couplings often have half-weights that are in the range of 1 to 4 percent of the total shaft weight of the driven machine. However, in high-power applications, coupling weights exceeding 10 percent do occur.

Figure 4-17 illustrates how a rotordynamic model of a flexible disk coupling may appear for use with the compressor of Fig. 4-2. Bear in mind that for large machines like turbine–generator sets, the coupling could be over 20 inches in diameter with a spacer over 10 ft long. There are four levels of complexity one may utilize in adding this coupling to the compressor model:

1. Place a concentrated mass at the end of the compressor shaft having weight, polar, and transverse inertia corresponding to one-half of

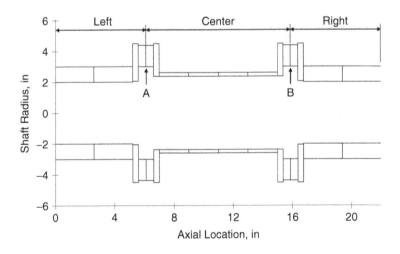

Figure 4-17 Sample model of a flexible disk coupling.

the coupling. This would be slightly conservative if the center of gravity of the coupling half were actually inboard of the end of the shaft.

2. Determine the actual location of the center of mass of the coupling half, and place the half-coupling concentrated mass properties on the compressor at that location. This would also require providing a station in the shaft model at that axial location.

3. Actually add the coupling model shown in Fig. 4-17 to the compressor model. For example, the two beam elements in hub B would go on top of a corresponding pair of elements of the compressor shaft, and hub A would be immobilized with either a rigid constraint or stiff springs to ground (see Fig. 4-18 discussed below).

4. Same as level 3, except hub A would go onto a shaft model of the machine to which the compressor is coupled. The entire model then contains two machine rotors joined by the flexible coupling.

It is often adequate to place the concentrated weight and inertia values at the very end of the shaft. This is method 1 listed above. Transverse inertia values are usually not provided in coupling catalogs, in which case it can be approximated as one half of the polar inertia value.

For cases where the coupling is heavy enough to have considerable influence on critical speeds, the concentrated coupling half-weight should be placed at its actual center of gravity, which is often slightly inboard of the end of the shaft. This is method 2 listed above. This is a situation in which is it often advantageous to specify a *reduced moment* style coupling hub.

If the center of gravity of the coupling half-weight is beyond the physical end of the shaft, then extending the shaft model to include the coupling may be warranted. This is method 3 listed above. Figure 4-18 illustrates the appearance of the compressor shaft model with the addition of the coupling model. A concentrated mass for the coupling half-weight is no longer needed because the beam elements of the coupling provide this. The connection of the leftmost coupling hub to ground is done as a rigid constraint to completely fix the position of that hub.

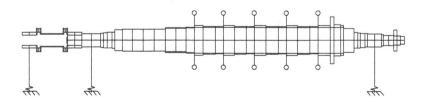

Figure 4-18   Sample compressor model with addition of flexible coupling.

Coupling spacer shafts are generally hollow and their length is kept short to avoid having their own critical speed problems, and so are generally just long enough to accommodate typical amounts of misalignment. Nonetheless, some applications require extra long spacer shafts to bridge the distance between machines. In this case the coupling spacer shaft may be relatively heavy, or it may be flexible enough in a lateral sense so that it does not behave a simple rigid shaft. This is method 4 listed above. This enables accurately accounting for the effect of the coupling on the critical speeds of each of the coupled machines. It also enables calculation of the critical speeds of the coupling itself in the final installation.

For a flexible disk coupling like that in Fig. 4-17, the coupling maker should be requested to provide lateral and angular spring rates for each flexible element. These spring rate values should be inserted into the model to form the connection between the two elements that meet at point A, and likewise at point B, in Fig. 4-17. This means the coupling model that looks like a single-shaft entity in Fig. 4-17 is actually three shaft entities (left, center, and right members joined by discrete springs). Typically, the lateral spring rate will be relatively high, and the angular spring rate will be relatively small (true for both disk pack and diaphragm couplings). Such a combination makes each flexible element behave much like a pin joint. So when spring rate values are not available from the coupling maker, the connections at A and B could be modeled as pin joints, and this would be a conservative modeling assumption from the standpoint of calculating critical speeds which are above maximum operating speed.

EXAMPLE SIMULATIONS

The remainder of this chapter presents examples of different types of rotordynamic computer simulations that can be performed. Each example falls into one of two categories: eigenanalysis or linear forced response. Transient response analysis would be a third important category, but this is beyond the scope of this book. Most of these examples can be distinguished by the type of graphical output they produce.

Quite often in rotordynamics the rather generic term "critical speed analysis" is used. The obvious meaning is to perform an analysis to identify the *critical speeds* of a machine. To some, this would be done with an eigenanalysis. To others, this would be done with a linear forced response analysis. By most definitions, a critical speed is any machine operating speed that should be avoided (with sufficient margin) because otherwise excessive vibration may occur. Excessive vibration is generally the result of a resonant condition. Resonance occurs when an excitation frequency (e.g., shaft speed) coincides with a natural frequency. Thus, it seems appropriate to use eigenanalysis to compute the natural frequencies of the rotor–bearing system. The calculated natural frequencies can then be used to state what the critical speeds are of the machine. This is complicated, however, because some natural frequencies (i.e., modes) may be damped well enough so they do not produce objectionable critical speeds. In addition, rotors generally have natural frequencies corresponding to both forward and backward whirling of the shaft. Backward whirling modes are known to be much less likely to produce critical speeds than forward whirling modes. So using eigenanalysis to identify critical speeds is sometimes straightforward, and sometime not.

As will be seen in the following examples, linear forced response analysis can be used to produce plots of predicted shaft vibration versus machine speed. Since such plots can be easily configured to closely mimic field measurements of shaft vibration, they can be used to identify critical speeds in a very direct fashion. However, plots of calculated response are very dependent on the excitation forces applied to the model. A critical speed can be identified only if a properly configured set of excitation forces are employed. In addition, not all critical speeds exhibit the same degree of severity (e.g., lightly damped versus heavily damped modes). So using linear forced response to identify critical speeds is also not always completely straightforward.

The net result is that a blend of eigenanalysis and response analysis is generally the most reliable way to predict critical speeds, and the severity of those critical speeds. This is done by first performing the eigenanalysis to compute natural frequencies and mode shapes. The natural frequencies essentially provide hints as to the probable locations of critical speeds. The mode shapes dictate how excitation forces should be configured in a subsequent forced response analysis to check if a particular mode will respond with enough severity to be deemed a critical speed, and also to identify the speed at which vibration response attains its maximum value (i.e., the critical speed). This overall approach to critical speed analysis is the same as that described in many API purchase specification documents (for example, see [16]).

Damped Natural Frequency Map (NDF)

Calculating and plotting a damped natural frequency map (DNF) is among the most common tasks in rotordynamics. When computing eigenvalues of a rotor system model, the speed of the shaft is a key input. The shaft speed is necessary to set the values of the gyroscopic terms that go into the system damping matrix. The stiffness and damping properties of bearing and seal elements also are often highly dependent on shaft speed. For these reasons the eigenvalues will vary with speed, and must be computed one speed at a time, typically, for a range of speed values from close to zero rpm or minimum operating speed up to some speed above maximum operating speed (say 10 or 20 percent above max speed). Five to ten speed values across this range are quite often sufficient. For each value of shaft speed the eigenanalysis produces one set of complex eigenvalues. The frequency part (i.e., imaginary part) of each individual eigenvalue can be displayed on a single plot, with speed as the abscissa, called a damped natural frequency map. Figure 4-19 shows an example DNF map. The points on the plot for each mode have been connected by lines, thereby forming a continuous *mode curve* for each mode. In actuality, the eigenanalysis is computed independently for each speed, and determining which points belong to which mode curve can at times be difficult to do. Information extracted from mode shapes (each point on the DNF has an associated mode shape) can be helpful in doing this. In the sample plot, the frequency of one mode is seen to decrease to zero at 4000 rpm, where it has become critically damped.

The dashed line on the plot has a slope of one, and represents a synchronous excitation (1×). This line is on the plot because nearly any rotor system will be subjected to a 1× excitation. When the frequency of a mode equals the frequency of an excitation, there is a possibility of resonance, or a critical speed. Any intersection of a mode curve with the 1× excitation line suggests a possible critical speed, and may warrant further

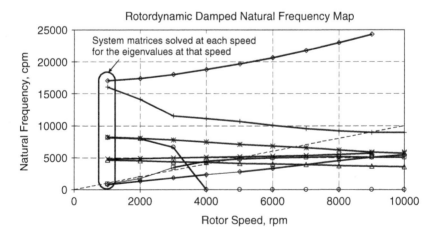

Figure 4-19 Example of a damped natural frequency map (DNF).

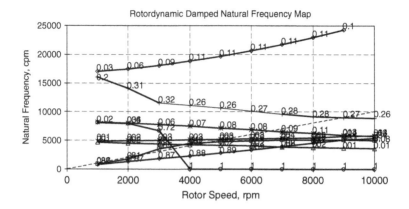

Figure 4-20 Same DNF with points labeled with corresponding damping ratio.

investigation such as reviewing the damping of the mode or whether it is forward or backward.

The DNF plot does not directly reveal any information about the damping of a mode, other than zero frequency implying when a mode is critically damped. Some or all of the points on the DNF plot can be individually labeled with their corresponding damping ratio or log decrement, as illustrated in Fig. 4-20.

The DNF plot also does not directly reveal any information about the whirl direction of a mode. Quite often, forward modes will have their frequency increase with speed, while backward mode frequencies decrease with speed. But this is not a general rule, as exceptions occur in heavily damped systems and in systems where bearing and seal coefficients vary greatly with speed. The only way to know if a mode is forward or backward is from its mode shape (discussed later). For any given mode, the whirl orbit direction could be forward at every station, or backward at every station, or be a mix (some forward and some backward). Mixed modes are frequently encountered with rotors supported on fluid film bearings (example below). Figure 4-21 shows the sample DNF with frequencies of backward modes being assigned negative values for display purposes only. The mode whose frequency drops to zero at 4000 rpm is a mixed mode in this particular example.

Modal Damping Map

The counterpart to the DNF is a modal damping map, which plots the damping part of each individual complex eigenvalue (i.e., real part). This plot may also be called a stability map. Figure 4-22 shows an example for the same rotor as above, but supported on 2-lobe sleeve bearings.

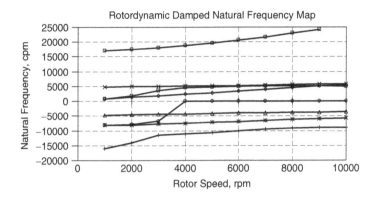

Figure 4-21 Same DNF with backward whirl modes plotted with negative frequency.

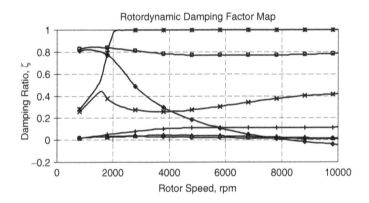

Figure 4-22 Sample damping factor map showing one unstable mode. The rotor is supported by 2-lobe sleeve bearings.

There is one mode that becomes critically damped at 2200 rpm. When looking for potential rotordynamic instability, this plot will show if any modes become unstable, and at what speeds. Usually, but not always, it is the first forward whirling mode that becomes unstable, if any. The speed where the damping ratio becomes negative is termed the *instability threshold speed*. In Fig. 4-22 this would be near 8000 rpm. The unstable mode is seen to be unstable for all speeds above 8000 rpm. For turbomachinery like pumps, compressors, and turbines, a predicted damping ratio of less than a threshold value like 0.05 (same as log decrement of 0.3) could be considered to mean that some amount of subsynchronous whirling of the mode may occur, although not necessarily of a destructive amplitude. The appropriate threshold damping value for stability purposes is best obtained from experience with similar machines, and will depend on how accurately the system has been modeled. Theoretically, unstable modes always diverge. In practice, however, system nonlinearities eventually limit the amplitude of the vibration. The vibration may or may not prevent machine operation. The vibration is termed *subsynchronous* whenever the frequency of the vibration (i.e., the frequency of the unstable natural mode) is less than shaft speed. This is by far the most common situation with unstable modes.

Root Locus Map

A root locus plot combines the DNF and damping factor maps into one plot. This plot type shows both the real and imaginary parts of each eigenvalue. As the rotor speed changes, these points move around the plot. This plot can take on different appearances, depending on the choice of frequency units (Hz, cpm, or rads/sec) and the type of damping factor (log decrement, damping ratio, damping exponent). Also, the axes may be interchanged. Regardless, the plot shows how the eigenvalues move around the complex plane as speed varies. Similar to Fig. 4-20, the points could be labeled with their corresponding rotor speed. And, similar to Fig. 4-22, backward modes could be plotted with negative frequency values. Instabilities can be identified, such as when a root crosses into the unstable half of the complex plane (i.e., negative log decrement or damping ratio, or positive damping exponent).

The plot in Fig. 4-23 shows how the eigenvalues move with speed. A root locus plot can also be used to show how the eigenvalues move as any system parameter is varied. For example, the stiffness or damping value of one or more supports, or the level of aerodynamic cross-coupling applied at an impeller.

Undamped Critical Speed Map

In the damped natural frequency map of Fig. 4-19, the intersections of forward mode curves with the $1 \times$ excitation line are sometimes all that is desired from that plot. By incorporating some additional assumptions into the eigenvalue problem formulation, these intersections can be computed in a single calculation of system eigenvalues (i.e., no need to sweep the shaft speed to find the intersections). The assumptions are necessary to enable such a calculation, but they also limit the applicability of the results. The assumptions are as follows:

  1. The imaginary part of the eigenvalue is equal to the rotor speed.

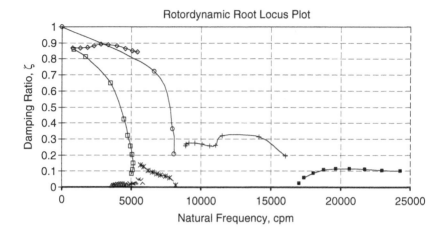

Figure 4-23 Sample root locus plot corresponding to Fig. 4-19.

2. The real part of the eigenvalue is zero.
3. The rotor whirls forward with circular orbits.

Assumption 1 stems directly from what we are trying to accomplish, that the eigenvalue frequency equal the rotor speed. Every place in the problem formulation where rotor speed is used, it is replaced with the eigenvalue frequency. Referring back to Eq. 4-7, the frequency of the eigenvalue is the imaginary part of the complex eigenvalue. So the frequency in radians/second is equal to $(-is)$ if the real part of s is identically zero. This leads to assumption 2, which is that the system be undamped and there must be no destabilizing cross-coupled stiffness. Assumption 3 is incorporated into the problem by setting $y = (-i)x$ so that y motion lags x motion by 90 degrees (also $ay = (-i)ax$). This assumption makes all calculated modes be forward modes (we only want critical speeds, so we exclude backward modes). This assumption results in the whirl orbits being circular.

To illustrate what is taking place in this analysis, let's start with a model of a rigid rotor. Figure 4-24 shows a rotor on simple bearings. The rigid rotor's dimensions can be seen in the figure, and the material density is 0.283 lb/in³. The rotor model can be seen to have 19 stations, and thus 76 dof. As many as 76 modes could be computed from this model. However, for this example the rotor has been rendered completely rigid, and any rigid rotor can be modeled as an equivalent rotor with a single station having just 4 dof. So any rigid rotor has exactly four natural modes. Figure 4-25 shows the DNF for this rotor when the bearing stiffnesses are each 31,600 lb/in in both horizontal and vertical planes. This is a rigid rotor, so there are exactly four modes in the DNF plot.

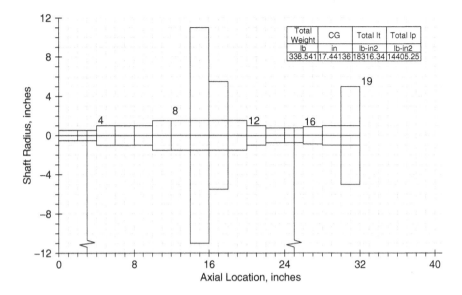

Figure 4-24   Example rotor for undamped critical speed analysis.

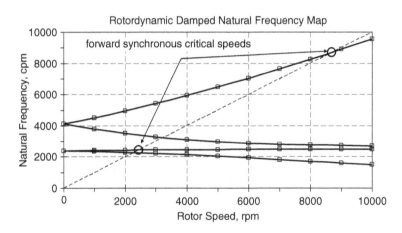

Figure 4-25   Eigenvalues (DNF) versus shaft speed for rigid rotor of Fig. 4-24 on 31.6e3 lb/in bearings.

The lowest two modes are the backward and forward counterparts of a bounce mode. The two higher modes are a rocking mode. At zero rpm there is no gyroscopic affect, so the backward and forward modes coalesce and have equal frequencies. Figure 4-25 points out the forward synchronous critical speeds appearing near 2400 and 8800 rpm. These two points are the only points that will be computed in an *undamped critical speed* (UCS) analysis. Figure 4-26 shows the UCS map for this rotor. The abscissa in this plot is bearing stiffness, since the UCS analysis

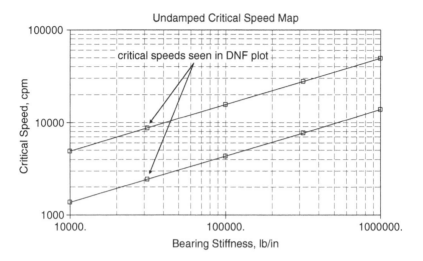

Figure 4-26 Undamped critical speed map for rigid rotor of Fig. 4-24.

is performed for a range of different bearing stiffness values. The plot shows how the undamped forward synchronous critical speeds vary as the bearing stiffness varies. The sample rotor is rigid, so there are just two forward mode curves to be displayed on this plot. Both curves plot as straight lines on the logarithmic scale because the critical speeds of a rigid rotor are directly proportional to the square root of the bearing stiffness (i.e., $\sqrt{k/m}$). The two critical speeds that are singled out in Fig. 4-26 at a bearing stiffness of 31,600 lb/in are an exact match for the two critical speed points in Figure 4-25.

Figure 4-27 shows the DNF for the example rotor when the rotor is not rigid ($E = 30e6$ psi). The rotor now has three natural modes below 20,000 cpm, and each spawns a backward and forward component. This undamped model produces 76 complex conjugate eigenvalue pairs (38 are forward, 38 are backward), but we'll look only at the lowest 6 frequencies in this discussion. The first two forward modes cross the 1× line near 2800 and 4200 rpm, as pointed out in the figure. These are the undamped synchronous critical speeds for the particular value of bearing stiffness used to generate this plot. The frequency of the forward component of the third mode is increasing rapidly due to strong gyroscopic effects of the two large disks on the shaft. It eventually will cross the 1× line at a very high speed.

The corresponding UCS map for the elastic version of the example rotor is in Fig. 4-28. Only the lowest six (out of 38) forward mode critical speeds are displayed in the plot. Ordinarily, a UCS analysis will be conducted to span three regions of bearing stiffness. To the left of the plot, where the bearings are very soft, the rotor is approaching a free–free

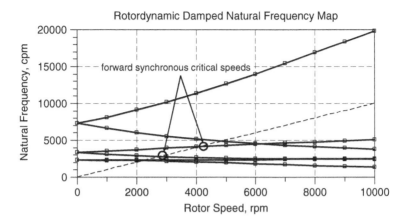

Figure 4-27 Damped natural frequency map for the nonrigid rotor of Fig. 4-24. Bearing stiffness is 31.6e6 lb/in.

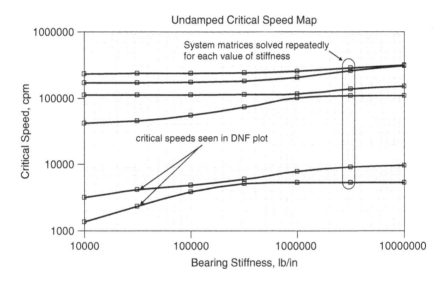

Figure 4-28 Undamped critical speed map for nonrigid rotor of Fig. 4-24.

condition, and there will be one or two rigid body modes sloping downward to the left, with the remainder of the modes being flexible modes having constant critical speed values. To the right of the plot, where the bearings are very stiff, the rotor is approaching a simply supported *rigid bearing* condition, and all modes level off and have constant values. In between these two regions the rotor transitions from unsupported to simply supported.

An interesting phenomenon occurs when $(T - P)$ is negative. T is the transverse inertia of the rotor, and P is the polar inertia of the rotor. This case arises when the rotor is shaped like a pancake as opposed to a pencil. When $(T - P)$ for a rotor is negative, the conical mode synchronous critical speed will not occur. This does in fact happen in practice where pancake-shaped rotors do not exhibit a conical mode critical speed. In our rigid rotor example case, increasing the diameter of the large disk, station 9, causes the conical mode critical speed to increase. As the diameter approaches 25 in., the conical mode critical speed increases toward infinity as the value of $(T - P)$ approaches zero. When $(T - P)$ eventually does become negative, the conical mode critical speed no longer appears in the UCS plot.

As a last example of a UCS plot, consider a case where the stiffness of the rotor's actual bearings has been calculated. A common case is a rotor supported on two nominally identical fluid film bearings. The direct stiffness values for this bearing, K_{xx} and K_{yy}, each vary with speed. The bearing stiffness data can be plotted on the UCS plot as illustrated in Fig. 4-29. The K_{xx} data curve intersects the first mode curve at a speed value of 4000 rpm, and intersects the second mode curve at a speed value of 5000 rpm. The K_{yy} data curve intersects the first mode curve at a speed value of 5300 rpm, and intersects the second mode curve at a speed value of about 7900 rpm.

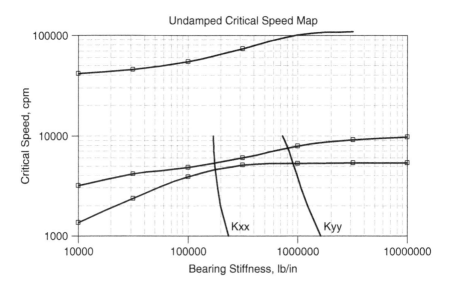

Figure 4-29 Undamped critical speed map with overlaid bearing stiffness data.

The two K_{xx} intersections suggest there may be two predominately horizontal critical speeds near 4000 and 5000 rpm. The two K_{yy} intersections suggest two vertical critical speeds may occur near 5300 and 7900 rpm. But the critical speeds on a UCS plot always assume circular orbits, and when $K_{xx} \neq K_{yy}$ the orbits will be elliptic and not circular. So the rotor's actual critical speeds on these particular bearings (5 tilting pads, load on pad) may be quite different than what the UCS map would indicate. This is a limitation of the UCS analysis, which is exacerbated in this example by the K_{xx} and K_{yy} values differing by over an order of magnitude. Also bear in mind that bearing damping is neglected in a UCS analysis, and the damping in fluid film bearings can exert a strong influence on a rotor's critical speeds.

Mode Shapes

The calculation of mode shapes will be our last example concerning eigenvalue analysis. Every eigenvalue has associated with it an eigenvector. Equation 4-9 is solved to find matched pairs of s and $\hat{q}$. The complex eigenvector $\hat{q}$ contains the mode shape in the form a complex amplitude for each dof in the model. An eigenvector describes only the *shape* of the rotor system when it vibrates in a single mode. The amplitudes can be scaled up or down arbitrarily. The simplest way to present this information is to simply plot the real and imaginary parts of each dof. Figure 4-30 shows a sample mode shape for the example rotor of Fig. 4-24. The modal amplitudes are plotted versus the axial coordinate of the model. The ordinate scale is entirely arbitrary, and this mode shape has been arbitrarily scaled so that the maximum plotted value is 1. Only the translational dof are plotted in Fig. 4-30. Another plot of the rotational dof could also be generated, but such plots are not normally presented since the same

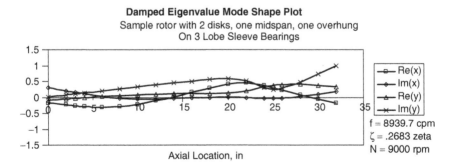

Figure 4-30   Example of a complex mode shape (only translational dof are shown).

information is conveyed in the slope of the curves in Fig. 4-30. Since the mode shape in Fig. 4-30 is for a particular eigenvalue (at 9000 rpm in Fig. 4-19), the specifics of that eigenvalue are annotated on the figure.

Equations 4-6 and 4-7 can be used to convert complex modal amplitudes to magnitude and phase for each dof. The magnitudes and phases can be used to plot whirl orbits (y versus x) for each station. Figure 4-31 shows how these orbits can be used to create a three-dimensional display of the mode shape. This particular style of displaying a mode shape is not actually a true 3D plot because each orbit is being displayed as it would appear in a conventional x versus y plot. This is done so orbits that are truly circular will display as true circles. If desired, the plot could instead be displayed as a true 3D plot (e.g., isometric).

Examination of the whirl orbits is the only way to tell if the rotor is whirling forward or backward, or both. In the 3D plot of Fig. 4-31 the whirl orbit direction is depicted graphically. Forward whirl could be either clockwise or counterclockwise, depending on which direction the shaft is rotating. In the example mode shape there are both forward and backward orbits, resulting in what is called a *mixed* mode. Mixed modes occur frequently with rotors supported on fluid film bearings. At an axial location where the whirl orbit direction transitions, the orbit must reduce to a straight line. This happens at three places in Fig. 4-31. It is also not uncommon, especially with fluid film bearings, that a mode shape may be a bit like a corkscrew, in which case there may be no place along the rotor where the shaft centroid intersects the centerline.

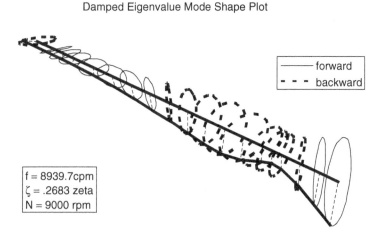

Damped Eigenvalue Mode Shape Plot

forward
backward

f = 8939.7cpm
ζ = .2683 zeta
N = 9000 rpm

Figure 4-31 Same mode as in Fig. 4-30 displayed in pseudo-three-dimensional form.

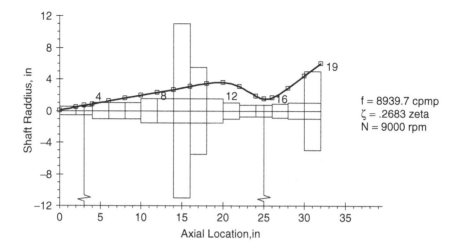

Figure 4-32 Same mode as in Fig. 4-30 overlaid on the model geometry.

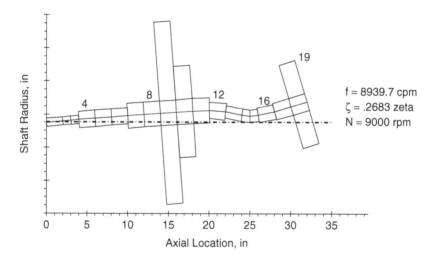

Figure 4-33 Same mode as in Fig. 4-30 displayed as a deformed geometry plot.

Two additional ways to display a mode shape are illustrated in Figs. 4-32 and 4-33. In the former, one selected component of the complex mode shape (here the real part of x) has been overlaid on the geometry of the model. In the latter, the same mode shape component has been used to produce a deformed geometry plot.

It is worth pointing out that for damped rotordynamic systems, animation of a mode shape like the one in Fig. 4-33 would be done for motion in one selected plane (e.g., either x or y), and should properly account for both real and imaginary parts of the mode shape (i.e., magnitudes *and* phases). All stations on the rotor then may *not* reach their peak deflections in that plane at the same instant, and also may not pass through zero at the same instant.

Bode/Polar Response Plot

We now turn our attention to linear forced response (LFR) analysis with Eq. 4-14. The same rotor system model used for the damped eigenanalysis is used for LFR, with the addition of a forcing function $F(t)$. It is worth emphasizing that this is a linear analysis, so doubling the strength of the applied forcing function will exactly double the computed responses. For this example we'll use rotating imbalance as the forcing function, which is specified as a value, u, in units such as g-in, oz-in, kg-mm, or similar. Circumferential location, β, in degrees or radians, would also be specified separately for each imbalance applied to the rotor. Each imbalance is assigned to a station, and produces a rotating force that rotates at the speed of the rotor Ω, and acts on the rotor at that station. The components of this force are given in Eq. 4-15. An appropriate amount of imbalance is normally selected relative to the tolerances to which the rotor is manufactured and/or shop balanced. The imbalance locations would generally be at the heaviest objects on the shaft, like impellers and turbine wheels. When there are multiple imbalances, the relative circumferential locations become important. A conservative approach to calculating response at a critical speed is to have the imbalance angular distribution follow the mode shape of the critical speed. Two documents that can help establish practical amounts of imbalance are balancing standards ISO 1940 for rigid rotors [17] and ISO 11342 for flexible rotors [18]. API purchase specifications recommend basing imbalance amounts on the value of $u = 4W/N$, where u is in oz-in, W is rotor weight in lbm, and N is the rotor's maximum operating speed in rpm. Mode shapes provide guidance as to how to configure the imbalances on the rotor. As an example, a LFR analysis might be performed specifically to show where the critical speed is for the mode shape shown in Fig. 4-33, and to quantify its severity. To specifically excite this mode, imbalances would be placed at stations 10 and 19, and these would be given the same phase angle:

$$F_x(t) = u\Omega^2 \cos(\Omega t + \beta)$$
$$F_y(t) = u\Omega^2 \sin(\Omega t + \beta)$$

(4-15)

With the definition in Eq. 4-15, rotor rotation is counterclockwise on an xy graph, and the angle β would be measured on the rotor in the counterclockwise direction. To apply this rotating force at a station in Eq. 4-14, we convert it to a complex representation using Eq. 4-7.

$$F_x(t) = \text{Re}\left(u\Omega^2 e^{i(\Omega i + \beta)}\right) \quad = \text{Re}\left(ue^{i\beta}\Omega^2 e^{i\Omega t}\right) \quad = \text{Re}\left(\hat{u}\Omega^2 e^{i\Omega t}\right)$$

$$F_y(t) = \text{Re}\left(-iu\Omega^2 e^{i(\Omega t + \beta)}\right) = \text{Re}\left(-i\hat{u}e^{i\beta}\Omega^2 e^{i\Omega t}\right) = \text{Re}\left(-i\hat{u}\Omega^2 e^{i\Omega t}\right)$$

$$(4\text{-}16)$$

where the complex version of the imbalance is $\hat{u} = u \cdot (\cos\beta + i\sin\beta)$ and contains both its magnitude and phase. An exercise for the reader is to prove that Eq. 4-15 is equivalent to Eq. 4-16 by using Euler's relation for a complex exponential and expanding the indicated multiplications.

In Eq. 4-14 $s = i\Omega$, and with the complex imbalance values assigned using Eq. 4-16, the set of complex linear equations is solved simultaneously for complex response amplitudes for every dof. For output we'll want the magnitude and phase at stations we're most interested in—for example, at bearing, seal, and impeller locations. To convert the complex response values to magnitude and phase, we use Eq. 4-17. The phase angle in this form of output is a *lagging* phase. For a leading phase angle the sign on ϕ would be reversed.

Calculated by program: $\hat{x} = x_r + ix_i$

Desired output: $x(t) = X\cos(\omega t - \phi)$ (4-17)

$$X = \left|\hat{X}\right| = \sqrt{x_r^2 + x_i^2} \qquad \tan\phi = \frac{-x_i}{x_r}$$

For our sample rotor from Fig. 4-24 supported on 3-lobe sleeve bearings, an example of magnitude and phase response presented in a Bode type of plot format is shown in Fig. 4-34. The magnitude portion of the plot clearly reveals a dominant critical speed near 5200 rpm, which is the forward component of this rotor's first mode (see Fig. 4-19). There is a minor critical speed near 4200, which is the backward component of the first mode. The stiffness properties of the bearing are asymmetric (see Fig. 4-29), which is what enables the backward mode to show up in the Bode plot. There is also a small critical near 6500 rpm, which is a well-damped "mixed" mode that has the overhung disk whirling forward.

Along with the x and y magnitudes in Fig. 4-34, there is also a major axis magnitude, which is the length of the major axis of an ellipse formed by the x and y responses. An example of this ellipse is shown below

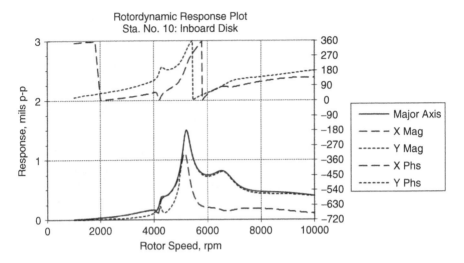

Figure 4-34 Sample imbalance response plot in Bode form (8 g-in at stations 10 and 18, 15 degrees apart).

in the orbit response plot example. The expression used to compute the major axis magnitude is derived in many texts (e.g., Childs [2]).

$$\text{Major axis} = \sqrt{\frac{x_r^2 + x_i^2 + y_r^2 + y_i^2}{2} \pm \sqrt{\frac{\left(x_r^2 - x_i^2 + y_r^2 - y_i^2\right)^2}{4} + (x_r x_i + y_r y_i)^2}}$$

(4-18)

$$\alpha = \frac{1}{2}\arctan\left(\frac{2\,(x_r y_r + x_i y_i)}{x_r^2 + x_i^2 - y_r^2 - y_i^2}\right)$$

(4-19)

For the ($\pm$) in Eq. 4-18, ($+$) yields the major axis length, which is always positive, and ($-$) yields the minor axis length. The minor axis can be either positive or negative. Positive means the whirl direction is forward (counterclockwise) and negative means backward. The expression for α gives the orientation of the major axis measured counterclockwise from the x axis. The major axis length of shaft displacement at a seal location could be compared to the available clearance in the machine to determine margin against rubbing.

The same magnitude and phase information in a Bode plot can be displayed in a polar plot, as shown in Fig. 4-35. For each transition through a critical speed in a Bode plot, the magnitude will peak and the phase angle will try to change 180 degrees. In a polar plot each transition through a critical speed will try to sweep out a circle. The phase data in both example figures have been plotted as lagging phase angles. When using lagging

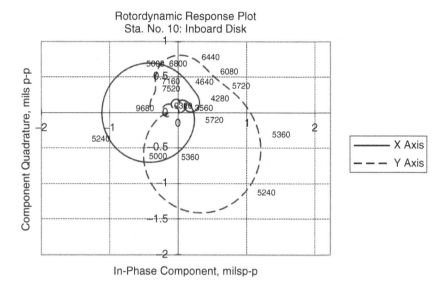

Figure 4-35 Polar format response plot of same data in Fig. 4-34.

phase, the phase angle *increases* 180 degrees while going through a critical speed. In the polar plot this means the circles go counterclockwise as the rotor speeds up. The *y* response is also seen to lag the *x* response by about 90 degrees, which implies the direction of rotation is counterclockwise on an *xy* graph (i.e., an orbit plot). The polar plot does not indicate what the speed of the rotor is, so some of the points have been labeled with their corresponding speed.

The use of a lagging phase angle convention is consistent with most textbooks on vibration and machine dynamics. Many computer-based data acquisition systems, however, will output phase angles as *leading* phase. In a plot of measured vibration data, check whether the phase angle increases (lagging) or decreases (leading) in either a Bode or polar plot as the rotor speeds up through a critical speed.

Orbit Response Plot

The Bode and polar plots show the magnitude and phase of both the x- and y-axis responses across an entire speed range. At any selected speed value, the *x* and *y* response can be used to construct a shaft orbit plot by using Eq. 4-7 to generate *x* and *y* time waveforms for one revolution. Figure 4-36 shows shaft orbit examples at three speeds: below the critical speed, at the peak of the critical speed, and above the critical speed. The • marker on the plot designates the start of the revolution, with rotation being counterclockwise. As the rotor traverses the critical speed, this plot

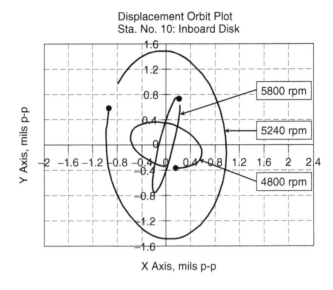

Displacement Orbit Plot
Sta. No. 10: Inboard Disk

X Axis, mils p-p

Figure 4-36 Polar format response plot of same data in Fig. 4-34.

shows both the rise in amplitude and the change in phase angle. The orbits are elliptic due to the asymmetric properties of the 3-lobe sleeve bearings supporting the rotor. These orbits could be compared directly to orbits measured with a pair of orthogonal displacement sensors. In situations where installed displacement sensors are not aligned with the x and y axes, either the measured or calculated responses can be rotated via a coordinate transformation to coincide with the other.

Bearing Load Response Plot

Another common form of output from a LFR analysis is the reaction force in a bearing. Figure 4-37 shows an example plot for our test rotor of Fig. 4-24 mounted on a pair of 3-lobe sleeve bearings. The applied imbalance is 8 g-in at each of stations 10 and 18. The rotating imbalance force for the total of 16 g-in at 5000 rpm is 25 lb, and at 10,000 rpm it is 100 lb. The applied imbalance loads are modified by the dynamics of the rotor, and shared by the two bearings that support the rotor. At the critical speed, the applied load in this bearing is being magnified by more than a factor of 7 (i.e., 90/12.5). At 10,000 rpm, where mass center inversion has taken place, the applied load is attenuated by more than a factor of 2. The load reduction that takes place above the critical is one of the key benefits of running rotors at supercritical operating speeds.

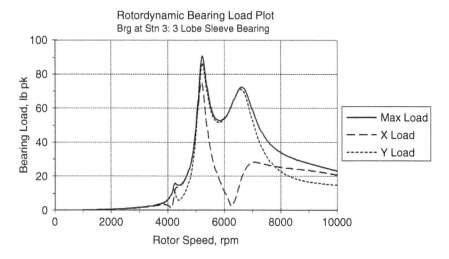

Figure 4-37 Plot of bearing reaction load.

A very good use of bearing load response output is judging how smooth or rough a machine will run with a given amount and distribution of imbalance. The force in each bearing inevitably will find its way to the ground or the structure on which the machine is mounted, and will be sensed not only by installed vibration instrumentation, but also by persons near the machine. It is often felt that if the bearing forces at operating speed are no more than 10 percent of the static journal load on the bearing, then the machine can be considered to be running "smooth," perhaps even very smooth. If the peak bearing force is greater than the static journal load, then (for a horizontal machine) there will be load reversal taking place in the vertical axis during every revolution. For many machines this level of load would result in the machine vibration being considered "rough," perhaps even very rough.

Operating Deflected Shape (ODS)

The Bode plot discussed above presents the response results for a single station for a range of speeds. The deflections for *all* stations at *one* speed can be used to produce an *operating deflected shape* (ODS) plot, which is very similar to a mode shape plot. Figure 4-38 shows a plot of the ODS of our example rotor when running at the critical speed of 5240 rpm. The ODS of a whirling rotor is difficult to display in a two-dimensional plot if the orbits are elliptic. The *x*-axis result in Fig. 4-38 displays the peak *x* displacement for each station, and these may not all occur at the same

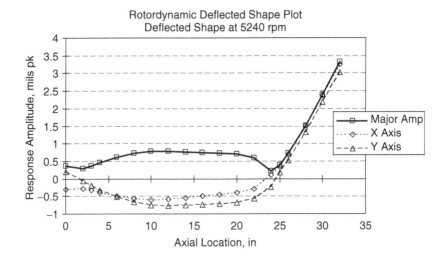

Figure 4-38 Example of an ODS plot displayed as a two-dimensional plot.

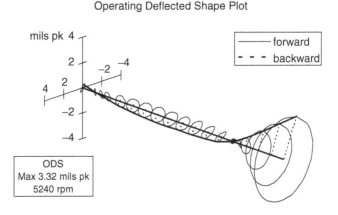

Figure 4-39 Example of a 3D ODS plot.

time due to damping. Analogous comments apply to the y-axis result. In the general case of elliptic orbits, the length of the major axis is the maximum rotor displacement at any station. So this value is also displayed in Fig. 4-38. The major axis amplitude could be compared to available clearances within the machine (although it will be necessary to also account for static deflections).

The same ODS is shown again in Fig. 4-39 in three-dimensional form. This is a true three-dimensional plot as opposed to the pseudo-3D mode shape plot seen earlier in Fig. 4-31. As was the case for mode shapes,

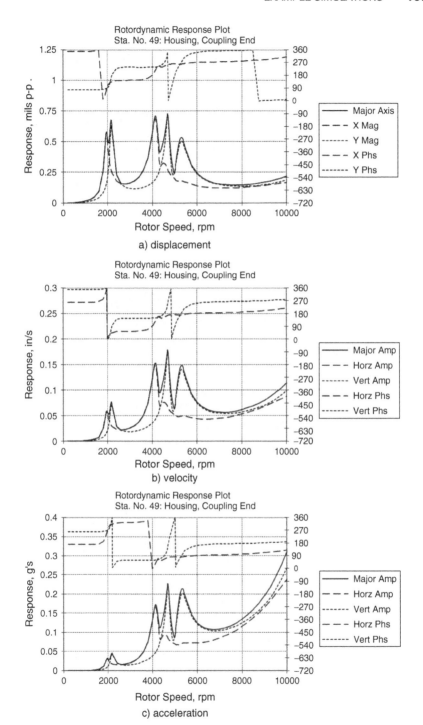

Figure 4-40   Sample housing responses for model shown in Fig. 4-4: (a) displacement, (b) velocity, and (c) acceleration.

an ODS can also have different parts of the rotor whirling in opposite directions (in the example plot the entire rotor whirls forward).

Housing Vibration (ips and *g*'s)

Figures 4-3 and 4-4 presented example models that contain nonrotating structures. It is common practice to have seismic transducers, such as accelerometers and velocity sensors, mounted on machine housing components. Outputs of these sensors are typically presented in their native units of *g*'s for acceleration, and inches/sec or mm/sec for velocity. The analytical result from a LFR analysis in the form of displacement is readily converted to equivalent velocity or acceleration units. This is conveniently done with the complex form of the solution by invoking Eq. 4-8, which results in the following relationships:

$$\text{Velocity: } \{\dot{\hat{q}}\} = s\,\{\hat{q}\} = (i\,\Omega)\,\{\hat{q}\}$$
$$\text{Acceleration: } \{\ddot{\hat{q}}\} = s^2\,\{\hat{q}\} = \left(-\Omega^2\right)\{\hat{q}\}$$

$$(4\text{-}20)$$

After performing the indicated complex multiplications, the real part is again taken to obtain the desired physical response. This is also often referred to as "differentiating" the displacement once for velocity or twice for acceleration. In addition to the magnitude change brought about by multiplying the displacement either once or twice by the frequency (in radians/sec), there is also a phase change. As is the general rule for harmonic motion, the corresponding velocity waveform leads the displacement waveform by 90 degrees. The acceleration leads by 180 degrees, and is effectively *out of phase* with the displacement.

Figure 4-40 shows sample responses computed for the housing shown in Fig. 4-4. The three plots demonstrate that differentiation magnifies response at higher frequencies, and also the phase changes mentioned above (*note*: plots are lagging phase). This partially explains why displacement is an effective means of vibration measurement at lower speeds, and accelerometers are more effective for very high speeds.

REFERENCES

[1] Vance, J. M. *Rotordynamics of Turbomachinery*. New York: Wiley, 1988.

[2] Childs, D. W. *Turbomachinery Rotordynamics*. New York: Wiley, 1993.

[3] Schweitzer, G., Bleuler, H., and Traxler, A. *Active Magnetic Bearings*, ZTH Books (Zurich) 1994.

[4] De Jongh, F. M. The synchronous rotor instability phenomenon—Morton Effect. Proceedings of the 37th Turbomachinery Symposium, Turbomachinery Laboratory, Texas A&M University, College Station, Texas, 2008.

[5] Someya, T. (ed.). *Journal Bearing Data Book*. New York: Springer-Verlag, 1989.

[6] Schwibinger, P., Neumer, T., Zurbes, A., and Nordmann, R. The influence of torsional–lateral coupling in geared rotor systems on its eigenvalues, modes and unbalance vibrations. IMechE, 1988, p. 279.

[7] Murphy, B. T., and Vance, J. M. Improved method for calculating critical speeds and rotordynamic stability of turbomachinery. *Journal of Engineering for Power, Transactions ASME* 105: 591–595 (1983).

[8] Press, W. H., Teukolsky, S. A., Vetterling, W. T., and Flannery, B. P. *Numerical Recipes: The Art of Scientific Computing*. New York: Cambridge University Press, 2007.

[9] Murphy, B. T. Improved unbalance response calculations using the polynomial method. 2nd Latin American Turbomachinery Conference, 1993.

[10] Lalanne, M., and Ferraris, G. *Rotordynamics Prediction in Engineering*. New York: Wiley, 1998.

[11] Genta, G. *Vibration of Structures and Machines*, 3rd ed. New York: Springer, 1999.

[12] Murphy, B. T., Ouroua, H., Caprio, M. T., and Herbst, J. D. Permanent magnet bias, homopolar magnetic bearings for a 130 kW-hr composite flywheel. 9th International Symposium on Magnetic Bearings, Lexington, KY, 2004.

[13] Darlow, M., Murphy, B., Elder, J., and Sandor, G. Extension of the transfer matrix method for rotordynamic analysis to include a direct representation of conical sections and trunnions. *Journal of Mechanical Design, Transactions of the ASME* 102: 122–129 (1980).

[14] Roark, R. J., Young, W. C., and Budynas, R. G. *Roarks Formulas for Stress and Strain*, 7th ed. New York: McGraw-Hill, 2001.

[15] Vance, J. M., Murphy, B. T., and Tripp, H. A. Critical speeds of turbomachinery computer predictions vs. experimental measurements, I: the rotor mass-elastic model. *Journal of Vibration, Acoustics, Stress, and Reliability in Design* 109 1–7 (1987).

[16] *API Recommended Practice 684, Tutorial on the API Standard Paragraphs Covering Rotor Dynamics and Balancing: An Introduction to Lateral Critical and Train Torsional Analysis and Rotor Balancing*, 2nd ed. American Petroleum Institute, August 2005.

[17] ISO 1940-1. Mechanical vibration—balance quality requirements for rotors in a constant (rigid) state, 1: specification and verification of balance tolerances. International Organization for Standardization, 2003.

[18] ISO 11342. Mechanical vibration—Methods and criteria for the mechanical balancing of flexible rotors. International Organization for Standardization, 1998.

[19] *Xlrotor Reference Manual*, Rotating Machinery Analysis, Inc., Austin, Texas, 2009.

[20] Bathe, K. J., *Finite Element Procedures in Engineering Analysis*, New Jersey: Prentice-Hall, 1982.

5

BEARINGS AND THEIR EFFECT ON ROTORDYNAMICS

Bearings constitute one of the most critical components in turbo and rotating machinery today. Their influence on the rotordynamic performance, life, and reliability of the machine cannot be ignored. Many of the problems we face with machinery today can be attributed to the design and application of the bearings. An understanding of how bearings work and some knowledge of the basic principles that underline their operation is therefore essential for making the proper choice for the particular design that best matches the service requirements of the machine in question. Even after the machine is designed and placed in operation, changes or modifications to the bearings constitute one of the most effective, direct, and economical means to alter and improve the machine's dynamic performance.

FLUID FILM BEARINGS

Fluid film bearings have been around for centuries, but the understanding of how they actually worked did not come about until 1882–1883. This is when an English inventor and railway engineer by the name of Beauchamp Tower [1] was appointed by the Institute of Mechanical Engineering in England to study high-speed railway engine bearings. He demonstrated in his test rig that with a suitable supply of lubricating oil, the surfaces of the bearing and the shaft were separated by a continuous film of lubricant, which prevented them from ever coming into contact. Tower was able to show that the load in the bearing was carried by the thin oil film through his measurements of the pressure profile in the bearing, but did not explain how and why this happens. Three years later (1886), Osborne Reynolds at the University of Manchester in England published a 77-page paper [2] in which he was able to predict and explain the experimental measurements conducted by Tower. Reynolds developed what is known as the classical Reynolds hydrodynamic lubrication equation through combining the simplified Navier Stokes equation (assumed Newtonian incompressible fluid, neglected inertia and variations of the viscosity across the film) and the continuity equation.

Fluid film bearings operating in the full hydrodynamic regime support the load on a very thin film and thus there is no contact between the shaft and the bearing. It is for this primary reason that fluid film bearings offer infinite life provided the lubricant is kept clean and the machine is operating in a safe dynamic range away from stability and critical speed problems. This is why when long term operation (2, 3, 5, and even 10 years of continuous operation) and a high online factor (90 percent and higher) are required, the fluid film bearings are the bearings of choice.

The primary requirement for hydrodynamic lubrication is that sufficient lubricant (typically mineral or synthetic oil, but it could be any fluid with the proper characteristics for the application) exists between the shaft journal and the bearing at all times. The formation of an oil wedge to lift the shaft journal is similar to hydro planing (controlled hydro planing in the case of bearings) and is dependent on the speed (relative speed between the shaft and the bearing), load (weight of the rotor or any additional side loads from process fluid, or gear loads, or side loads due to misalignment), and the oil viscosity of the lubricant. These parameters are combined and presented by the *ZN/P* curve shown in Fig. 5-1. This is also known as the

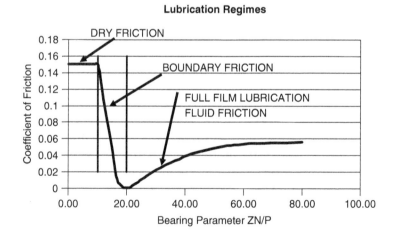

Figure 5-1   The *ZN/P* curve and the three lubrication regimes.

Stribeck curve [3], after the German engineer who introduced it in 1902. The symbol Z represents viscosity, N is the speed in rpm, and P is the unit loading in lb/in^2. This curve describes the three regimes of operation a bearing passes through while the machine accelerates to operating speed or decelerates from the operating speed to stand still conditions. These regimes are *dry friction*, where contact between the asperities of the shaft and the bearing exist; mixed lubrication regime, or *boundary friction*; and full hydrodynamic lubrication, *fluid friction*, where a thin film exists between the shaft and the bearing, which supports the static and dynamic loads in the rotating shaft. These three regimes are also shown graphically with an exaggerated clearance in Fig. 5-2 as the shaft accelerates from standstill to full operating speed. The discussion in this section of the book is limited to the full hydrodynamic regime. This regime has the most influence on the rotordynamic characteristics of the machine as it speeds up from stand still to possibly traverse one or more critical speeds on its way to reaching the design operation speed.

Fluid film bearings develop pressure in the converging wedge, which supports the radial load on a thin oil film. Most fluid film bearings are of the *hydrodynamic* type, where the film pressure is produced by the shaft rotation dragging the oil into the converging wedge formed by the shaft and the bearing surface. In hydrodynamic bearings the oil pressure is just enough to keep the bearing supplied with oil and to remove the frictional heat generated by the viscous shear in the thin film. A less common type of fluid film bearing is the *hydrostatic* type, where the film pressure is provided through external pressurization and feeds a recess in the bearing that supports the load. Such a bearing requires pressurization that is orders of magnitude higher than that needed for a hydrodynamic bearing, and a significant power is required to drive the high-pressure oil

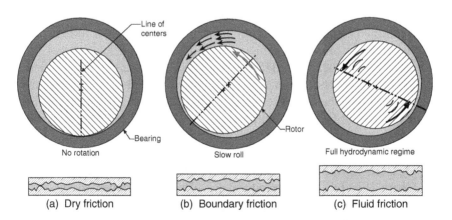

Figure 5-2   Three lubrication regimes in fluid film journal bearings.

pump. This purely hydrostatic bearing is sometimes applied based on the premise of providing a very stiff bearing through external pressurization of the bearing. This may be needed in some instances if the hydrodynamic generated film is not sufficient to build such stiffness due to unfavorable conditions such as low speed or inadequately sized bearings. Hydrostatic bearings are also used to provide a very stiff support and to allow rigid rotors to operate subcritical. This approach of providing very stiff bearings is successful only in limited applications. The discussion in Chapter 3 and further discussion on squeeze film damper supports show that a softer bearing approach is better for both stability and synchronous response. The hydrostatic bearing approach is sometimes utilized to complement or in combination with the hydrodynamic bearing. This is usually the case in large equipment and in conditions where the start-up torque is high. The hydrostatic features described above may be employed just for starting and when shutting the machine down. The hydrodynamic action is relied on once the shaft reaches the full hydrodynamic regime.

From the standpoint of rotordynamics, the main advantage of fluid film bearings is their inherent damping characteristics. Fluid film bearings exist in a variety of configurations depending on the application, space availability, and the rotordynamic requirements. The trend over the last few decades has favored the variable geometry tilting pad type bearings over the fixed-geometry sleeve type bearings, particularly in high-speed, supercritical machinery due to their inherent stability characteristics. However, these variable-geometry tilt pad bearings are more complex, contain more parts, and generally have lower damping than fixed-geometry bearings.

FIXED-GEOMETRY SLEEVE BEARINGS

A fixed-geometry or fixed-profile sleeve bearing supports the weight of the rotating shaft by developing a hydrodynamic pressure in the converging wedge formed by the shaft and the bearing surface, shown in Figure 5-3. The asymmetric pressure profile in the oil film is characteristic of fixed-geometry journal bearings. This gives rise to an attitude angle formed by the line of centers (line connecting the center of the shaft and the center of the bearing) and the load vector. This characteristic, present in all fixed-geometry journal bearings, is indicative of the presence of cross-coupling in the bearing. The load, which in this case is due primarily to the weight of the rotor, is acting directly downward on the journal bearing. Integrating the pressure profile results in a net force that balances the weight of the rotor. This force generated in the film is accompanied by a shaft displacement with a component that is along the load direction (direct) and a displacement that is orthogonal to the load direction and is shifted in the direction of rotation (cross). This cross-coupling characteristic is

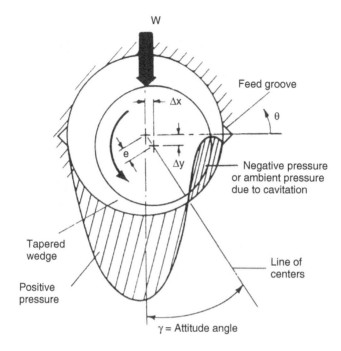

Figure 5-3 Pressure profile and attitude angle in a sleeve bearing.

present in all rotating machinery where a fluid or gas is rotating with the shaft in a small annulus.

This phenomenon is unique to rotating machinery and is not present in other nonrotating structures or mechanisms. It is this cross-coupled stiffness force that promotes self-excited, subsynchronous vibrations and instability in rotating machinery. In fluid film bearings this is referred to as *oil whirl*, where the shaft will whirl at a frequency that is equal to a fraction of the running speed frequency. This subsynchronous component tracks the running speed until it latches on or locks on a natural frequency where its amplitude will increase rapidly and results in what is commonly referred to as *oil whip*. The frequency of the subsynchronous vibrations in the oil whirl region is dependent on the *L/D* ratio of the bearing. The higher that ratio is, the closer the subsynchronous oil whirl is to half running speed. This gives rise to the thought that the subsynchronous frequency is directly related to the average circumferential flow of the lubricant between the shaft and the bearing. This circumferential flow is closer to half running speed as the length of the bearing increases and the effect of the side leakage is relatively smaller.

It is also worth noting that the cross-coupling in the journal bearing tends to always move or push the shaft to the opposite side of the converging wedge. Therefore, this cross-coupling effect is always biased by the

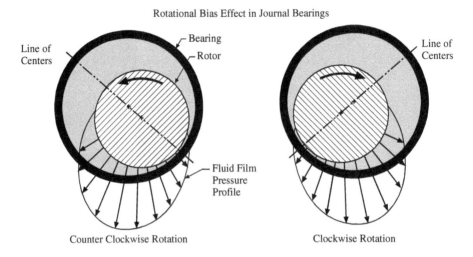

Figure 5-4 Rotational bias caused by cross-coupled stiffness coefficients.

direction of rotation as shown in Fig. 5-4. The term rotational bias is often used to describe this phenomenon, but what is also relevant is the fact that it is forward driving. Therefore, when diagnosing subsynchronous vibrations it is important to note if the whirl of the subsynchronous vibrations is forward or backward to help distinguish oil whirl and whip from rub and other anomalies that produce backward whirl.

The cross-coupling force vector shown in Fig. 3-15 provides an excellent graphic representation of the effects of the cross-coupled stiffness coefficients in the bearings and how they act on the rotor. The cross-coupled stiffness forces $K_{XY}Y$ and $K_{YX}X$ combine vectorially to produce a force that is tangential to the circular whirl orbit and in the direction of rotation. This is why it is often referred to as a follower force (following the direction of rotation), forward driving force (pushing the rotor to whirl forward in reference to the direction of rotation), or destabilizing force (tending to add energy to the system, which is destabilizing). All these terms are often used to describe this destabilizing force (generated from the cross-coupled stiffness), which acts in the same manner as direct damping (colinear with the instantaneous velocity) but in the opposite direction to direct damping, thus the term negative damping. Positive damping dissipates energy and in this manner it reduces vibrations and whirl amplitude, while the negative damping adds energy to the dynamic system and therefore increases vibrations and amplitude of motion. What may determine the final stability of the rotor–impeller–bearing–seal dynamic system is often dependent on the net effect of these two opposing forces—direct damping and the destabilizing cross-coupled stiffness (negative damping) present in the bearings, seals, and impellers.

The degree of the destabilizing cross-coupled stiffness in a fixed-geometry bearing is mostly influenced by the circular geometry in the bearing and the fluid rotation. Although load and eccentricity also play a role in the stability of these bearings, the circular geometry and fluid rotations are directly related to the bearing design and configuration. These two parameters can be changed and/or modified to help an unstable or marginally stable bearing become more stable. Therefore, fixed-geometry bearings can achieve a higher stability threshold by adding axial grooves as shown in Fig. 5-5 to reduce the fluid rotation. The more grooves that are added, the more the net fluid rotation in the bearing is reduced, and the higher the stability threshold. The increase in the stability threshold is demonstrated in Fig. 5-6, which shows an increase in the log dec for the first forward mode (the mode most likely to go unstable) as a function of the number of axial grooves. There is a limit to the addition of grooves in the bearing where it can become counterproductive. This is because as we add more grooves, we are effectively reducing the load area, increasing the bearing temperature, and eventually reducing the direct damping. There is also the possibility of the load vector being directly in line or very close to the groove, which will affect the flow of oil into the bearing. Thus, it is important to note that improving one aspect of the bearing performance may have deleterious effects on other operating parameters. Therefore, a complete bearing design must examine all aspects of the bearing performance and limits.

Additional and more practical gains in stability with fixed-geometry bearings can be achieved by altering the circular geometry. This is

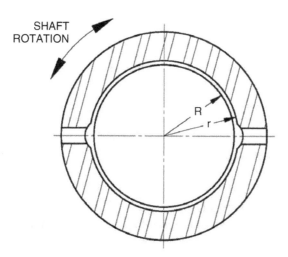

Figure 5-5   Sleeve bearing with two axial grooves.

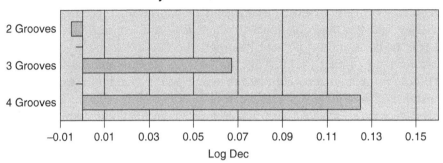

Figure 5-6   Stability as a function of axial grooves.

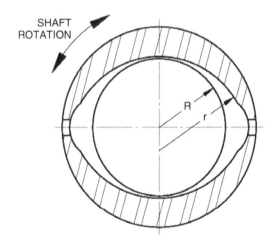

Figure 5-7   Elliptical or lemon bore journal bearing.

achieved by using preloaded lobes, often referred to as elliptical or lemon bore in the case of 2-lobe bearings (Fig. 5-7), and preloaded 3-lobe or 4-lobe bearings as shown in Figs. 5-8, and 5-9. The elliptical 2-lobe bearing and the preloaded 3- and 4-lobed bearings are bidirectional and this feature makes them more commonly used. The elliptical bore essentially provides a geometric preload on the shaft journal. Typical preloads used in elliptical bearings can range from 0.25 to 0.5. A preload of 0.5 means that the machined bore (bore machined with shims at the split line) will have a clearance twice of the assembled bore (vertical bore once the shims are removed from the split line). Thus, the horizontal clearance in this case is twice as large as the vertical clearance and provides a significant asymmetry in the fluid film direct stiffness coefficients. The plot in Fig. 5-10 shows the effect preload has on the stability of a

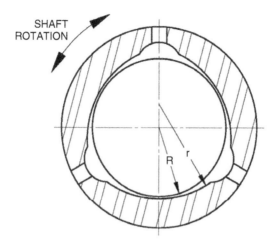

Figure 5-8   Three-lobe preloaded journal bearing.

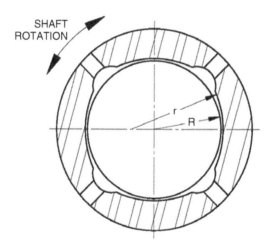

Figure 5-9   Four-lobe preloaded journal bearing.

rotor–bearing system. The plot shows that there is an optimum value for preload and as we keep increasing the preload the stability will start to degrade. This is because the stiffness in the bearing increases with increasing preload and, as a result of that, the damping becomes less effective. Another important characteristic of the preload optimization plot shown in Fig. 5-10 is the slope of the curve below and above the optimum point. In the lower preload range (zero preload to 0.25), the curve is steeper and any wear will cause the stability to drop significantly. Above the optimum point (0.25 to 0.5), the slope appears to be more

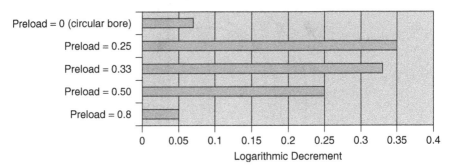

Figure 5-10 Influence of preload on stability.

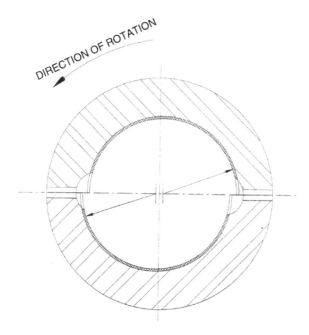

Figure 5-11 Offset half sleeve bearing.

gradual and suggests that a proper design for long-term operation should target a preload value slightly above the optimum point.

The offset half bearing shown in Fig. 5-11 also produces a more stable bearing in comparison to a circular sleeve bearing. The offset half can be a circular bearing with the upper and lower halves offset or it can be an elliptical bearing with an offset between the upper and lower halves. Both the elliptical and offset half bearings produce a significant asymmetry in the fluid film direct stiffness coefficients. Tripp and Murphy [4]

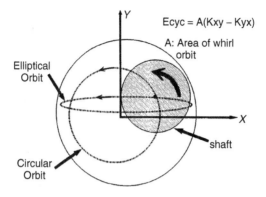

Figure 5-12 Schematic showing the stabilizing effects of asymmetry.

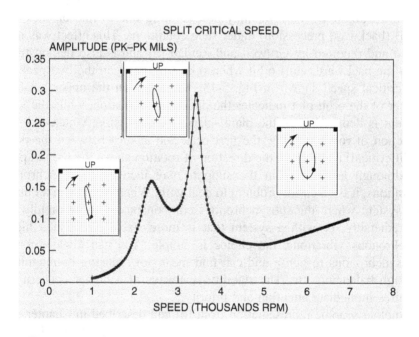

Figure 5-13 Split critical and backward precession or backward whirl.

have shown how this asymmetry combines with the cross-coupled stiff-ness coefficients to influence the total energy added to the dynamic system. Energy added is destabilizing, while energy dissipated due to direct damp-ing is stabilizing. The net effect on stability can be physically explained by the schematic shown in Fig. 5-12 and the following equation:

$$E_{cyc} = A\,(K_{XY} - K_{YX}) \tag{5-1}$$

The energy added to the system is calculated by integrating the force due to the bearing cross-coupled coefficients over the displacement around the closed curve of the whirl orbit. Therefore, the energy added (destabilizing) to the dynamic system is equal to the product of the whirl orbit area (A) times the net value of $(K_{XY} - K_{YX})$. The more asymmetric the whirl orbit, the smaller the orbit area and the less destabilizing energy is added to the dynamic system. As is true with most engineering applications, there is a penalty associated with this approach. Asymmetry, which tends to reduce the destabilizing influence of cross-coupled stiffness, results in higher synchronous vibration amplitudes along the axis with the lower stiffness. The asymmetry may also cause a split critical speed as shown in Fig. 5-13, which, depending on its proximity to the operating speed, may reduce the safe operating range of the machine. Operating between the peaks of the split critical speed can also make the machine very sensitive to rub, since the rotor may be executing line to line or backward whirl (backward precession) under such conditions. This effect was measured and reported by Gruwell and Zeidan [5] and is also demonstrated with the backward whirl orbit when operating between the two peaks of the critical speed shown in Fig. 5-13. The arrow in the upper left-hand corner of the orbit plot indicates the direction of rotation, while the whirl motion is deduced from the blank–dark dot precession, which is in the direction of rotation below the first peak and against between the peaks (split critical) and then in the direction of rotation above the second peak.

Although asymmetry in the support may increase the synchronous vibrations, it is an easier problem to deal with. When faced with an unstable system where the subsynchronous component can grow rapidly and unpredictably, or with a system that is more stable but with a higher synchronous vibrations, the choice is simple. You can always control the synchronous response and can find means of reducing its magnitude through balancing, etc. The stability is always of a higher concern and requires immediate attention and priority.

Another graphic representation from [6] and described in Chapter 3 on the effect of orbit ellipticity is worth mentioning here as well. The follower force or destabilizing force resulting from the cross-coupled stiffness coefficients tends to be in line with the instantaneous velocity of the shaft only when the orbit is circular. This is when the follower force is most effective and, in fact, most destructive as it transfers the destabilizing energy into the rotor dynamic system. When the orbit is elliptical, the follower force and the instantaneous velocity are no longer in line, and the effectiveness of the follower force in adding destabilizing energy to the rotor is diminished.

The canted bore bearing is a special type of the 3-lobe preloaded bearings with a taper or a fixed tilt. The lobes in this bearing configuration are continuously converging across each segment in the direction of rotation. A schematic of such a bearing configuration is shown in Fig. 5-14 and it resembles a preloaded pad with a fixed pretilt about a large offset pivot location in the direction of rotation. This bearing is used in some high-speed integrally geared compressors. They offer very high stiffness, which is not necessarily a good thing for most rotordynamic issues. Cavitation is a major problem with these bearings when used in high-speed applications.

The pressure dam bearing, shown in Fig. 5-15, is similar to a plain two-groove sleeve bearing with one relief track in the bottom half in the midsection of the bearing or two relief tracks at the ends. In the upper half it has a relief track in the midsection of the bearing and extends from the split line to around 120–135 degrees in the direction of rotation. The relief track in the upper half comes to an abrupt sharp edge or *dam*. The sharp edge was thought to be critical to the proper operation of the pressure dam. Childs et al. [7] showed that not to be true and, in fact, a radius was determined to provide better stability than an abrupt or

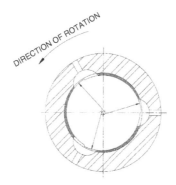

CANTED BORE JOURNAL BEARING

Figure 5-14   Canted bore journal bearing.

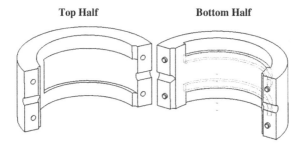

Figure 5-15   Pressure dam bearing.

sharp edge at the bottom of the dam. There are two main features that make the pressure dam bearing more stable. The first and most important is due to the pressure from the fluid inertia effects hitting the dam and generating an additional downward load on the journal. This forces the shaft into a position of greater eccentricity and, consequently, greater stability. The other aspect is due to the relief in the bottom half of the bearing, which essentially reduces the effective length of the bearing. This serves to increase the operating eccentricity and thus further enhances the stability of the bearing. The important parameters that influence the effectiveness of the pressure dam bearing are the optimum depth and width of the dam, the angular position of the dam, and the width of the relief track in the bottom half. These parameters must be optimized to achieve higher stability values while still maintaining good synchronous response characteristics. A good starting value for the optimization process for the pressure dam depth is a value five times the radial clearance based on Someya's layout [8]. Nicholas and Allaire [9] performed extensive theoretical and experimental calculations and found this value to be 2–3 times the radial clearance for stability. Often improving the stability through the use of a pressure dam bearing results in higher amplification factors when traversing the critical speed, as shown by Carlson and Zeidan [10].

All the bearings discussed thus far fall in the category of what is commonly referred to as fixed-geometry or fixed-profile journal bearings. The geometry of the bearing is fixed (no moving parts) and cannot be changed or adjusted to different loading conditions. These bearings are simple and therefore very economical to manufacture. They are very limited in stability when used in machinery operating supercritical (above their first critical speed) and are very susceptible to variations in operating parameters such as load, alignment, and increase in wear and clearance. A general guide on the relative stability of the different fixed-geometry bearings can be assessed from the plot shown by Zeidan and Paquette, [11].

One of the major limitations of the fixed-geometry bearings in general is the fact that they are designed for a certain load magnitude and direction. In speed-increasing and speed-reducing gears, as shown in Fig. 5-16, the load magnitude and direction can cover a wide angle, making it very difficult for these bearings to perform well under such conditions. This is also true in integrally geared compressors and expanders where the radial load is the net effect of the weight of the rotor, which must be added vectorially to the gear reaction forces that are a function of the torque or power applied. In pumps, the side load from the pump volute must also be taken into account. In power generation and process steam turbines, the partial steam admission can also change the magnitude and direction of the load as shown in Fig. 5-17. A bearing that is designed for a load in the bottom lobe could end up with a load in line with the oil

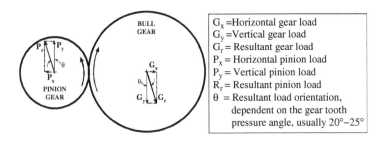

G_x	=Horizontal gear load
G_y	=Vertical gear load
G_r	= Resultant gear load
P_x	= Horizontal pinion load
P_y	= Vertical pinion load
R_r	= Resultant pinion load
θ	= Resultant load orientation, dependent on the gear tooth pressure angle, usually 20°–25°

Figure 5-16   Gear reaction forces in a speed-increasing gear box.

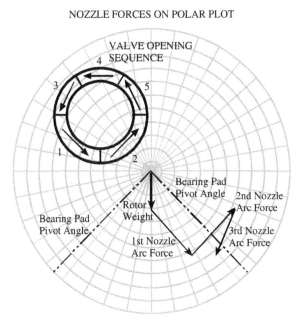

Figure 5-17   Partial steam admission effects on bearing load magnitude and direction.

feed groove. These are some of the drawbacks of fixed-geometry journal bearings and the reasons why in such equipment fixed-geometry bearings have a distinct limitation.

VARIABLE-GEOMETRY TILTING PAD BEARINGS

Variable-geometry tilt pad bearings are characterized by the inherent stability that arises from the low or negligible cross-coupling present in these bearings. The pads tilt or rotate about their pivot in response to the radial load applied by the shaft journal, and will always produce a reaction

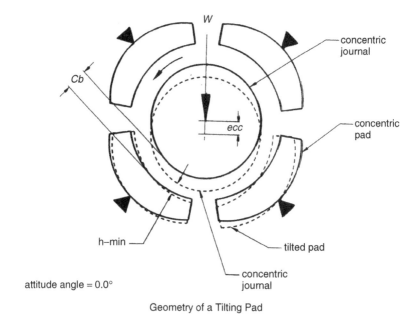

Geometry of a Tilting Pad

Figure 5-18 Schematic of a tilt pad bearing.

force that is in line with the shaft centers, as shown in Fig. 5-18. There is no attitude angle (attitude angle is zero) and therefore no corresponding cross-coupled stiffness. The attitude angle is almost zero provided the pad inertia and the friction in the pivot are low or negligible.

Conventional tilt-pad bearings use a point or a line contact for the pivot and achieve the tilt motion through this mechanism. These bearings came about as a natural progression from the invention of tilt-pad thrust bearings by Albert Kingsbury [12] in 1907 and independently by A.G.M. Michell [13], a highly respected Australian engineer in England, in 1905. In these bearings the tilt is achieved by rocking over the pivot and therefore the term rocker back tilt-pad bearing is used to describe them. The contact stresses can be very high on these bearings particularly when used in integrally geared compressors or when the radial load or weight is high. Under these circumstances, the pivot wear and brinneling will accelerate and cause a loss in the preload factor and a drop in stiffness and damping. Depending on the proximity of the critical speed to the operating speed, this may result in a loss of the separation margin and a further degradation in the bearing pivot leading to further increase in synchronous vibrations. The pivot wear can increase the bearing set clearance, and in many instances could significantly reduce the preload and cause the rotor–bearing dynamic system to become unstable.

The spherical pivot tilt pad bearing shown in Fig. 5-19 is often referred to as a ball-in-socket tilt pad bearing, was pioneered by CentriTech and

Figure 5-19   Ball-in-socket tilt pad bearing.

its founder Bernie S. Herbage [14]. This bearing tends to reduce the contact stresses in the pivot and allow the pad to tilt in the axial direction as well as in the circumferential direction. The former (axial tilt) allows the bearing to avoid edge loading caused by the shaft sag or misalignment between the rotor and the bearing. These bearings are advantageous for retrofitting older equipment since they provide a higher tolerance to deteriorating field conditions. They work very well when the radial loads applied are moderate. However, when used in high-speed integrally geared compressors, the highly loaded pads can lock in a fixed position under load. This can cause the bearing to behave like a fixed-geometry bearing resulting in subsynchronous vibrations.

The conventional variable-geometry tilt pad bearings offer greater stability than the fixed-geometry sleeve bearings, but this benefit comes at a cost and brings some additional drawbacks that we need to be aware of. By virtue of having multiple parts, the tilt pad bearings are more costly to make and require longer lead time. The unloaded pads are inherently susceptible to pad instability often referred to as pad flutter. The phenomenon is limited to the unloaded pads and often leads to Babbitt fatigue at the leading edge of the pad, as demonstrated by Adams [15]. In some applications, the pad instability can lead to significant wear in the pivots of the unloaded pads, increasing the clearance and significantly influencing the performance of the bearing.

The multipiece assembly in conventional tilt pad bearings, where each major component is made with a set of manufacturing tolerances, can lead to a significant tolerance stack-up during assembly. The stack-up can be a significant percentage of the bearing clearance, particularly for the smaller bearings (5 inches in diameter and lower). This has a significant effect on the range of preload that can be achieved and subsequently on the dynamic bearing coefficients. It is now a requirement (refer to API specifications 617 for compressors and 612 for turbines) to evaluate the rotordynamic stability and synchronous response at minimum, nominal, and maximum conditions due to the stack-up effect on the bearing coefficients.

The drawbacks with the conventional tilt-pad bearing have given rise to what is known as the Flexure Pivot tilt pad bearing shown in Fig. 5-20. This bearing achieves the tilt necessary for bearing stability through flexure of the web supporting the pad as shown in the schematic of Fig. 5-21. The construction of the Flexure Pivot tilt pad bearing eliminates the multiple piece features inherent with conventional tilt pad bearings, thus eliminating the relative movement between parts that produces wear. The fact that the pad, pivot, and bearing shell are all made from the same piece of material in the Flexure Pivot bearing also reduces the tolerance stack-up, as illustrated in the schematic of Fig. 5-22. The fact that the unloaded pads cannot drop from their pivots, and the higher rotational stiffness in these pivots, are two factors that are inherent in Flexure Pivot bearings. These features in the Flexure Pivot bearings either prevent pad flutter from taking place or move the threshold speed at which this takes place to a much higher speed than is attainable with conventional tilt pad bearings.

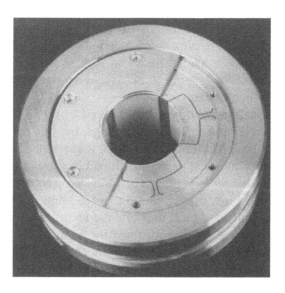

Figure 5-20 Flexure Pivot tilt pad bearing. Courtesy of KMC, Inc.

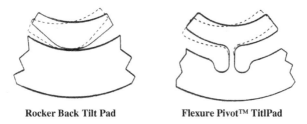

Rocker Back Tilt Pad   Flexure Pivot™ TitlPad

Figure 5-21   Schematic of rocker back and Flexure Pivot tilt pad bearings.

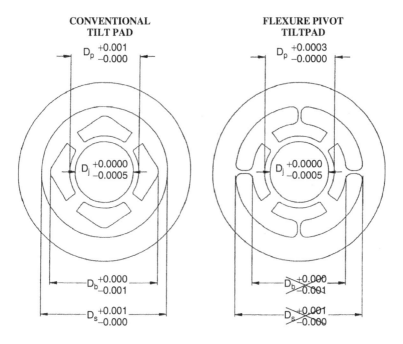

Figure 5-22   Tolerance stack-up as a result of manufacturing tolerances in multi components forming typical tilt pad bearings.

The advantages in reducing the tolerance stack-up with the Flexure Pivot bearings are highlighted in Chen et al. [16]. This reference shows a comparison between the conventional and the Flexure Pivot preload calculations due to tolerance stack-up. The possibility of achieving the optimum preload is greatly enhanced with the Flexure Pivot bearing due to the narrow range for preload variation. Furthermore, the manufacturing process that utilizes computer numerical controlled (CNC) combined with state of the art wire electric discharge machining (EDM) ensures a very high repeatability. Every part performs in the same manner and there is no variation in the dynamic coefficients between different batches of manufactured bearings.

FLUID FILM BEARING DYNAMIC COEFFICIENTS AND METHODS OF OBTAINING THEM

The Reynolds equation is derived from combining Newton's 2nd law for fluids (momentum transport equation) and the continuity equation (conservation of mass). The momentum transport equation is basically the Navier-Stokes equation, which has been simplified using the assumptions of incompressible, Newtonian, constant viscosity (isoviscous), laminar, isothermal, and neglecting the fluid inertia effects. These assumptions lead to an elliptical partial differential equation for the pressure distribution. This is what is commonly referred to as the Reynolds classical lubrication equation which can be solved by assuming certain boundary conditions.

The solution to the Reynolds equation provides the pressure distribution. The pressure can then be integrated around the journal to provide the fluid film reaction force required to counter the applied load. Since the pressure distribution is not symmetric in the case of journal and fixed-geometry bearings, this means that the film reaction force is not collinear with the applied load. Therefore, a component of the force is in line with the load and the other component is perpendicular to it. This vector relationship defines the eccentricity and the attitude angle shown in Fig. 5-3. Now that we have the eccentricity and attitude angle, the minimum film thickness and its circumferential location on the bearing surface can be determined. This analysis defines the static equilibrium position for the shaft journal in the bearing.

Once the static position is defined, the dynamic coefficients can then be obtained, assuming that the shaft journal vibrations or whirl has sufficiently small amplitude in reference to the clearance. This assumption is vital for the oil film dynamic coefficients to be assumed linear. Mathematically, the components of the oil film force are expressed as functions of coordinates X and Y of the journal center. These forces are not linear in each variable, but can be approximated with sufficient accuracy by the Taylor series of the first order as shown by Eqs. 3-18 and 3-19. This assumption is valid when vibration of the journal occurs in the vicinity of the static equilibrium position and both amplitude and velocity are small. Experimental determination of the dynamic coefficients follows a similar approach in terms of the journal bearing being subjected to a static loading and a hydraulic shaker is used to impart small displacements and velocities about this static position and the force coefficients are thus measured [7].

Sommerfeld in 1904 [17] proposed a solution to Reynolds equation that simplified it and solved for one of its limits. He used the infinitely long bearing limit and assumed the axial pressure gradient to be zero. In the process he defined a dimensionless parameter, known as Sommerfeld number, which is given as

$$S = \frac{\mu N L D}{W} \left(\frac{R}{C} \right)^2 \qquad (5\text{-}1)$$

The Sommerfeld number is helpful to designers, because it includes the following design parameters; bearing dimensions radius R and radial clearance C, viscosity μ, speed of rotation N, and the inverse of the bearing unit loading or pressure (LD/W); but it does not include the bearing arc. Therefore, the functional relationship can be obtained for bearings with different arcs, such as $360°$, $60°$, etc.

Raimondi and Boyd in 1958 [18, 19] gave a methodology for computer-aided solution of Reynolds equation using an iterative technique. For L/D ratios of 1, $\frac{1}{2}$, and $\frac{1}{4}$ and for bearing angles of $360°$ to $60°$ extensive design data are available. Charts have been prepared by Raimondi and Boyd for various design parameters, in dimensionless form and plotted with respect to the Sommerfeld number.

Someya et al. [8], through the activity of the Research Subcommittee on Dynamic Characteristics of Journal Bearings and Their Applications, which was established and organized in June 1979 through May 1982 within the Japan Society of Mechanical Engineers (JSME), compiled one of the most complete and useful collections of data for fluid film bearing coefficients. These coefficients, together with the static characteristics, have been calculated and also measured on a number of test rigs. A condensed sample of these tables extracted from [8] is shown in Tables 5-1 through 5-4 for illustrative purposes. The dimensionless stiffness parameters can be made dimensional by multiplying the dimensionless stiffness by the load W and divided by the pad bore radial clearance. Likewise the dimensionless damping parameter is multiplied by the load W and divided by the product of the pad bore radial clearance and journal velocity in radians per second. The value of the Sommerfeld number is first calculated using the bearing operating parameters that defines this number. Then interpolation or extrapolation takes place for the different L/D and the different preload tables shown for the particular bearing geometry and configuration. This process requires good experience in determining the proper temperature and viscosity to use and an extensive double interpolation between tables. The second and third author collaborated on automating this process, which the third author further developed into a more complete and comprehensive computer program, XLJRNL. This process provides a quick extraction of the dynamic coefficients without

Table 5-1   Dynamic coefficients for 4-pad bearing with LBP, $L/D = 0.5$, $m = 0.5$

S	e	Stiffness dimensionless coefficients				Damping dimensionless coefficients			
		K_{XX}	K_{XY}	K_{YX}	K_{YY}	C_{XX}	C_{XY}	C_{YX}	C_{YY}
0.976	0.1	14.7	0	0	14.7	16.4	0	0	16.4
0.474	0.2	8.03	0	0	8.03	8.49	0	0	8.49
0.301	0.3	6.13	0	0	6.13	5.96	0	0	5.96
0.208	0.4	5.39	0	0	5.39	4.75	0	0	4.75
0.150	0.5	5.13	0	0	5.13	4.08	0	0	4.08
0.109	0.6	5.2	0	0	5.20	3.69	0	0	3.69
0.0797	0.7	5.48	0	0	5.48	3.48	0	0	3.48
0.0571	0.8	6.01	0	0	6.01	3.40	0	0	3.4
0.0398	0.9	6.85	0	0	6.85	3.47	0	0	3.47
0.0264	1.0	8.16	0	0	8.16	3.64	0	0	3.64
0.0162	1.1	10.3	0	0	10.3	3.95	0	0	3.95
0.00868	1.2	14.4	0	0	14.4	4.74	0	0	4.74

Table 5-2   Dynamic coefficients for 4-pad bearing with LBP, $L/D = 0.5$, $m = 2/3$

S	e	Stiffness dimensionless coefficients				Damping dimensionless coefficients			
		K_{XX}	K_{XY}	K_{YX}	K_{YY}	C_{XX}	C_{XY}	C_{YX}	C_{YY}
0.419	0.1	25.2	0	0	25.2	18.5	0	0	18.5
0.204	0.2	13.4	0	0	13.4	9.51	0	0	9.51
0.13	0.3	9.87	0	0	9.87	6.64	0	0	6.64
0.0914	0.4	8.41	0	0	8.41	5.29	0	0	5.29
0.067	0.5	7.84	0	0	7.84	4.58	0	0	4.58
0.0499	0.6	7.8	0	0	7.80	4.22	0	0	4.22
0.0371	0.7	8.14	0	0	8.14	4.03	0	0	4.03
0.0273	0.8	8.84	0	0	8.84	3.99	0	0	3.99
0.0195	0.9	10	0	0	10	4.07	0	0	4.07
0.0134	1.0	11.9	0	0	11.9	4.26	0	0	4.26
0.00848	1.1	15	0	0	15	4.59	0	0	4.59
0.00473	1.2	21	0	0	21	5.17	0	0	5.17

the convergence problems associated with the more complex computer codes, but is limited to the bearing geometries these tables were developed for. Furthermore, it does not allow for loads that are not on pivots or between pivots, and for different bearings arcs or different bearing pivot offset positions.

Table 5-3 Dynamic coefficients for 4-pad bearing with LBP, $L/D = 1$, $m = 0.5$

S	e	Stiffness dimensionless coefficients				Damping dimensionless coefficients			
		K_{XX}	K_{XY}	K_{YX}	K_{YY}	C_{XX}	C_{XY}	C_{YX}	C_{YY}
0.617	0.1	14	0	0	14	18.7	0	0	18.7
0.3	0.2	7.65	0	0	7.65	9.61	0	0	9.61
0.192	0.3	5.85	0	0	5.85	6.74	0	0	6.74
0.134	0.4	5.14	0	0	5.14	5.32	0	0	5.32
0.097	0.5	4.93	0	0	4.93	4.54	0	0	4.54
0.0718	0.6	4.97	0	0	4.97	4.08	0	0	4.08
0.0529	0.7	5.13	0	0	5.13	3.78	0	0	3.78
0.0385	0.8	5.56	0	0	5.56	3.64	0	0	3.64
0.0275	0.9	6.32	0	0	6.32	3.64	0	0	3.64
0.0187	1.0	7.51	0	0	7.51	3.76	0	0	3.76
0.0119	1.1	9.46	0	0	9.46	4.07	0	0	4.07
0.00669	1.2	13.2	0	0	13.2	4.58	0	0	4.58

Table 5-4 Dynamic coefficients for 4-pad bearing with LBP, $L/D = 1$, $m = 2/3$

S	e	Stiffness dimensionless coefficients				Damping dimensionless coefficients			
		K_{XX}	K_{XY}	K_{YX}	K_{YY}	C_{XX}	C_{XY}	C_{YX}	C_{YY}
0.285	0.1	25.2	0	0	25.2	21.3	0	0	21.3
0.139	0.2	13.3	0	0	13.3	10.9	0	0	10.9
0.0884	0.3	9.63	0	0	9.63	7.51	0	0	7.51
0.0624	0.4	8.07	0	0	8.07	5.91	0	0	5.91
0.046	0.5	7.45	0	0	7.45	5.04	0	0	5.04
0.0347	0.6	7.37	0	0	7.37	4.56	0	0	4.56
0.0262	0.7	7.63	0	0	7.63	4.29	0	0	4.29
0.0196	0.8	8.25	0	0	8.25	4.20	0	0	4.2
0.0143	0.9	9.28	0	0	9.28	4.25	0	0	4.25
0.01	1.0	10.9	0	0	10.9	4.42	0	0	4.42
0.00659	1.1	13.7	0	0	13.7	4.73	0	0	4.73
0.00383	1.2	19.5	0	0	19.5	5.27	0	0	5.27

The procedures described above provide a quick approximation for obtaining the dynamic coefficients and are often referred to as "look-up" tables. This may not be adequate for a detailed and accurate analysis, particularly when the thermal effects that have a major impact on the viscosity are not taken into consideration. A review of how the Energy and Reynolds equations are solved to provide a more accurate and detailed analysis is shown next.

The more comprehensive computer codes provide a solution to the Reynolds equation for a given bearing geometry and loading. This involves choosing (in fact, guessing) an operating position (eccentricity), finding pressures (related to the cube of film thickness and to its derivative), integrating the pressures to find loads, evaluating trigonometric relationships for the film thickness and its position, and updating the operating position or eccentricity. This procedure is repeated iteratively until the operating position generates pressures that balance the applied loads. The convergence characteristics of this iterative procedure suggest that there will be convergence problems, and this is often the case.

The convergence problems and lengthy iterative schemes in bearing codes have been improved by further simplification of the Reynolds equation. The formulation for pressure in two dimensions is reduced to a variation for pressure in only the circumferential direction by assuming a parabolic pressure profile in the axial direction. This assumption has been validated by the work of Shelly and Ettles [20] and by Jones and Barrett [21] to not have a large impact on the accuracy of the solution.

The most difficult step in analyzing fluid film bearings continues to be the determination of the equilibrium position for the shaft in the bearing. As discussed above this position is determined by balancing the hydrodynamic pressure forces with the applied shaft loads. Since the Reynolds equation for hydrodynamic pressure is dependent on the fluid film viscosity, a form of the energy equation is used to establish the film temperature profile, and related viscosity profile. However, since the Energy equation is dependent on the fluid film pressures, the Reynolds and Energy equations are coupled. This coupling further adds to the convergence problems in the analysis of both fixed-geometry and tilt pad bearing programs. The third author often uses the previous procedure defined as look-up tables to establish a value for the eccentricity. This value for eccentricity is then input into the more sophisticated bearing analysis codes as a first guess, a step that tends to help convergence.

Dynamic analysis of bearings requires the determination of the position and velocity derivatives of the integrated bearing pressure profile. In the case of fixed pad bearings, this produces a set of eight stiffness and damping coefficients. In tilt pad bearings, the additional pad degrees of freedom generate a large number of stiffness and damping coefficients depending on the number of pads and the degrees of freedom for each pad. The full set of coefficients is reduced to a frequency dependent set of eight stiffness and damping coefficients by an explicit or implicit reduction. The pad assembly method, following the work of Lund [22] and Nicholas [23], involves an implicit reduction to the synchronous coefficients (synchronous speed). Others such as Parsell et al. [24], and Wilson and Barrett [25] have extended the pad assembly method to allow for full eigenvalue reduction of these coefficients. Their work indicates that eigenvalue reduction is most important for low preload, high Sommerfeld number bearings. Branagan [26] extended the development of the approximate thermal analysis algorithms to consider cross-film viscosity in fixed-geometry and tilting pad bearings and also considered the effects of deformations (thermal, mechanical, and pivot) on tilting pad bearings. The analysis provided by him allows for synchronously reduced coefficients as well as nonsynchronous reduced coefficients. The latter tends to produce stability predictions that are closer to those actually measured. However, the latest API specifications (API 617, 7th ed.) have adopted the synchronously reduced coefficients as the universal standard in order to achieve conformity.

Now that we have been able to review the methods used in obtaining the dynamic coefficients, it would be beneficial to see how these influence the rotordynamics of the rotor–bearing system. Although tilt pad bearings, like fixed-geometry sleeve bearings, come in many variations and configurations, they have many parts and therefore more variations are available. These variations allow greater flexibility for the designer and rotordynamicist to tailor the bearing for the application. These variables include the orientation of the bearing for a load on pivot (LOP) versus a load between pivots (LBP), the number of pads used, the pad pivot offset angle, axial length of the pad, clearance, lubricant viscosity, and preload. They are all factors that can modified to achieve the desired thermal and rotordynamic characteristics. These factors will be examined further to provide a better understanding of their use in the bearing design and influence on the rotordynamics of modern turbomachinery.

LOAD BETWEEN PIVOTS VERSUS LOAD ON PIVOT

The orientation of a tilt pad bearing with respect to the radial load plays an important role as this directly relates to the stiffness and damping characteristics of the bearing. The schematic shown in Fig. 5-23 illustrates the orientation with respect to the load vector for a LOP and a LBP tilt

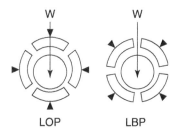

Figure 5-23   Schematic representation of LOP versus LBP in a 4-pad bearing.

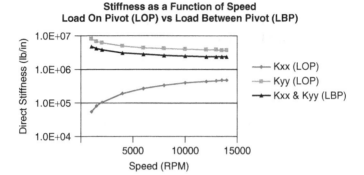

Figure 5-24 Direct stiffness coefficients for a 4-pad bearing in LOP versus LBP configuration.

pad bearing. The vertical and horizontal stiffness shows a large stiffness asymmetry for the LOP orientation, as shown in the plot in Fig. 5-24, whereas the LBP shows very symmetric characteristics and the vertical and horizontal stiffness curves are virtually identical and cannot be discerned in the plot. Furthermore, the LBP configuration tends to allow both bottom pads to share the load and thus this configuration offers higher load capacity than the LOP configuration. This is apparent in the plot shown in Figs. 5-25 and 5-26 for the LBP and LOP, respectively. In the LBP configuration the peak pressure in either of the two loaded pads is around 800 psi. In the LOP case the peak pressure reaches 1140 psi, which exceeds some of the conservative limits often employed for long term operation.

While it is clear that the LBP offers higher load capacity and most likely better synchronous response characteristics, there are applications in which the LOP offers a better solution. The benefits of the LOP configuration become more apparent when coupled with a high degree of aero cross-coupling. The stability analysis for an industrial high-pressure centrifugal compressor showed that the basic log dec (the log dec evaluated with no aero cross-coupling or any destabilizing seal effects) was higher for the LBP configuration, as expected. This is shown in the eigenanalysis and mode shape plots for the first forward mode in Figs. 5-27 and 5-28 for the LBP and LOP, respectively. However, when the aero cross-coupling effects were included in the model, the log dec for the LOP dropped, as expected, but not at the same rate as for the LBP case. The stability for the first forward mode with the LOP bearing configuration resulted in a more stable system than with the LBP. This is mainly due to the beneficial asymmetry effects introduced by the bearings, which more than compensate for the lower damping. The mode shape for the first forward mode with the LBP and LOP with aero cross-coupling included is shown in Figs. 5-29 and 5-30. One can clearly see the difference in the orbit

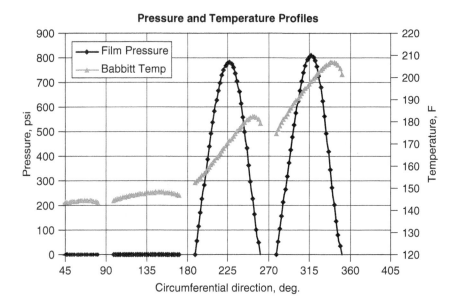

Figure 5-25   Pressure and temperature profile on a 4-pad LBP bearing.

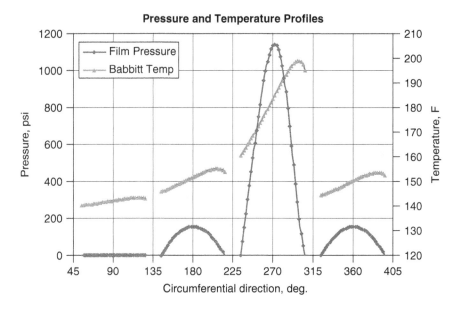

Figure 5-26   Pressure and temperature profile on a 4-pad LOP bearing.

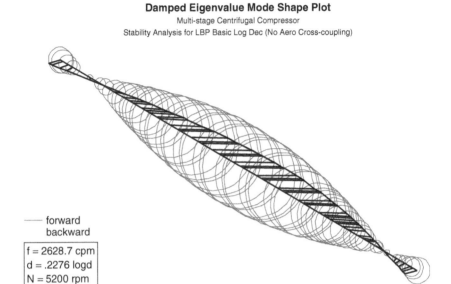

Figure 5-27 Natural frequency for first forward mode with basic log dec for LBP bearings.

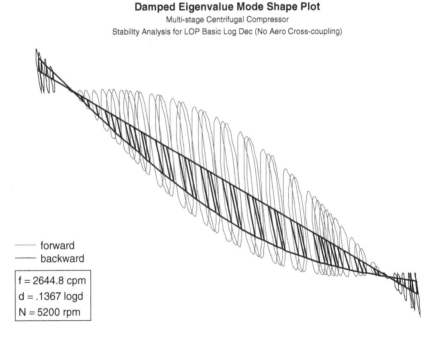

Figure 5-28 Natural frequency for first forward mode with basic log dec for LOP bearings.

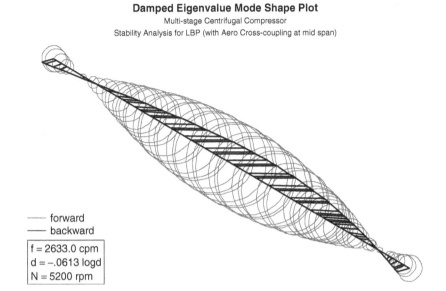

Figure 5-29 Natural frequency for first forward mode with aero cross-coupling for LBP bearings.

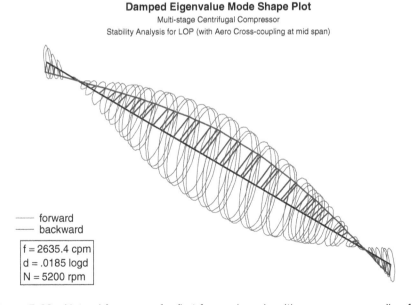

Figure 5-30 Natural frequency for first forward mode with aero cross-coupling for LOP bearings.

shape for each of the two bearing configurations and the high degree of ellipticity introduced by the LOP configuration.

However, when we examine the response to imbalance, it is clear there is a distinct advantage for the LBP configuration. The response plot in Fig. 5-31 clearly shows a more damped and symmetric response exhibited by the identical amplitudes predicted for the vertical and horizontal response. The amplification factor is also lower for the LBP with 11.08 as opposed to the 17.58 for the LOP configuration shown in Fig. 5-32. The LOP response also shows a split critical which is introduced by

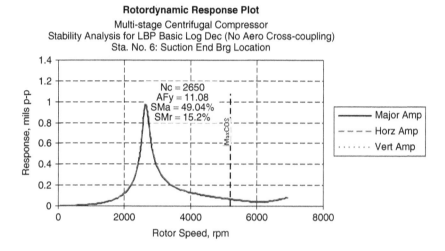

Figure 5-31 Unbalance response with LBP bearings.

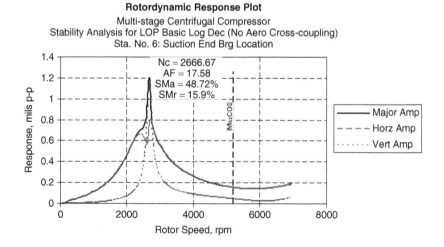

Figure 5-32 Unbalance response with LOP bearings.

Table 5-5   Stability and imbalance response summary for LBP and LOP 4-pad Bearing

| Bearing configuration | Logarithmic decrement | | Amplification factor |
	Basic log. dec. (no aero cross-coupling)	Log. dec. (with aero cross-coupling)	
4-pad LBP	0.2276	−0.0613	11.08
4-pad LOP	0.1367	0.0185	17.58

the asymmetric stiffness characteristics for this bearing configuration. A comparison of the two bearings configuration is summarized in Table 5-5 and shows that the synchronous response is better with the LBP. The LOP is advantageous only when relatively high cross-coupled stiffness values are present.

INFLUENCE OF PRELOAD ON THE DYNAMIC COEFFICIENTS IN TILT PAD BEARINGS

The geometric preload in tilt pad bearings is a critical and effective parameter to alter the magnitude of the force coefficients. Preload can be explained in the graphic shown in Fig. 5-33. A positive preload is essential also from the standpoint of providing a larger inlet for the oil at the leading edge of the bearing. In contrast, zero preload due to manufacturing tolerance can quickly lead to a negative preload. This will reduce the effectiveness of the oil entering the leading edge of the bearing and will cause pad flutter or pad instability.

Increasing the preload in the bearing, as shown in Fig. 5-34, will result in higher stiffness particularly at the higher Sommerfeld numbers. Increasing the stiffness in the bearing may be desirable from the standpoint of shifting a critical speed located just above operating speed further up in frequency to provide a better separation margin. However, it is also important to note that while increasing the preload will increase the stiffness, it will result in a lower effective damping. As discussed earlier in Chapter 3 and as will be shown later when we discuss squeeze film dampers, the lower stiffness is essential for the damping to be effective. There have been several incidents when one would attempt to provide a higher separation margin from synchronous vibrations by increasing the preload and stiffness, and the end result is a subsynchronous instability due to the loss of effective damping. Therefore, the total rotordynamic consequences of a change should address both stability (eigenanalysis) and synchronous response (forced analysis). The latter determines the critical

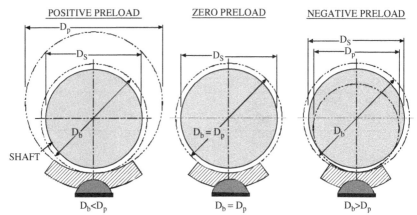

POSITIVE PRELOAD ZERO PRELOAD NEGATIVE PRELOAD

$D_b < D_p$ $D_b = D_p$ $D_b > D_p$

D_b = ASSEMBLED (SET BEARING DIAMETER C_b = BEARING SET CLEARANCE = $D_b - D_s$
D_p = PAD (MACHINED) DIAMETER C_p = PAD BORE CLEARANCE = $D_p - D_s$
D_s = SHAFT DIAMETER

$$\text{PRELOAD} = 1 - \frac{C_b}{C_p} = 1 - \frac{D_b - D_s}{D_p - D_s} = \frac{D_p - D_b}{D_p - D_s}$$

*D_b CAN BE CHANGED IN CERTAIN DESIGNS TO ALTER THE PRELOAD ONCE
THE BEARING HAS BEEN MANUFACTURED. OTHER FACTORS THAT MAY
INFLUENCE THE PRELOAD IN OPERATION ARE THERMAL DISTORTION OF
THE PAD, AND MECHANICAL DEFORMATION OF THE PAD AND PIVOT SUPPORT.

Figure 5-33 Schematic representation of positive, zero, and negative preload in tilt pad bearings.

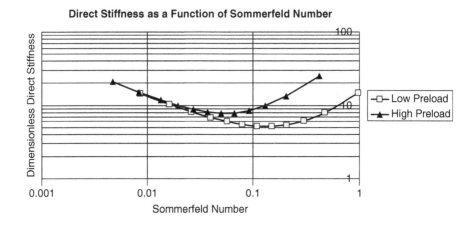

Direct Stiffness as a Function of Sommerfeld Number

Dimensionless Direct Stiffness

Sommerfeld Number

- Low Preload
- High Preload

Figure 5-34 Effect on preload on the direct stiffness.

speed location and the safe separation margin between the critical speed and the operating speed.

INFLUENCE OF THE BEARING LENGTH OR PAD LENGTH

Although manufacturers tend to standardize on a certain L/D ratio for their bearings, field conditions and varying process requirements may push the standard bearing design close to its limit. One of the parameters that can be changed to impart a desirable change in the stiffness and damping coefficients in a bearing is the pad length. This approach has been utilized in addressing stability problems as well as critical speed problems. One such field problem that the second author experienced, which left a lasting impression in his mind, was the redesign of a bearing to help shift the critical speed on a high-speed compressor. A meeting was held to address this critical speed that was just above operating speed. This meeting was attended by the chief compressor engineer and the chief turbine engineer representing the two product lines offered by this manufacturer. While everyone agreed that the problem (critical speed separation) should be addressed by increasing the bearing stiffness, the two chief engineers offered opposing views on the effects the length has on the bearing and its stiffness characteristics. The compressor chief engineer, who is experienced with relatively light rotors running at high speeds and thus high Sommerfeld numbers, suggested that increasing the length will increase the stiffness. The turbine engineer whose experience focused on heavier rotors running at lower speeds, and therefore generally lay in the lower range for the Sommerfeld numbers, stated that reducing the length would result in higher eccentricity and therefore higher stiffness. What was unique about this dilemma was the fact that both chief engineers were correct in their assessment based on the range of equipment they were experienced with. The turbine engineer was more used to dealing with driving equipment, which tend to be relatively heavy and runs at lower speeds at least in process and conventional power generation equipment. A heavier rotor and lower speed means that the range of Sommerfeld number covers the left side of the dimensionless stiffness curve shown in Fig. 5-35. When operating in that region, decreasing the length in the expression for the Sommerfeld number will result in a lower value for the Sommerfeld number and moves it further to the left increasing the stiffness. On the other hand, driven equipment, such as compressors in this case, tend to be relatively light and run at higher speeds, and therefore lie in the higher range of the Sommerfeld number. In this case, increasing the length in the Sommerfeld expression will increase the Sommerfeld number and shift the value to the right in the right part of the curve shown in Fig. 5-35, thus resulting in a higher stiffness.

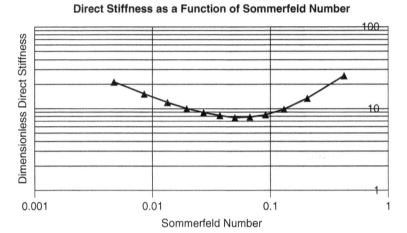

Figure 5-35   Dynamic stiffness as a function of Sommerfeld number.

INFLUENCE OF THE PIVOT OFFSET

The pivot offset is generally expressed in dimensionless form as the ratio of the angle from the leading edge to the pivot divided by the pad arc. The influence of the pivot location or offset position has been established in the use of tilt pad thrust bearings to increase load capacity and also reduce operating temperatures. The schematic in Fig. 5-36 shows that with the offset pivot, the pad rotation will result in a larger gap at the leading edge and thus more cool oil or lubricant will enter the pad. Although both of these effects—load capacity and cooler temperatures—are also true when applied to journal (radial) bearings, the offset can also be beneficial from the distinct rotordynamic characteristics such a parameter has on the dynamic coefficients.

Integrally geared compressors, such as the one shown in Fig. 5-37, tend to have a relatively large overhung mass (the impellers) and run at very high speeds and very high loads imparted by the gear reaction forces. The bearings designed for these compressors often utilize the offset pivot feature to provide lower temperatures at these speeds and loads as well as higher stiffness. The compressor in reference [16] operated above two rigid modes and above a flexible mode with a conventional center pivot bearing. The use of the Flexure Pivot offset pivot bearings allowed operation above the two rigid modes but below the flexible mode, as shown in the undamped critical speed map presented in reference [16]. This is significant in terms of allowing the unit to tolerate more imbalance and eliminating the necessity for trim balancing. The other important advantage offered with the offset pivot is the reduced sensitivity to bearing

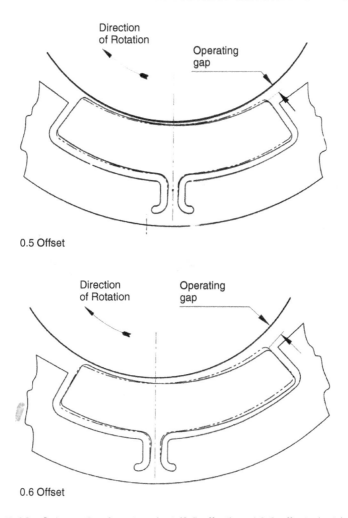

Figure 5-36 Schematic of center pivot (0.5 offset) and 0.6 offset pivot bearing.

clearance when evaluating stability. The unit had a larger range of stability with increased clearance, while it had a much narrower stability range with the center pivot pads.

INFLUENCE OF THE NUMBER OF PADS

The optimum number of pads in the case of thrust bearings has been established and centers on having an optimum aspect ratio. This more or less determines the optimum number of pads for thrust bearings. This is not the case for journal (radial) bearings as there are other aspects that need to be considered, in particular, their influence on the rotordynamics

Figure 5-37 Typical integrally geared compressor.

of the system. The predominant number of pads used in tilting pad journal bearing has been five pads in the past, but this trend is migrating toward the use of 4-pad bearings today.

The 4-pad bearings tend to provide a larger unit load area than a 5-pad bearing and thus can carry a higher load. They also tend to provide better synchronous response characteristics, and these two advantages are the main reason why more of the new equipment uses 4-pad bearings. The plot shown in Fig. 5-38 shows the advantages offered with the 4-pad in a high-load and high-speed integrally geared compressor. The x axis has been normalized to provide the temperature profile for the pad in both a 4-pad and a 5-pad bearing, since they have different arc length. The 4-pad bearing runs at roughly $30°F$ lower temperature, although this is partly due to the offset pivot configuration as well.

There are also applications, particularly in gears or integrally geared compressors, where operation dictates different loading direction and magnitude. The bearing dynamic stiffness coefficients will see a significant variation in the direct stiffness as the load changes directions from LOP to LBP. This variation in the direct stiffness can be minimized as the number of pads is increased, as illustrated in Fig. 5-39. Some gears use 7-pad tilt pad bearings for this same reason. In some instances the limited radial space dictates the use of more pads to reduce the pad deflections. The narrower radial space may result in very thin pads with long arcs

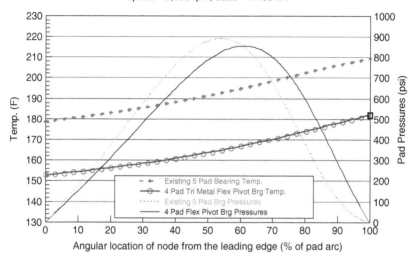

Figure 5-38 Temperature profile from leading to trailing edge of a 5-pad and a 4-pad bearing.

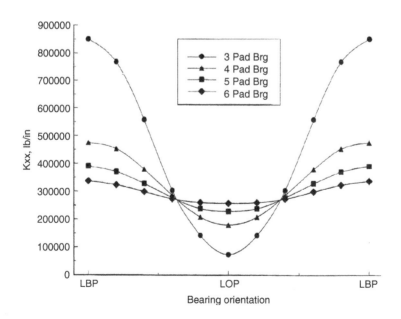

Figure 5-39 Variation in stiffness for different numbers of pads as the load swings from LBP to LOP.

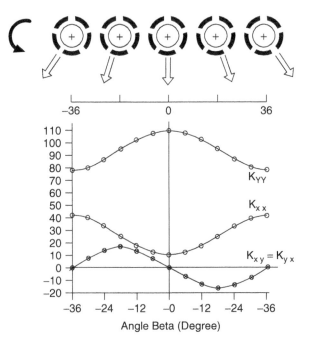

Figure 5-40 Cross-coupled coefficients in a tilt pad bearing when the load is between LOP and LBP.

and these may experience excessive thermal and mechanical deformations. Another reason for using more pads or in effect a shorter pad arc is the potential in the unloaded pads to experience pad flutter. The larger the number of pads and the subsequent shorter pad arc, the higher the threshold for pad flutter or pad instability. Another aspect that is not commonly considered when selecting the number of pads, but should be accounted for in some applications, is the increase in startup torque as the number of pads is reduced.

The swing in load direction with tilt pad bearings does produce cross-coupled stiffness coefficients even when the pad inertia and pivot friction are neglected. This is shown in Fig. 5-40 and is attributed to the coordinate transformation. The cross-coupling in this case is not destabilizing, since the K_{XY} and K_{YX} are of the same magnitude and sign. These cross-coupled coefficients do not produce any destabilizing force since they will always sum to zero.

BALL AND ROLLING ELEMENT BEARINGS

Ball bearings are used in many aerospace and aircraft applications due to their capacity for large transient overloads, slow and safe failure modes with premature warning, and low friction. The angular contact types are designed to support combined loading from both axial and radial directions by incorporating a contact angle between the balls and races that usually varies between 0 and 40 degrees. Ball bearings can tolerate high speeds, which are generally limited by a *DN* number <1,200,000, where *D* is the pitch diameter in mm and *N* is the speed in rpm. Ball bearings are also used in many land-based rotating machines because of their capacity for large transient overloads, ready "off-the-shelf" availability from standardized catalogs, and ease of installation in certain applications.

From the standpoint of rotor dynamics, the most important characteristics of ball bearings are their high radial stiffness and their very low damping. Neither characteristic is generally desirable. Chapter 3 showed the advantages of low support stiffness, and the desirability of an optimum amount of support damping for flexible rotors. In addition to high stiffness, another undesirable characteristic of angular contact ball bearings is that the stiffness is dependent on a number of variables, including speed, thrust, and misalignment. These undesirable characteristics can be overcome by mounting the ball bearing on a soft support with damping. If the structure supporting the bearing is soft enough, the ball bearing stiffness becomes irrelevant. This, along with the requirement for damping, is why all modern aircraft turbine engines have squeeze film dampers, as described in a following section.

Nevertheless, many rotating machines have been built with ball bearings mounted in stiff housings, due either to lack of understanding of the benefits of support flexibility or to other special considerations. For example, cryogenic turbopumps for rocket engines have been developed in the United States with ball bearings on stiff supports due to a belief that significant damping is not attainable in the cryogenic environment. Under that paradigm, precise knowledge of the bearing stiffness becomes very important, since it controls the critical speeds. With no significant damping, the critical speeds should be kept above the maximum running speed. This is usually not possible in high-speed turbopumps, so a critical speed "margin" at the running speed is mandatory. The following example is adapted from references [27] and [28].

CASE STUDY: BEARING SUPPORT DESIGN FOR A ROCKET ENGINE TURBOPUMP

An experimental and analytical study was carried out in the Turbomachinery Laboratory at Texas A&M University to determine if ball bearing stiffness values could be determined with adequate precision to accurately predict the critical speeds of a liquid hydrogen turbopump, assuming that the bearings would not be mounted on soft supports. The study went further, to determine the effect on rotordynamics of mounting the bearings on soft wire mesh dampers. Wire mesh dampers for these types of pumps had been explored earlier at Rocketdyne [29], and implemented by the Japanese in their pumps [30]. These devices are discussed in a following section in this chapter.

The type of turbopump that was studied is shown in Fig. 5-41. It has two hard-mounted, single-row, angular contact symmetric split inner race ball bearings, one at the pump end and one at the turbine end. Detailed views of the bearing mounting configurations are shown in Fig. 5-42. The

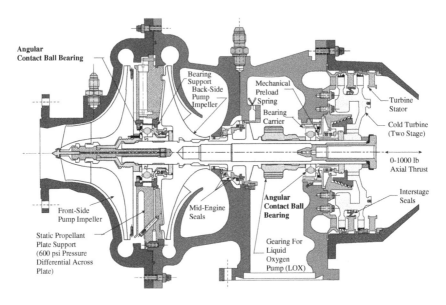

Figure 5-41   Turbopump cross section. From Ertas and Vance [27].

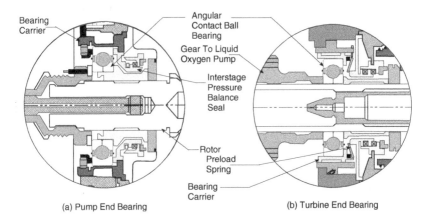

Figure 5-42   Turbopump bearing mounts. From Ertas and Vance [27].

pump end bearing is used to locate the rotor axially and the turbine end bearing is allowed to float axially by the use of an axial preload spring.

This turbopump was known to have a critical speed close to the maximum operating speed of 36,000 rpm. The lack of damping produces very narrow-banded critical speeds, so accurate (±5 percent) knowledge and control of the bearing stiffness could possibly allow avoidance of the critical speed by a "safe" margin. An XLROTOR computer model was developed, which has critical speed characteristics similar to those observed on the existing turbopump. This model was used to predict critical speeds, rotordynamic stability, response to unbalance, and dynamic bearing loads, with the bearings mounted on the stiff supports, as shown in Fig. 5-42.

Figure 5-43 is a computer-drawn sketch of the rotor–bearing model. The bearing supports are seen to be at stations 5 and 11. See Chapter 4 for a detailed description of XLROTOR models.

The bearings and supports are first modeled with a small amount of structural (hysteretic) damping, as no bearing dampers are incorporated into the original design. A small amount of internal damping was also included in the rotor assembly, which can be destabilizing. The exact

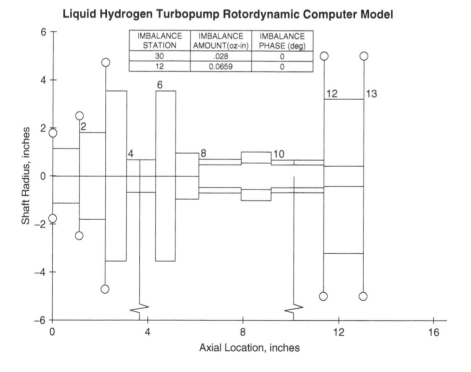

Figure 5-43 Turbopump computer model.

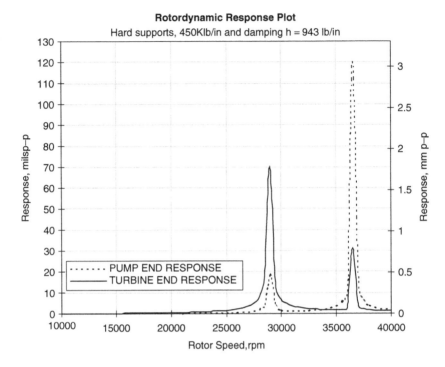

Figure 5-44 Computed synchronous response to unbalanced, hard supports.

amount depends on the tightness of the rotor assembly fits. Figure 5-44 shows the computer-predicted response to unbalance with 0.001" C.G. offset at the pump wheel and the turbine wheel.

Note the second critical speed just above the running speed of 36,000 rpm. The dynamic bearing load at the second critical speed is predicted to exceed 25,000 lb. Both eigenvalues associated with the two critical speeds are predicted to be unstable at the running speed primarily by the destabilizing internal rotor friction. Due to the narrow-banded response, it might be possible to operate safely with this design, assuming that the rotor fits are tight enough and assuming that the bearing stiffness values are accurate and will hold constant. In fact, the stiffness of angular contact ball bearings is known to be a function of speed, thrust load, and misalignment. A test apparatus was designed and built to measure the stiffness of the ball bearing in this turbopump as a function of those variables. See [27], [31], and [32] for details of the test rig and all of the measured results. A brief summary of a few of the experiments and results follows here.

Ball Bearing Stiffness Measurements

Figure 5-45 shows a cross section of the test apparatus. This apparatus was designed to make its critical speeds and nonsynchronous eigenvalues highly sensitive to the test bearing stiffness, so that those two measurements could be used as surrogates to the stiffness. Bearing misalignment, axial thrust, imbalance, and speed could all be varied in this test rig. Experimental measurements of the forward eigenvalues, backward eigenvalues, backward critical speeds, and forward critical speeds were matched with computer model simulations to determine the stiffness values for each of the tested cases.

Figure 5-46 shows the measured radial stiffness as a function of axial thrust for one of the tested misalignment cases. Considering all the cases tested, the radial stiffness of the tested ball bearing was shown to vary between 44,880 kN/m (0.1 million lb/in) and 186,850 kN/m (1.067 million lb/in) for a range of axial preloads of 0–1,148 N (258 lb). The measured radial stiffness varied by a ratio of 10:1. The applied range of thrust loads is realistic for this type of turbopump. The study showed conclusively that ball bearing radial stiffness cannot be reliably determined within the ±5-percent range for realistic variations in axial loading, radial dynamic loading, and relative race misalignment. However, the ball bearing stiffness can be made irrelevant by mounting the bearing on a support much softer than the bearing itself. All of the rotordynamic characteristics of

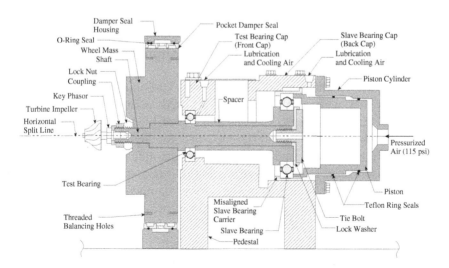

Figure 5-45   Cross section of the ball bearing test rig. From Ertas and Vance [27].

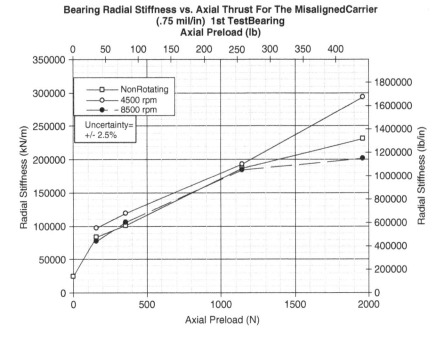

Figure 5-46   Measured stiffness versus axial thrust for one case. From Ertas and Vance [27].

the machine will be improved, especially if the support also has some damping.

Wire Mesh Damper Experiments and Computer Simulations

Several years of intensive research at the Turbomachinery Laboratory has shown that wire mesh rings can be used as an effective vibration damper for turbomachines. Zarzour and Vance [33] tested stainless steel wire mesh for a range of interference fits, for high operating temperatures, and in lubricated conditions. Their results indicated that metal mesh could be used in aircraft gas turbine engines as a substitute for squeeze film dampers. Further testing of wire mesh dampers, conducted by Al-Khateeb and Vance [34], also confirmed the presence of useful amounts of damping for both stainless steel wire mesh and copper wire mesh. They also investigated the behavior of the wire mesh element in parallel with a squirrel cage. Vivek and Vance [35] developed a refined hysteretic model for the damper in the form of spreadsheets. Although this research revealed significant amounts of damping, the effect of cryogenic temperatures was unknown. The possible application of wire mesh damper elements in liquid fuel rocket engines depends on the amount of damping retained at cryogenic temperatures and the overall impact on turbopump rotordynamics. Ertas and Al-Khateeb [36] measured damping and stiffness of steel and copper wire mesh rings at liquid nitrogen temperatures and found both coefficients to be slightly higher than at ambient conditions for most of the tested samples.

The same rotordynamic model from Fig. 5-43 above was modified with a set of metal mesh bearing supports replacing the hard supports. All rotor parameters were exactly the same as before. The wire mesh stiffness and damping coefficients were taken from the measurements in [36]. The simulation shows the two whirl modes to be rendered stable and with no rotor bending. The response to imbalance at both bearings is shown in Fig. 5-47.

Due to the damping available at the soft bearing supports, and the lower stiffness, the peak vibration response shows a decrease in major amplitude by a factor of 20 (compare with Fig. 5-44). Also note that the critical speeds are now well below the operating speed of the machine. The maximum dynamic bearing loads are predicted to be reduced by a factor

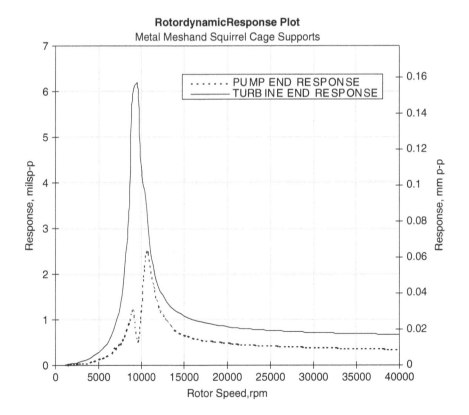

Figure 5-47 Computed synchronous response to unbalance, wire mesh supports.

of 250. This can be attributed to (1) soft supports, with lower stiffness whereby the lower critical speeds produce lower rotating imbalance loads, and (2) direct damping at the bearing supports.

This study showed the likelihood of achieving the following improvements in rotordynamic performance of a cryogenic fuel turbopump by mounting the bearings on metal wire mesh supports:

1. Reducing peak vibration amplitudes by a factor of 20 at the bearings
2. Reducing dynamic loads at the bearings by several orders of magnitude
3. Removing critical speeds from the normal operating range of the turbopump
4. Reducing rotor bending in any potential subsynchronous whirl modes
5. Stabilizing subsynchronous whirl modes that might otherwise be unstable from internal friction.

SQUEEZE FILM DAMPERS

Squeeze film dampers are not bearings by the strict definition of the word "bearing", which defines elements designed to carry or bear the static load and the dynamic load imparted by the rotating shaft journal and rotor. However, since they are always associated with bearings, a discussion of their design and application fits appropriately in this chapter. The historical development of squeeze film dampers and their relevance is described and discussed in Chapter 7.

Squeeze film dampers have traditionally been used in conjunction with a bearing to overcome stability and vibration problems that are not adequately handled with a conventional bearing alone. One of the key design attributes that the damper brings is the introduction of support flexibility and damping in the bearing/support structure. This translates to lower transmitted forces and longer bearing life, particularly for machinery designed to operate at supercritical speeds. Machinery that runs above the first critical speed is classified as supercritical and constitutes an increasing number of the new high-performance machinery manufactured today. The advantages of running supercritical were discussed theoretically in Chapter 3 and become apparent when computing the forces transmitted to the bearings. Lower stiffness in the supports will generally result in lower transmitted forces. Lower damping will also lower the transmitted forces, but some damping is required to go through the critical speed. Furthermore, damping is required to counter the destabilizing effects of cross-coupling. As discussed in Chapter 3 and shown later in this chapter through an example, there is an optimum amount of damping that needs to be quantified to achieve the most desirable stability characteristics. Some of the most commonly used dampers and their advantages and limitations will be discussed next.

Squeeze Film Damper without a Centering Spring

The simplest form of a squeeze film damper is what is commonly referred to as a "loose," "uncentered," or squeeze film damper without a centering spring. A schematic for such a configuration is shown in Fig. 5-48, where the outer race of the rolling element bearing or the outer shell in the case of a fluid film bearing is allowed to float, translate, and whirl, but is prevented from rotation or spinning. This loose anti-rotation mechanism allows the damper journal (outer race) to squeeze the oil or lubricant in the small clearance space (exaggerated here for illustrative purposes) formed by the outer race and the bearing housing or stator. This squeeze motion will build pressure in the film and generate a damping force.

The absence of a mechanical centering mechanism in this damper configuration means that the damper journal will be bottomed out at start-up. As the speed increases and the shaft starts to whirl, the damper's journal (bearing shell outer surface) will lift off. The oil film in a squeeze film damper does not produce stiffness (i.e., support a static

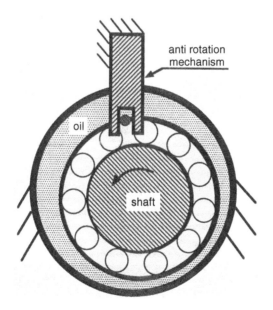

Figure 5-48   Schematic of a simple squeeze film damper without a centering spring.

load) like conventional fluid film bearings discussed above. However, the squeeze film damper does develop stiffness-like behavior. This stiffness is due to cross-coupled damping coefficients, which exhibit stiffness-like (spring) characteristics when the damper operates in a cavitated mode.

The uncentered damper is one of the most nonlinear of the squeeze film damper designs. There are two basic mechanisms that are responsible for this nonlinear behavior. The first of the two nonlinear mechanisms is attributed to the nonlinear characteristics produced by the cross-coupled damping coefficients. These are thought to produce what is referred to as the nonlinear "jump" phenomenon caused by the hardening spring effect, depicted in the response plot in Fig. 5-49. The second source of nonlinear behavior present with this type of damper comes as a direct consequence of the bottoming out of the damper journal. This generally occurs at high side loads or due to excessive unbalance forces. The bottoming out of the damper journal, which is very likely with this design due to the absence of a centering or restraining spring, will result in a bilinear spring behavior, as shown in Fig. 5-50. This nonlinear behavior can be inferred

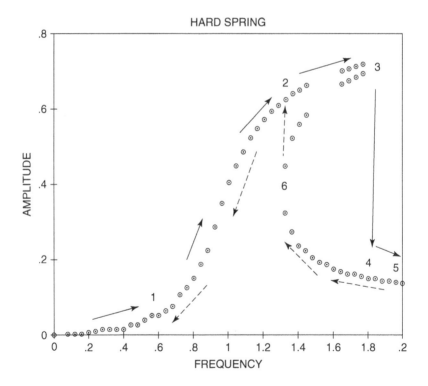

Figure 5-49 Response due to hardening spring effect leading to the "jump" phenomenon.

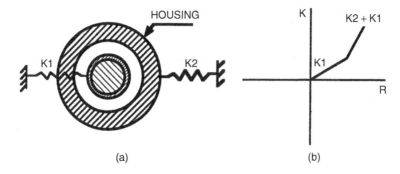

Figure 5-50 Bilinear spring effect when the damper journal makes contact with the stator/housing.

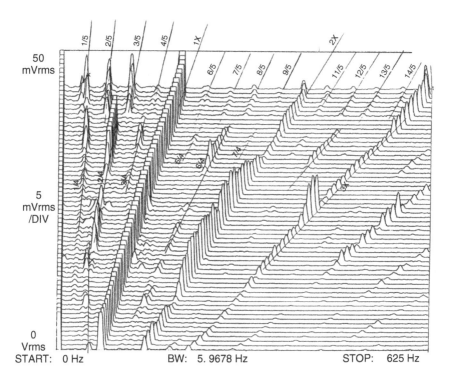

Figure 5-51 Waterfall plot showing the effect of the bilinear spring effect on the vibrations.

from the subsynchronous and supersynchronous vibration characteristics often noted on this type of damper, shown in Fig. 5-51. In some cases, the impact force generated when the damper journal bottoms out excites the lowest natural frequency of the rotor. In the case of flexible casings and support structures, the resonance frequencies of the structure can also be excited. The fact that this produces subsynchronous and supersynchronous vibrations at exact fractions of the synchronous running speed is not to be confused with subsynchronous vibrations due to instabilities. In this case these vibrations are stable but demonstrate nonlinearity in the system. It is also worth noting that the fraction of the subsynchronous is related to the ratio of the first natural frequency to the operating speed.

The uncentered dampers are commonly used on earlier generation aircraft gas turbines, lightweight process compressor rotors, and automotive turbocharges. In aircraft engines, their use has been limited to the smaller engines and to engines where the use of a conventional style centering spring (squirrel cage spring) is difficult to implement due to space limitations.

O-ring Supported Dampers

The simplest means of providing a centering spring in a squeeze film damper is through the use of elastomeric O-rings. An illustration of this damper design is shown in Fig. 5-52. The advantages of this design stems from its simplicity, ease of manufacture, and the ability to incorporate the damper into small radial envelopes. The relatively low radial space required makes it the preferred method to retrofit existing problem machines in the field. The O-ring doubles as a good end seal, which helps increase the effectiveness of the damper by reducing side leakage.

Some of the disadvantages with this design are attributed to the limited range of stiffness that can be achieved with elastomers. Predicting the stiffness with a good degree of certainty is difficult in elastomeric materials, due to the material variance from one batch to another, and due to the influence of temperature and time on its properties. The elastomer stiffness and damping is strongly influenced by temperature and the amount

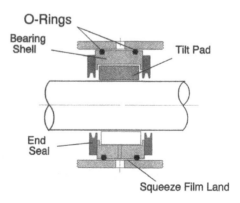

Figure 5-52 Schematic of an O-ring-supported squeeze film damper.

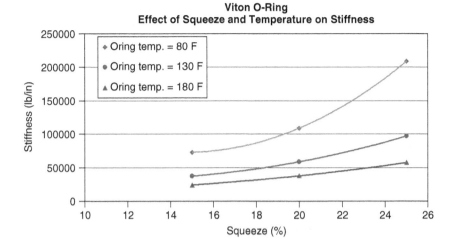

Figure 5-53 Effect of squeeze and temperature on O-ring stiffness.

of squeeze, as shown in Fig. 5-53. The O-ring design is also susceptible to creep, causing the damper to bottom out, which, as discussed above, may lead to a bilinear spring behavior. Furthermore, O-ring dampers are not capable of taking thrust loads, and cannot be easily manipulated for centering of the damper journal within the damper clearance space. One means of achieving some centering capability is through making the O-ring groove eccentric. This limitation makes them suitable only for use with relatively lightweight rotors.

High-speed and high-pressure centrifugal compressors prone to stability problems frequently have been fitted with these O-ring dampers. The damper is installed in series with tilting pad fluid film bearings to enhance the stability of the compressor. Although tilt pad bearings have significant damping, in many cases their stiffness is so high (relative to the rotor bending stiffness) that there is not enough motion at the bearing to dissipate energy. In Fig. 5-52 the oil films in the tilt pad bearing and the squeeze film are in series, so the net resultant damping coefficient is less than that from the tilt pad bearing. But the *system* damping is increased due to the soft support of the O-rings that allows motion across the squeeze film. Although most of their use has been primarily aimed at improving stability, they have also been used to reduce the synchronous response due to imbalance, or to shift the peak unbalance response outside the operating speed range.

The O-rings provide a form of internal friction damping (hysteretic damping), in addition to the squeeze film damping (viscous damping) produced by the oil or lubricant in the damper. The elastomer material in the O-rings limits the use of such dampers to mostly low-temperature applications. The hysteretic damping from the O-rings has been utilized in series with rolling element bearings, and gas bearings without any oil in the clearance space (dry O-ring damper). High-speed dentist drills are an example where this configuration is commonly used. These drills are composed of air turbines running on ball or gas bearings, with elastomer O-ring supports for increased damping and improved stability. Although elastomers are not suitable for extreme temperature applications like the cryogenic example above or in high-temperature gas turbines, they are readily available with a wide range of properties and configurations for temperatures over the range $50-140°$F.

All solid elastomeric materials have some amount of *hysteretic* or *structural* damping. Some have a lot more than others, e.g. elastomers (rubber). Hysteretic damping can be modeled as the imaginary part of a complex stiffness coefficient K, as follows:

$$F = -Kx = -k(1 + i\varepsilon)x = -(k + ih)x \qquad (5\text{-}2)$$

The stiffness force is then $-kx$ and the damping force is $-ik\varepsilon x = -ihx$. Recall that the effect of i ($\sqrt{-1}$) on a harmonic function is to make it lead real variables by $90°$, so the damping force leads the stiffness force by $90°$ as it should. The parameter ε is dimensionless and is called the *loss coefficient*. Typically, its value is <1. The product $k\varepsilon = h$ is called the *hysteretic damping coefficient*. It has units lb/in or N/m.

Recall that a viscous damper in parallel with a linear stiffness produces a force given by

$$F = -kx - c(dx/dt) \qquad (5\text{-}3)$$

If we assume harmonic motion by substituting $x(t) = Ae^{i\omega t}$ into Eqs. 5-2 and 5-3 it will show that hysteretic damping can be expressed as an equivalent viscous damping coefficient. The equivalent viscous coefficient C_{eq} has the proper units for viscous damping, and it is frequency dependent as follows:

$$C_{eq} = k\varepsilon/\omega = h/\omega \qquad (5\text{-}4)$$

Equation 5-4 shows that elastomeric damping is much more effective at low frequencies, which is advantageous for ball bearing applications where high dynamic loads at high frequencies would reduce bearing life.

Darlow and Zorzi [37] carried out extensive research and tests to measure the performance of elastomeric bearing dampers. Figure 5-54 shows the configuration they favored, using elastomeric "buttons" to support the dynamic loads.

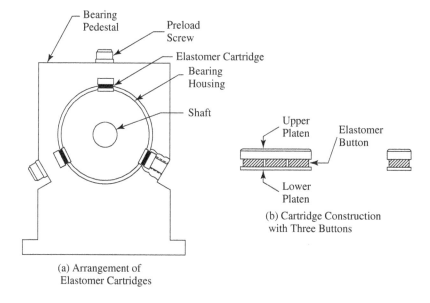

(a) Arrangement of
Elastomer Cartridges

Figure 5-54   Bearing supported on elastomeric buttons. From [37].

Squirrel Cage Supported Dampers

Squirrel cage-supported dampers are one of the most commonly used squeeze film damper designs, particularly in modern aircraft engines where its use is widespread. Most large aircraft gas turbine engines employ at least one, and, in many instances, two or three of these dampers in one engine. A schematic of this damper is shown in Fig. 5-55. A distinctive

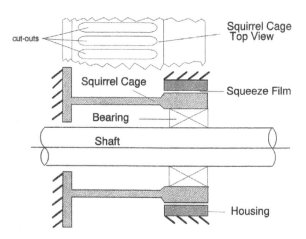

Figure 5-55   Schematic of squirrel cage support with damper journal on the right.

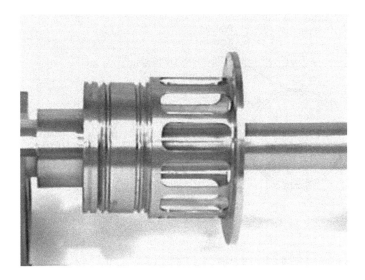

Figure 5-56 Actual squirrel cage damper with the damper journal on the left end.

feature necessary with such a design (and apparent from the schematic as well as the actual hardware shown in Fig. 5-56) is the relatively long axial space required in comparison to the actual damper length. This is one of the major drawbacks of this damper design. The squirrel cage forming the centering spring for the damper quite often requires three to four times as much axial space as the damper itself.

Assembling the squirrel cage spring and centering the journal within the clearance space requires special tools and skills and is difficult to achieve. The squirrel cage spring also complicates the damper end seal design and assembly. It is also difficult to offset the spring assembly, in order to account for the gravity load due to the rotor weight. Maintaining parallelism between the damper journal and the housing or stator is another factor that adds uncertainty and complications to this design.

Integral Squeeze Film Dampers

The integral squeeze film damper design [41, 42] shown in Fig. 5-57 depicts how the damper in general is used to complement an existing bearing that cannot on its own meet the application requirements. In this configuration the outer ring contains both the S- or L-shaped springs and a relatively narrow clearance space (shown exaggerated here for illustrative purposes) between the springs forming the squeeze film land. This design integrates the springs into the squeeze film space and addresses many of the drawbacks described with the conventional dampers cited above. The use of S-shaped springs tends to distribute the stresses evenly in the spring

Figure 5-57   Schematic of original integral squeeze film damper.

elements. The part can be also integrated with the bearing itself to form one integral part for ease of assembly, as shown in Fig. 5-58. High loads can be accommodated with this design and the part can also be made offset to account for the deflections caused by the static loading imparted by the rotor weight or gear reaction forces. The ability to manufacture this part using wire electric discharge machining (EDM) technology also makes it possible to make the clearance variable or offset to ensure that the damper operates in a centered position under load. This eliminates many of the cavitation problems and nonlinearities associated with the conventional dampers.

Squeeze Film Damper Rotordynamic Force Coefficients

The squeeze film force coefficients in dampers are obtained after solving the Reynolds equation for squeeze flow to obtain the pressure profile. Integrating the pressure and solving for the force coefficients is very similar to what is done for journal bearings. The solution to the damper pressure profile can be simplified using the same limiting cases discussed above

Figure 5-58   Integral squeeze film damper and flexure pivot bearing combination. Courtesy of KMC Inc.

for journal bearings. Most dampers in practice are of relatively short axial length ($L/D < 0.25$), and therefore the assumptions for the short bearing theory applies. To increase the damping with these dampers, some form of axial seals is employed to reduce the pressure gradient in the axial direction. Numerical analysis for the prediction of squeeze film pressures and computation of dynamic squeeze film forces are available in the open literature (Marmol and Vance [38], Szeri et al. [39], and Kinsale and Tichy [40]).

APPLICATIONS OF SQUEEZE FILM DAMPERS

The following examples demonstrate the process involved in the application of dampers to solve specific vibration and rotordynamic problems. These examples emphasize the principles presented in Chapter 3 and use the numerical techniques discussed by the third author in Chapter 4.

Optimization for Improving Stability in a Centrifugal Process Compressor

The design of aircraft gas turbines and other standard equipment offers the designer the opportunity to build prototypes and verify all aspects of the design. In process-type compressors, this is difficult since process conditions vary and the luxury of building a complete prototype is prohibitively expensive. This is one of the major reasons why process compressors tend to be more suceptible to experience stability problems when installed in the field. In many instances the field and process conditions also change making it very difficult to provide a design capable of meeting all these variables. The squeeze film damper bearings in these conditions provide a solution through widening and increasing the stability margin.

The rotor model for such a process compressor is shown in Fig. 5-59 with the bearings shown as springs attached to ground. This is a relatively heavy rotor and with a long bearing span, two conditions known to result in possible stability problems due to the lower natural frequency and a relatively flexible rotor. The first forward mode, the mode most likely to become unstable, is shown in Fig. 5-60. This mode, as expected, has a low natural frequency that lies below the operating speed, and the other more relevant characteristic in terms of stability is that the nodes for this mode are very close to the bearing locations. This tends to make the bearings ineffective in providing the damping required for a stable system and is evident in the relatively low basic log dec predicted for this mode.

In this case the use of a squeeze film damper in series with the existing tilting pad bearings will introduce flexibility and damping to the bearing supports. Quite often it is the flexibility in the squeeze film damper supports springs that is the most valuable and relevant contribution by virtue of moving the node away from the bearings and making their damping more effective. This is evident in the mode shape shown in Fig. 5-61 for the rotor with the squeeze film damper. The bearings are no longer nodes and the motion at the bearings makes the damping much more effective. As discussed in Chapter 3, there is an optimum amount of damping for any particular rotor–bearing system. Too much damping can lock

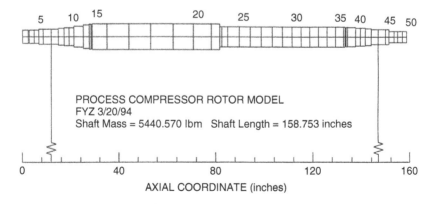

Figure 5-59 Rotor model for process compressor.

ROTORDYNAMIC MODE SHAPE PLOT
PROCESS COMPRESSOR STABILITY ANALYSIS
FYZ 3/20/94 EXISTING BRGS W/O SFD
SHAFT SPEED = 4152.0 rpm
NAT FREQUENCY = 1523.41 cpm, LOG DEC = 0.0369
STATION 25 ORBIT FORWARD PRECESSION

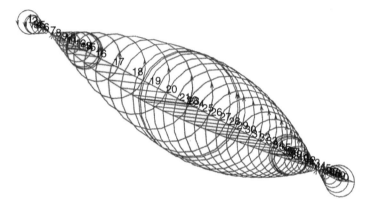

Figure 5-60   First forward mode with tilt pad bearings.

ROTORDYNAMIC MODE SHAPE PLOT
PROCESS COMPRESSOR STABILITY ANALYSIS
FYZ 3/20/94 WITH OPTIMIZED SQUEEZE FILM DAMPER
SHAFT SPEED = 4152.0 rpm
NAT FREQUENCY = 923.72 cpm, LOG DEC = 1.0804
STATION 24 ORBIT FORWARD PRECESSION

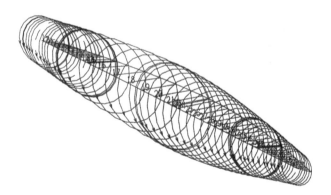

Figure 5-61   First forward mode with squeeze film damper.

the supports and result in a node at that location negating the benefits from the damped supports. The optimization process can be determined through a parametric variation of the damping and computing the log dec for the first forward mode, as shown in Fig. 5-62. Note that the curve has a steep slope on the left end and a much more gradual one on the right

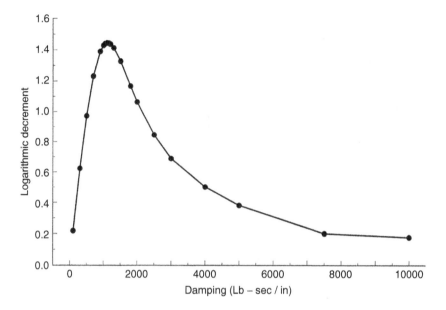

Figure 5-62   Squeeze film damper optimization process for stability.

end. This suggests that the design should target a nominal damping value to the right of the optimum point so the actual operation would not be too sensitive to variations in the actual damping value.

In discussing the stability of fluid film bearings, it was recognized that bearings tend to become unstable when the load is light or the operating eccentricity is low. While this fact is not in question, as it has been demonstrated both theoretically and through field experience, there are conditions where the higher loading can cause stability problems. This is due to the fact that at higher loads the stiffness in the bearings can become large enough to make the damping ineffective. Furthermore, the higher load can stiffen the bearing and move the node for the mode of concern closer to the bearing location as illustrated by the following example.

The rotor model for this high-speed, integrally geared expander is shown in Fig. 5-63. The overhung weights represent the impeller weight and the weight of the rotating components in the dry gas seal. The use of the dry gas seals in this case further increased the overhung weight of the rotor due to the longer axial space required by the dry gas seals. The vibration plot shown in Fig. 5-64 illustrates the load dependence of these subsynchronous vibrations, which took off once the load reached 35 percent of design conditions and tripped the machine. This condition was not detected on the test stand since the machine could not be loaded until it was placed in the field. The load is intuitively stabilizing as explained above. However, in this case with the integrally geared compressor the

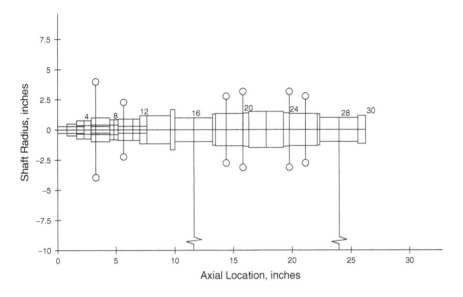

Figure 5-63 Rotor model for an integrally geared expander.

POINT: VEX-203 ∠45° Right
MACHINE: GEARBOX HS NDE
From 20NOV2002 12:20:24.1 To 20NOV2002 12:22:32.2 Startup 12:22:06.9
WINDOW: None SPECTRAL LINES: 400 RESOLUTION: 1500 CPM

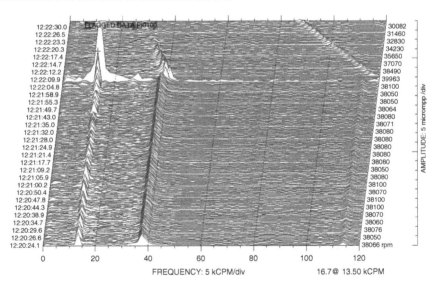

Figure 5-64 Frequency spectrum showing the subsynchronous instability when loading increased.

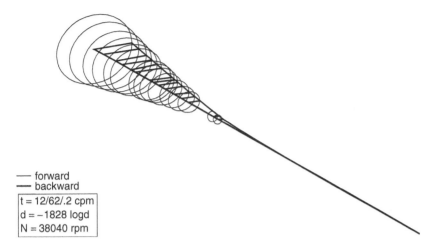

— forward
— backward

| t = 12/62/.2 cpm |
| d = −1828 logd |
| N = 38040 rpm |

Figure 5-65   Mode shape without the damper showing a node at the impeller bearing.

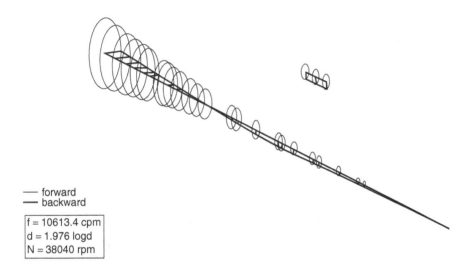

— forward
— backward

| f = 10613.4 cpm |
| d = 1.976 logd |
| N = 38040 rpm |

Figure 5-66   Mode shape with the ISFD showing motion in the bearing and the damper on the impeller end.

gear separation forces which increase with the increase in process loads caused the bearing stiffness to increase. This in turn moved the node very close to the bearings and made the damping ineffective, as shown in Fig. 5-65. The use of an integral squeeze film damper served to soften the support and allow motion on the impeller end bearing, as shown in Fig. 5-66. This is evident by the motion in the bearing and in the damper exhibited in the plot for the first forward eigenvalue.

Using Dampers to Improve the Synchronous Response

The examples above showed how dampers are used to improve stability for the first forward mode. In the following two examples, the damper is used to improve the synchronous response characteristics and allow for a greater tolerance to imbalance. Much of the driving or driven equipment today is required to operate within a wide range of speeds. This operational demand makes it very difficult to meet the API requirements in maintaining adequate separation margin between the critical speed and the minimum and maximum operating speed. The damper in these cases offers greater flexibility in its ability to critically damp the first mode or reduce its amplification factor, thus eliminating the requirement for the separation margin altogether. The picture in Fig. 5-67 shows such a damper used to increase the operating range for a power turbine. This damper is integrated with a rocker back tilt pad bearing, but what is also important to note is the damping gap. A significant part of the damper is cut out (large cut gaps) for the sake of keeping the damping value close to optimum for synchronous response in a similar manner to the previous case for stability. Extensive testing was done on this damper to ensure that the stiffness characteristics are linear and match the desired stiffness

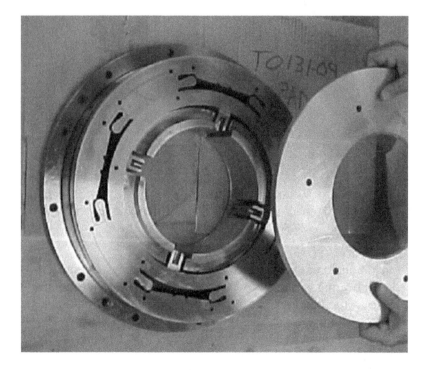

Figure 5-67   Integral squeeze film damper with a rocker back tilt pad bearing.

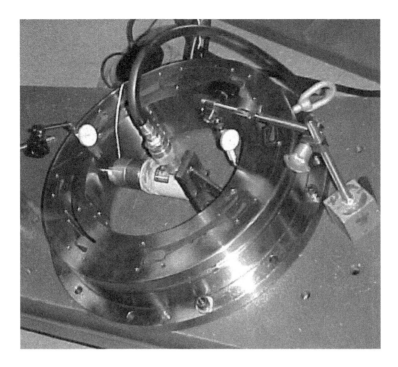

Figure 5-68   Setup used to measure the load/deflection curve for the damper.

specified. The picture in Fig. 5-68 shows the test setup to load the damper and measure the deflection in order to generate the spring or stiffness values. The plot in Fig. 5-69 shows the load/deflection curve and shows the total spring value to be linear and within 1 percent of the desired value. This linear stiffness is a very valuable characteristic not attainable with the conventional dampers.

In some applications, the balance quality will degrade with time, or the process in which the machine is applied will increase the imbalance force, causing damage and shortening the life of the bearings. The following example shows how the incorporation of the damper with a rolling element bearing served to reduce the dynamic loads transmitted to the bearing and allowed the machine to withstand excessive amounts of unbalance inherent in the design of this machine. The second and third authors collaborated on the design for such a damper used to support the generator shown in Fig. 5-70. The response and prediction to imbalance are shown in Fig. 5-71. This plot shows that the predictions for the critical speed matched the actual location predicted by the analysis showing good correlations with the stiffness. The shape of the curve also matched the predictions showing good correlation for the damping as well.

Machinery can often be exposed to process conditions that tend to increase the imbalance on the rotor. This can be dramatic, particularly

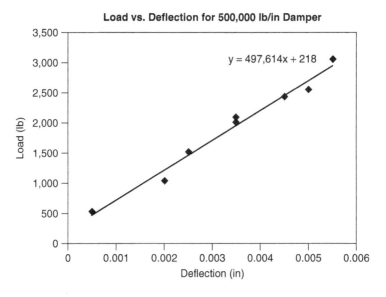

Figure 5-69 Load/deflection measurement for integral squeeze film damper.

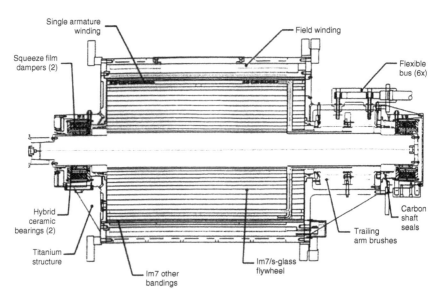

Figure 5-70 Schematic of the generator rotor with rolling element bearings and an integral squeeze film damper [43].

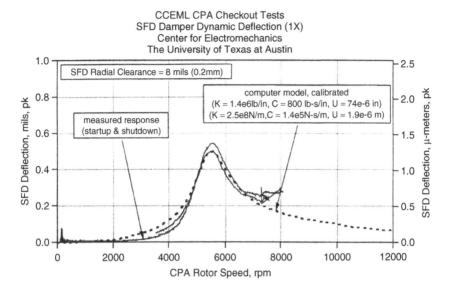

Figure 5-71   Measurements and predictions for an integral squeeze film damper [43].

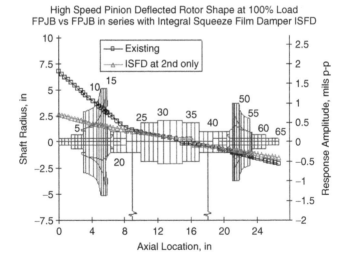

Figure 5-72   Deflected rotor shape at operating speed with and without an integral speed film damper.

in overhung machines, as shown in Fig. 5-72. The rotor deflections at the impeller can become excessive, leading to rub and eventual failure of the compressor. The use of the integral squeeze film damper bearing in this case allows more motion at the bearing and reduces the bending or deflection of the overhung unit, thus allowing much greater tolerance

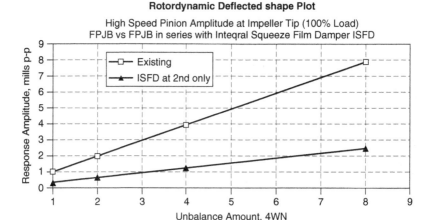

Figure 5-73   Response at the impeller location as a function of increased imbalance.

to imbalance. The difference is magnified as the amount of unbalance is increased, as presented in Fig. 5-73, and clearly shows the advantages offered by such an element.

Using the Damper to Shift a Critical Speed or a Resonance

The damper supports can also be tuned to shift a critical speed that is too close to the operating speed, or to shift a resonant frequency that tends to coincide with an excitation frequency. The generator model shown in Fig. 5-74 is presented to illustrate such a situation. In the case of such a heavy rotor and the limitation of what can be done with the bearing or the relatively soft pedestal support, the squeeze film damper offers a good solution. The fluid film bearing coefficients are in series with the squeeze film damper, which in turn are in series with the pedestal support. In this class of machinery the assumption of rigid supports is not adequate and the flexibility of the supports must be included in the analysis. The plot in Fig. 5-75 shows the response with the original elliptical bearing and compared to the response after incorporating a squeeze film damper. The resonance that was located at two times the line frequency is now shifted with the damper model.

The same approach was applied to the integrally geared compressor whose response without and with the damper is shown in Fig. 5-76. The peak without the damper was located too close to the operating speed, and the incorporation of a squeeze film damper critically damped the

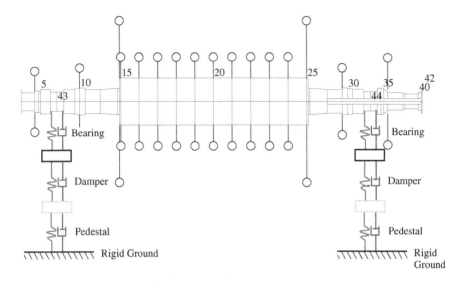

Figure 5-74   Generator rotor model with bearing–damper–pedestal supports.

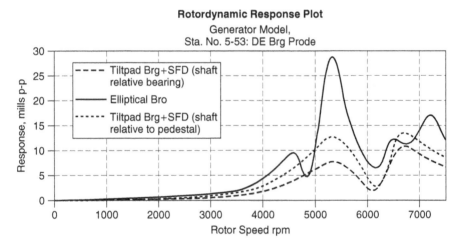

Figure 5-75   Response with and without the squeeze film damper.

response and eliminated the peak. What is more relevant in this case is the significant reduction in the dynamic loads transmitted through the bearings and is shown in Fig. 5-77. This integral squeeze film damper configuration allowed the machine to run for 10 years without any damage or degradation to the bearings.

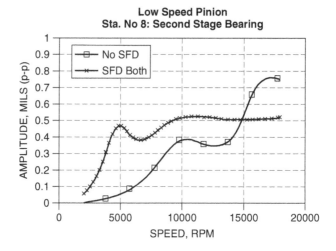

Figure 5-76 Response with and without the damper.

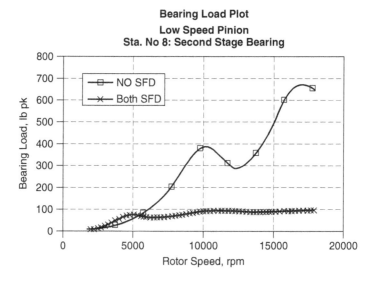

Figure 5-77 Dynamic load transmitted with and without the damper.

INSIGHTS INTO THE ROTOR–BEARING DYNAMIC INTERACTION WITH SOFT/STIFF BEARING SUPPORTS

One of the most basic principles to grasp in rotordynamics is the influence of the bearing stiffness on the critical speeds and mode shapes associated with a rotor–bearing system. While the focus in this chapter so far has been on bearings and their stiffness and damping characteristics, these properties cannot be fully understood without knowing how they interact with the rotor stiffness and flexibility. The best means of illustrating these effects is through the use of a simple example. For the sake of simplicity, the cross-coupled stiffness and damping coefficients in the bearings are neglected. The direct damping is set to a relatively small value of 2 lb-sec/in throughout this analysis.

The direct stiffness and direct damping coefficients are assumed symmetric for the purpose of generating simpler circular orbits. The rotor used in this example is shown in Fig. 5-78 and a listing of the rotor model is shown in Table 5-6 in the appendix. In this case a relatively high value for the bearing stiffness was used. The undamped critical speed map is shown in Fig. 5-79. This map is generated through a parametric variation of the bearing support stiffness and the numerical analysis to calculate it was discussed in Chapter 4. The undamped critical speed (UCS) map provides an overview of how the critical speeds and mode shapes will change with variation of the bearing support stiffness. The undamped critical speed map is a relatively simple approach and has been used since the early 1960s as one of the primary tools in analyzing the rotordynamic characteristics of a rotor–bearing system. As the stability, damped critical speed, and unbalance response analysis programs have become more readily available, the importance of the undamped critical speed analysis has lessened. However, in spite of the presence of more sophisticated tools and programs, the undamped critical speed map provides insights into some of the rotordynamic characteristics not readily seen with other analysis forms.

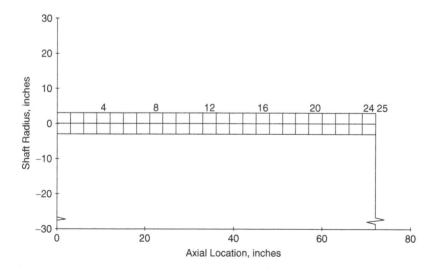

Figure 5-78 Simple rotor model with bearings at ends of shaft.

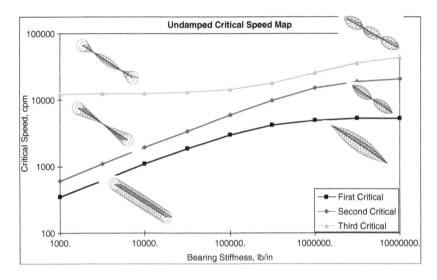

Figure 5-79 Undamped critical speed map with soft and stiff bearings.

INFLUENCE ON NATURAL FREQUENCIES
WITH SOFT/STIFF BEARING SUPPORTS

The first three critical speeds are typically the only speeds of interest for most machines and are presented in the plot shown in Fig. 5-79. The first two modes have a sloping section in the left half of the plot. This indicates that the critical speed can be shifted up or down as the stiffness in the bearing supports are varied. In the right section of the plot, the shaft stiffness dominates, and varying the bearing stiffness does very little to change the critical speed. This characteristic is often used to check and compare analysis results between two different programs or analysts. The value in the flat section resembles an infinitely rigid support and provides an upper limit to the critical speed. It eliminates differences in the computation of the bearing coefficients and allows a correlation for the bearings span and the stiffness or flexibility of the rotor.

The first three forward modes and corresponding natural frequencies are shown in Fig. 5-80. These were obtained using a bearing stiffness of 10 million lb/in (very stiff supports). Stiffness in this range falls in the horizontal (flat) section of the undamped critical speed map shown in Fig. 5-79 (the right half of the plot). The eigenvalue analysis was repeated with bearing stiffness coefficients in the range of 100,000 lb/in (relatively soft supports). This locates the rotor–bearing critical speeds in the sloping section of the undamped critical speed map lines (first and second modes). The natural frequencies and mode shapes for the first three forward modes with soft supports are shown in Fig. 5-81. Note that the mode shapes show

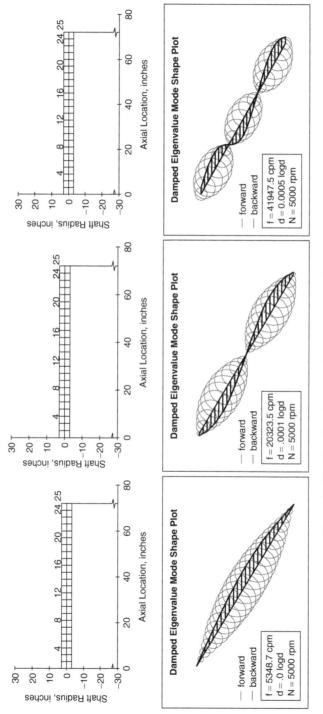

Figure 5-80 The first three modes with very stiff bearings.

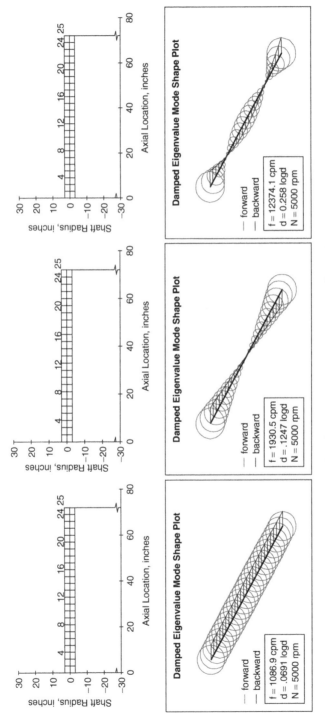

Figure 5-81 The first three modes with very soft bearings.

more motion at the bearings for the softer bearing supports, and that the shaft is relatively more rigid. The frequencies as expected are lower with the soft bearing supports. In both, the stiff and soft bearing supports analysis, the damping was held constant at 2 lb-s/in. In the case of the softer support, the motion at the bearings allows the damping to be more effective and can explain the relatively higher logarithmic decrements obtained.

EFFECTS OF MASS DISTRIBUTION ON THE CRITICAL SPEEDS WITH SOFT/STIFF BEARING SUPPORTS

To gain additional insights into the effect of mass distribution on the rotor mode shapes and critical speeds, a disk is placed at midspan. The analysis was made for the disk at midspan and repeated with the midspan disk split into two equal disks and placed at the two quarter-spans of the rotor. The effect of the mass at midspan and quarter-spans on the first natural frequency is shown in Fig. 5-82. The natural frequency will drop from 5348 to 3056 cpm when the disk is placed at midspan. The first natural frequency will also drop for the disks at the quarter-spans (4411 cpm), but not to the same extent as when the weight was placed midspan. This is because in the midspan weight case, the mass was located at an anti-node and thus increases the influence of the mass for the first mode. This simple analysis shows that the mass will lower the natural frequency, but of more importance is the distribution of the mass on the rotor and its location relative to the particular mode that has the greater influence. This becomes even clearer when we look at the effect of the added mass on the second natural frequency for all three mass distribution cases, as shown in Fig. 5-83. The midspan mass happens to fall at a node for the second mode, and thus does not have any significant influence on the second natural frequency. The frequency is virtually unchanged compared to the case with no mass. On the contrary, the quarter-span masses are located at anti-nodes and thus have a large influence on the second natural frequency, dropping it from 20,323 to 16,149 cpm. The influence of the mass distribution on the third natural frequency is shown in Fig. 5-84. In this case the midspan mass is at an anti-node and results in a larger drop in the natural frequency as compared to the quarter-span masses. The addition of mass to the rotor will result in lowering of the natural frequencies, but the magnitude of the drop is very much influenced by the distribution of the mass relative to the particular mode.

The influence of the mass distribution for the first mode with soft bearing supports is shown in Fig. 5-85. In the case of lower bearing stiffness relative to the shaft stiffness, the first mode is characterized by a relatively rigid shaft (no significant bending along the length of the shaft). The difference in motion at midspan and quarter-span is not as big

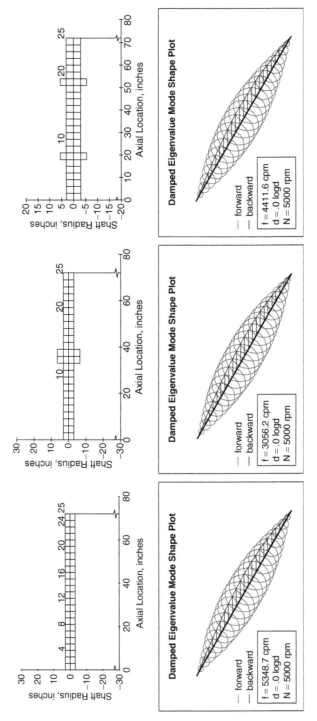

Figure 5-82 Influence of mass distribution on the first mode with relatively stiff bearings.

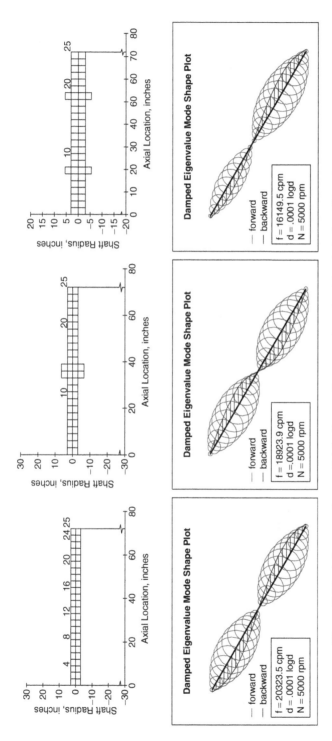

Figure 5-83 Influence of mass distribution on the second mode with relatively stiff bearings.

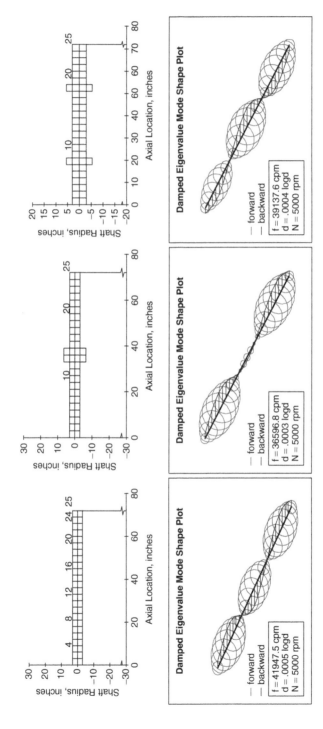

Figure 5-84 Influence of mass distribution on the third mode with relatively stiff bearings.

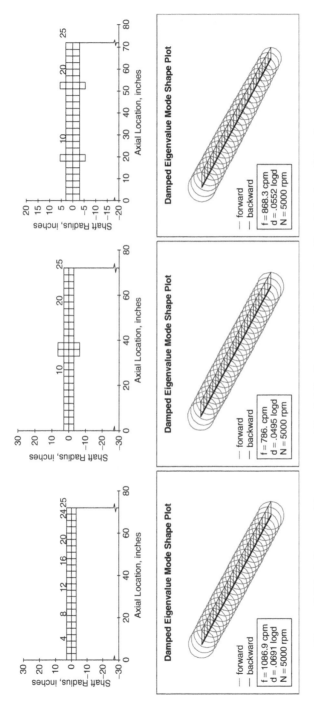

Figure 5-85 Influence of the mass and its distribution on the first mode with relatively soft bearings.

as with the rigid bearing supports. This explains why the difference in the natural frequencies between the midspan and the quarter-span mass is not as large as it was in the stiff bearing case. For the second natural frequency with soft supports, the midspan mass is located at a node and therefore has no influence on the natural frequencies when compared with the diskless shaft. The quarter-span masses have a bigger influence on this mode and result in lowering of the natural frequency from 2008 to 1745 cpm as shown in Fig. 5-86. The effect on the third mode is shown in Fig. 5-87. The midspan mass has the most influence since it is located at an anti-node for this mode. The quarter-span masses are very close to a node and therefore have negligible influence on the third mode.

The undamped critical speed map for the shaft without and with the midspan mass is shown in Fig. 5-88. The dashed lines correspond to the shaft without the midspan mass. As noted above, the midspan mass tends to reduce the first and third critical speeds with no effect on the second critical speed. The mass is located at a node in this case and does not influence the natural frequency corresponding to that mode shape. The undamped critical speed map is also generated for the shaft with two masses at the quarter-span locations. Superimposed on this map is the undamped critical speed generated without the added masses, indicated by the solid lines shown in Fig. 5-89. In this case the two masses at the quarter-span will reduce the first and second critical speeds, but have no effect on the third critical speed. Again, this is because the two masses tend to fall at the node locations for the third critical speed.

To evaluate the effect of increasing the shaft diameter from 6 to 8 inches on the first three modes, the undamped critical speed map was generated using the 8-inch-diameter shaft with the same length of 72 inches. The undamped critical speed map for the 6-inch-diameter shaft is superimposed and indicated by the dashed lines shown in Fig. 5-90. In the lower bearing stiffness region (the left half of the figure), the bearing stiffness dominates, since the bearing stiffness and the shaft stiffness add in series. This means that the lower stiffness of the bearings dictates what the natural frequency will be. The 8-inch-diameter shaft will result in a heavier shaft (increase in the mass), and since it is moving as a rigid body, there is no significant increase in effective stiffness. The increase in mass due to the 8-inch diameter will reduce the first two critical speeds in the lower bearing stiffness range. On the contrary, the right side of the undamped critical speed map (rigid bearings/flexible shaft), where the bearing stiffness is very high and the shaft stiffness, being lower, dominates (the bearings and the shaft are springs in series); the critical speed is higher with the 8-inch-diameter shaft. The increase in the shaft stiffness overcomes and offsets the effects resulting from the increase

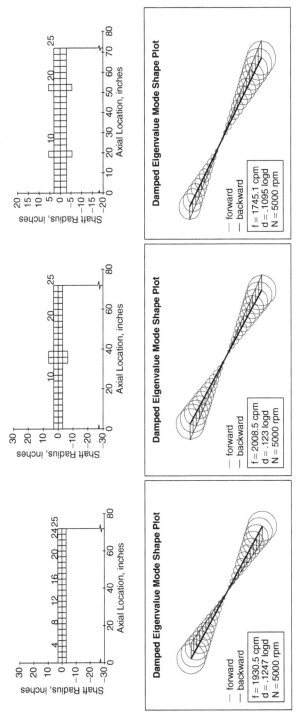

Figure 5-86 Influence of the mass and its distribution on the second mode with relatively soft bearings.

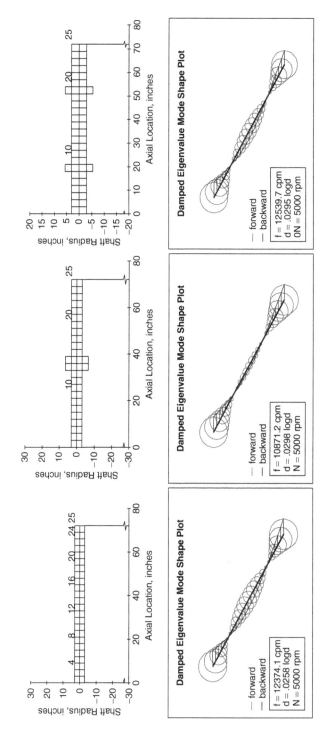

Figure 5-87 Influence of the mass and its distribution on the third mode with relatively soft bearings.

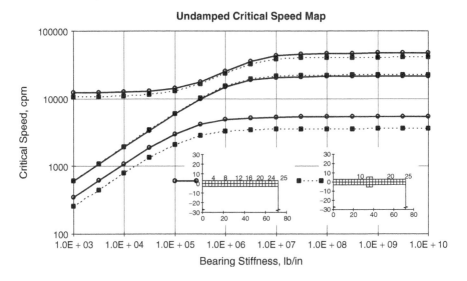

Figure 5-88 Influence of mass at midspan on UCS map.

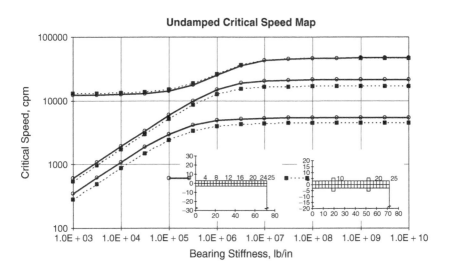

Figure 5-89 Influence of mass at quarter-span on UCS map.

in the rotor mass. This results in a net increase in the critical speeds for the first two modes. Note that the third mode is flat on the left side of the undamped critical speed map and therefore we see the same effect as we see for the first two modes on the right side. That is an increase in the critical speed as a result in the higher stiffness with the 8-inch shaft.

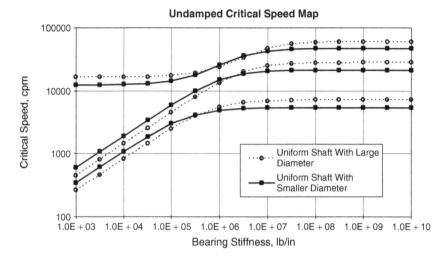

Figure 5-90 Influence of shaft diameter with soft and stiff bearings.

INFLUENCE OF OVERHUNG MASS ON NATURAL FREQUENCIES WITH SOFT/STIFF SUPPORTS

There is a large class of turbomachinery in which a significant portion of the rotor mass is located outboard of the bearing supports in an overhang configuration. Power turbines, gas expanders, turbochargers, fans, pipeline compressors, and many single-stage compressors fall into this category. Furthermore, the introduction of gas seals as a replacement to the conventional oil film seals often requires more axial space, which requires increasing the shaft overhang section for overhung compressors and expanders. It is therefore important to recognize the different mode shapes that result from such a configuration. We first examine the influence of the overhang on the natural frequencies with stiff bearings. The influence on the first natural frequency for an overhang length of 6, 12, and 18 inches is shown in Fig. 5-91. Although one expects that the natural frequency will drop as the overhang increases, the natural frequency for the 12-inch overhang is actually higher than that of the 6-inch overhang. The mode shape can help explain why this is so. The mode shape in the 12-inch overhang case results in larger orbits at the overhang disk. This increases the gyroscopic stiffening effects, which compensates for the loss in stiffness as a result of the increase in the overhang length. As the overhang is further increased to 18 inches, the shaft flexibility dominates and results in a decrease in the natural frequency from 7305 to 6478 cpm. The effects on the second mode are shown in Fig. 5-92. There are no surprises in this case—the natural frequency drops as the

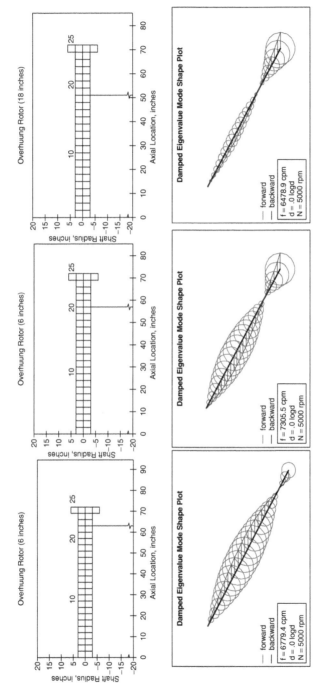

Figure 5-91 Influence of overhang mass on the first natural frequency with stiff bearings.

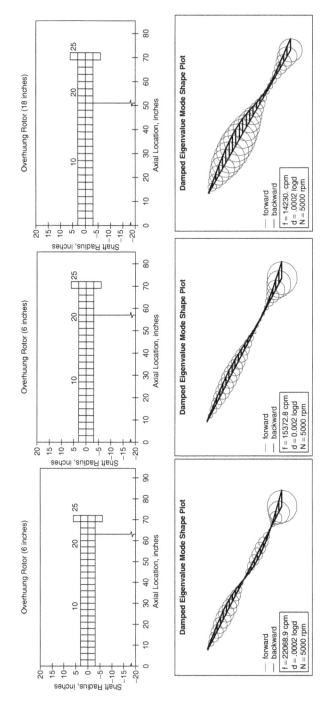

Figure 5-92 Influence of overhang mass on the second natural frequency with stiff bearings.

overhang is increased. The influence on the third natural frequency is shown in Fig. 5-93. As the overhang length increases in this case, the natural frequency increases, and the mode shapes tend to show why. Note that the third mode corresponds to a node to the right end of the shaft as the overhang is increased. The motion is mostly between the bearing spans in this mode and therefore the shaft stiffness between the two bearings dictates what the natural frequency will be. As the overhang is increased the distance between the bearings (bearing span) is reduced, thus increasing the stiffness in that section of the rotor. Since the third mode is governed by the section of the shaft between the bearings, its natural frequency will increase as this span is reduced. Therefore, the natural frequency for this mode will increase as the bearing span is reduced and the overhang is increased.

The influence of the overhang with soft bearing supports is shown in Figs. 5-94 through 5-96. The comments stated above for the third natural frequency result with stiff bearings apply to the soft bearing supports too.

INFLUENCE OF GYROSCOPIC MOMENTS ON NATURAL FREQUENCIES WITH SOFT/STIFF BEARING SUPPORTS

To gain some more insights into the influence of the gyroscopic moment on the natural frequencies and mode shapes, the disk overhang diameter was increased from 12 to 36 inches. Since this increase also increases the overhang mass and makes it difficult to separate the gyroscopic effects from the mass effects, the density of the 12-inch disk was increased so the weights of the two rotors matched. This eliminates the mass effects, and the only difference between the two rotors is due to the larger disk diameter, which tends to generate larger gyroscopic effects. The first natural frequency, which corresponds to a backward mode for the two models with different disk diameters but equal weights, is shown in Fig. 5-97. As expected, the gyroscopic effects, which are larger with the larger disk, tend to lower the backward mode. The backward mode drops from 2511 to 1582 cpm due to the larger gyroscopic moment with the larger disk.

The natural frequency that corresponds to the first forward mode, is shown in Fig. 5-98. As expected, the gyroscopic effects, which decreased the natural frequency for the backward mode, increased the natural frequency of the forward mode. The gyroscopic effect, which is higher in the case of the 36-inch disk, results in a higher frequency for the forward mode. In the forward mode case, the gyroscopic effect results in a moment on the shaft, which tends to increase its effective stiffness.

In the case of the second mode, the disks fall at a node and, therefore, as shown in Fig. 5-99, have very little influence on that natural frequency.

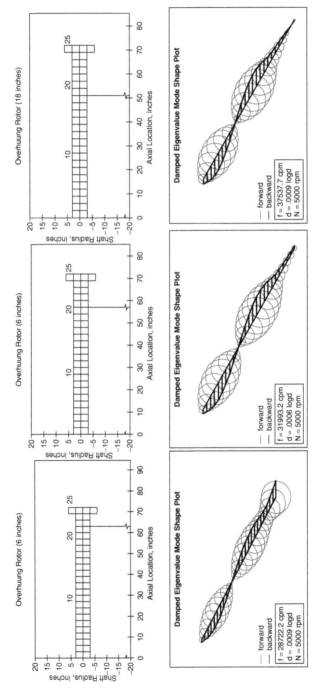

Figure 5-93 Influence of overhang mass on the third natural frequency with stiff bearings.

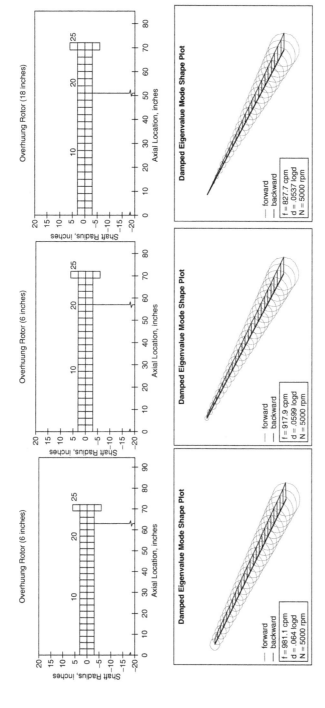

Figure 5-94 Influence of overhang mass on the first natural frequency with soft bearings.

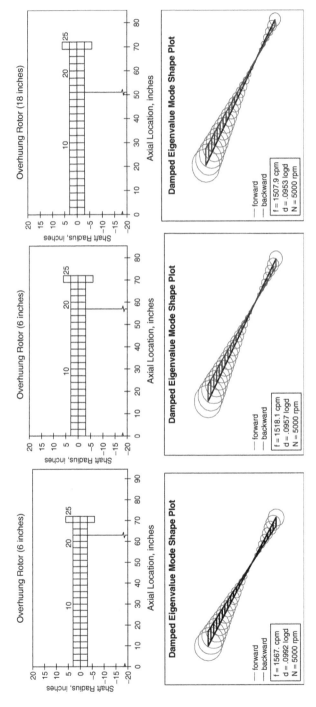

Figure 5-95 Influence of overhang mass on the second natural frequency with soft bearings.

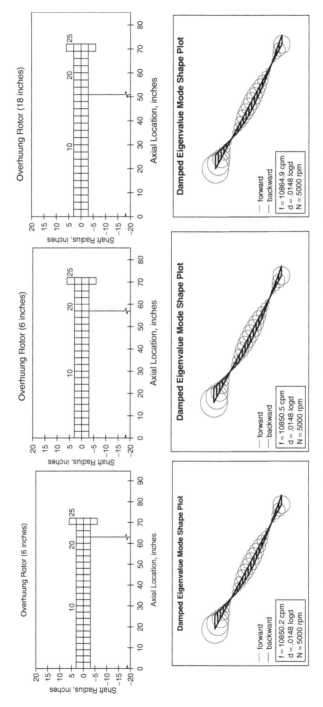

Figure 5-96 Influence of overhang mass on the third natural frequency with soft bearings.

259

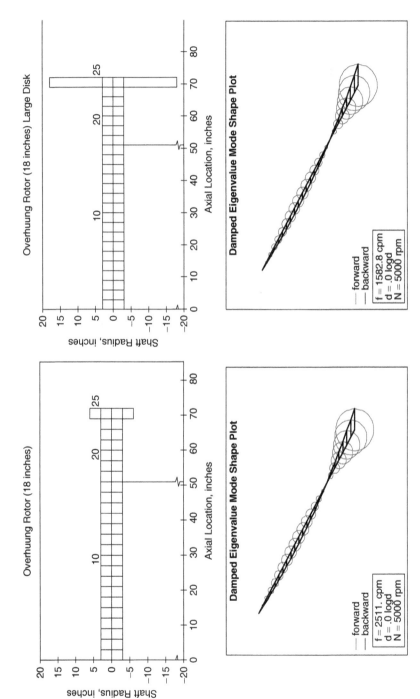

Figure 5-97 Influence of gyroscopic effects on the first backward mode.

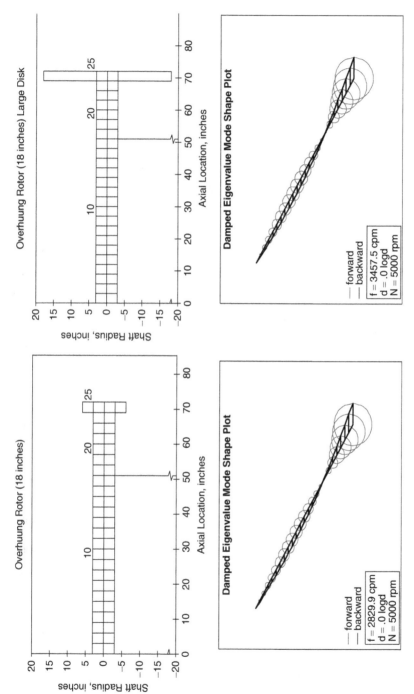

Figure 5-98 Influence of gyroscopic effects on the first forward mode.

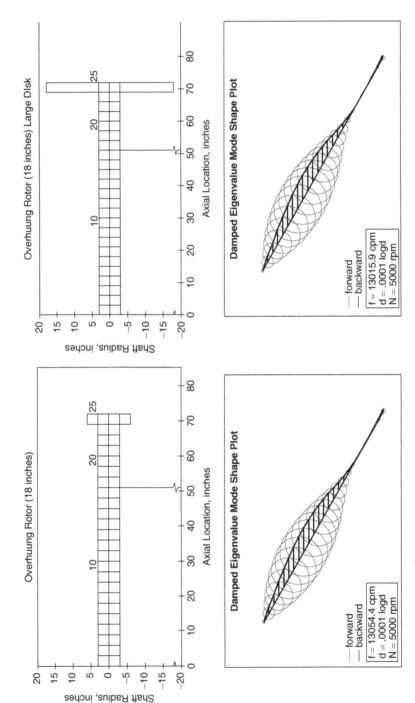

Figure 5-99 Influence of gyroscopic effects on the second forward mode.

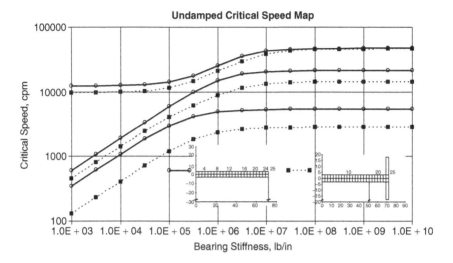

Figure 5-100   Influence of gyroscopic effects on the UCS map.

The effect of the overhang mass on the critical speed map can be seen in Fig. 5-100. Superimposed on this map is the critical speed map for a shaft of the same length and diameter, but without the added mass. Note also that the bearing span is different for the two rotors because of the overhung shaft section. The presence of the overhung mass can result in a significant reduction in all of the first three critical speeds. This is in spite of the fact that the rotor without the overhang has a longer bearing span and therefore would be relatively more flexible. The overhang mass overcomes the influence of shaft flexibility and results in a net reduction in the natural frequency of all three critical speeds.

To evaluate the gyroscopic effects without the influence of the overhang disk mass, the undamped critical speed map is generated for the 36-inch overhang disk. Superimposed on this plot is the undamped critical speed map for the 12-inch disk with a mass equivalent to the 36-inch disk. The critical speeds corresponding to the 12-inch disk are denoted by the dashed lines in Fig. 5-101. There is no difference for the first mode in the flexible bearings range (left half of figure), because, in this range, the mode is dominated by motion in the bearings and there are no deflections in the shaft. The shaft moves as a rigid body (translatory motion) and there is no significant stiffening moment generated by the overhang mass. In the higher bearing stiffness range (the right half of the undamped critical speed map), there are significant shaft deflections and the gyroscopic effects are larger for the 36-inch disk, resulting in a higher critical speed.

The second critical speed in the soft bearing support region will result in a bouncing mode. This tends to slightly increase the stiffness with the 36-inch disk, and thus results in a slightly higher critical speed. In

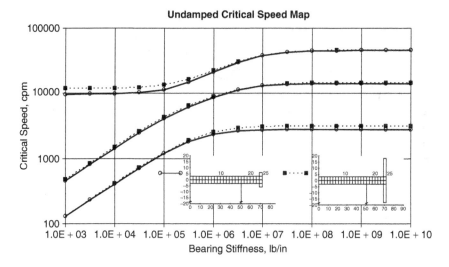

Figure 5-101 Influence of the gyroscopic effect on the UCS without the overhang mass effect.

the higher bearing stiffness region, the mode results in the disks being very close to a node, and therefore there is no difference in the critical speeds. The same characteristics noted for the second mode apply to the third mode.

The gyroscopic effects, although they appear to be minor on the undamped critical speed map (log scale), are generally very large and cannot be ignored. Overhung rotors generally tend to make the critical speed and stability predictions more difficult to analyze.

REFERENCES

[1] Tower, B. First report on friction experiments. *Proceedings of the Institution of Mechanical Engineers*, 632–666; 2nd report, *ibid.*, 58–70 (1885); 3rd report, *ibid.*, 173–205 (1888); 4th report, *ibid.*, 111–140 (1891).

[2] Reynolds, O. On the theory of lubrication and its application to Mr. Beauchamp Tower's experiments, including an experimental determination of viscosity of olive oil. *Philosophical Transactions* 177: 157–234 (1886).

[3] Stribeck, R. Die wesentlichen Eigenschaften der Gleit- und Rollenlager (The key qualities of sliding and roller bearings). *Zeitschrift des Vereines Seutscher Ingenieure*, 46H38,39I: 1342–1348, 1432–1437 (1902).

[4] Tripp, H., and Murphy, B. Eccentricity measurements on a tilting-pad bearing. *ASLE Transactions* 28: 217–224 (1985).

[5] Gruwell, D. R., and Fouad Y. Zeidan, F. Y. Vibration and eccentricity measurements combined with rotordynamic analyses on a six-bearing turbine generator. Proceedings of the 27th Turbomachinery Symposium, 1998, pp. 85–92.

[6] Vance, J. M., and Kar, R. Sub-synchronous vibrations in rotating machinery—methodologies to identify potential instability. GT2007-27048, Proceedings of ASME Turbo Expo 2007: Power for Land, Sea and Air, May 14–17, 2007, Montreal, Canada.

[7] Childs, D., Schaible, A., and Al Jugahaiman, B. Static and dynamic performance of pressure dam bearings with dam steps that are (I) square and (II) filleted. Proceedings of ASME Turbo Expo 2008: Power for Land, Sea and Air, GT2008, June 9–13, 2008, Berlin, Germany.

[8] Someya et al. *Journal Bearing Data Book*. Springer-Verlag, 1989.

[9] Nicholas, J., and Allaire, P. Analysis of step journal bearings—finite length and stability. *ASLE Transactions* **22:** 197–207 (1980).

[10] Carlson, T., and Zeidan, F. Subsynchronous vibrations: temporary fix and permanent solution. *Turbomachinery International* (Sept./Oct. 1991).

[11] Zeidan, F. Y., and Donald J. Paquette, D. J. Application of high speed and high performance fluid film bearings in rotating machinery. 23rd Turbomachinery Symposium, 1994, pp. 209–233.

[12] Kingsbury, A. US patent No. 947242 for the tilting pad thrust bearing in 1910.

[13] Michell, A.G.M. Improvements in thrust and like bearings. British Patent No. 875 (1905).

[14] Herbage, B. S. High speed journal and thrust bearing design. Proceedings of the 1st Turbomachinery Symposium, 1972, pp. 56–61.

[15] Adams, M. L., and Payandeh, S. Self-excited vibration of statically unloaded pads in tilting-pad journal bearings. *Journal of Lubrication Technology* 105: 377–384 (July 1983).

[16] Chen, W. J., Zeidan, F.Y., and Jain, D. Design, analysis, and testing of high performance bearings in high speed integrally geared compressor. Proceedings of the 23rd Turbomachinery Symposium, 1994.

[17] Sommerfeld, A. Zur Hydrodynamischer Theorie der Schmier mettleibung, *Zeit schrift fur Mathematische and Physik* 40: 97–155 (1904).

[18] Raimondi, A. A., and Boyd, J. Applying bearing theory to the analysis and design of pad type bearings. *Journal of Applied Mechanics, Transactions of the ASME*, 77: 287–309 (1955).

[19] Raimondi, A. A. and Boyd, J. A solution for the finite journal bearing and its application to analysis and design, II. ASLE Preprint No. 56LC-12, 1956.

[20] Shelly, P., and Ettles, C.M.M. A tractable solution for medium-length journal bearings. *Wear* 12: 221–228 (1970).

[21] Jones, W. R., and Barrett, L. E. Rapid solutions to Reynolds equation with applications to fluid film bearings. UVA/643092/MAE81/177, January 1981.

[22] Lund, J. W. Spring and damping coefficients for the tilting-pad journal bearing. *ASLE Transactions* 7: 342–352 (1964).

[23] Nicholas, J. C. *A Finite Element Dynamic Analysis of Pressure Dam and Tilting Pad Bearings*. Thesis, University of Virginia, May 1977.

[24] Parsell, J. K., Allaire, P. E., and Barrett, L. E. Frequency effects in tilting-pad journal bearing dynamic coefficients. *ASLE Transactions* 26: 222–227 (1982).

[25] Wilson, B. W., and Barrett, L. E. The effect of eigenvalue-dependent tilt pad bearing characteristics on the stability of rotor–bearing systems. UVA/643092/MAE85/321, January 1985.

[26] Branagan, L. A. *Thermal Analysis of Fixed and Tilting Pad Journal Bearings Including Cross-film Viscosity Variations and Deformations*. Thesis, University of Virginia, May 1988.

[27] Ertas, B.H., and Vance, J.M. Effect of static and dynamic misalignment on ball bearing radial stiffness. *AIAA Journal of Propulsion and Power* 20: 634–647 (2004).

[28] Ertas, B.H., Al-Khateeb, E., and Vance, J.M. Rotordynamic bearing dampers for cryogenic rocket engine turbopumps. *AIAA Journal of Propulsion and Power* 19: 674–682 (2003).

[29] Childs, D.W. Space shuttle main engine high-pressure fuel turbopump rotordynamic instability problem. *ASME Journal of Engineering for Power* 100: 48–57 (1978).

[30] Okayasu, A., Ohta, T., Azuma, T., Fujita, T., and Aoki, H. Vibration problems in the LE-7 liquid hydrogen turbopump. Proceedings of the 26th AIAA Joint Propulsion Conference, 1990, pp.1–5.

[31] Ertas, B. *The Effect of Static and Dynamic Misalignment on Ball Bearing Radial Stiffness for Various Axial Preloads*. Thesis, Texas A&M University, May 2001.

[32] Schmidt, B. *The Experimental Determination of the Dynamic Radial Stiffness of an Angular Contact Ball Bearing*. Thesis, Texas A&M University, May 2001.

[33] Zarzour, M., and Vance, J. M. Experimental evaluation of a metal mesh bearing damper. *ASME Journal of Engineering for Gas Turbines and Power*, April 2000.

[34] Al-Khateeb, E., and Vance, J. M. Experimental evaluation of a metal mesh damper in parallel with a structural support. ASME Paper No. 2001-GT-0247, Proceedings of the 46th International Gas Turbine and Aeroengine Congress, New Orleans, Louisiana, June 4–7, 2001.

[35] Vivek, V. C., and Vance, J.M. Design equations for wire mesh bearing dampers in turbomachinery. GT2005-68641, Proceedings of ASME Turbo Expo 2005, Reno–Tahoe, Nevada, June 6–9, 2005.

[36] Ertas, B., and Al-Khateeb, E., Cryogenic temperature effects on metal mesh dampers and liquid hydrogen turbopump rotordynamics. AIAA Paper 2002-4164, Proceedings of the AIAA–ASME–SAE Joint Propulsion And Power Conference, Indianapolis, Indiana, July 2002.

[37] Darlow, M. and Zorzi, E. Mechanical design handbook for elastomers. NASA Contractor Report, 1981.

[38] Marmol, R., and Vance, J. M. Squeeze film damper characteristics for gas turbine engines. *ASME Journal of Mechanical Design* 100: 139–146 (1978).

[39] Szeri, A., Raimondi, A., and Giron-Durate, A. Linear force coefficients for squeeze film dampers. *ASME Journal of Lubrication Technology* 105: 326–334 (1983).

[40] Kinsale, I., and Tichy, J. Numerical and experimental study of a finite submerged squeeze film damper. Proceedings of the 1989 ASME Vibrations Conference, Machinery Dynamics: Applications and Vibration Control Problems, DE-18 (2), 1989.

[41] Zeidan, F., San Andres, L. and Vance, J. Design and application of squeeze film dampers in rotating machinery. Proceedings of the 25th Turbomachinery Symposium, Houston, Texas, September 17–19, 1996, pp. 169–188.

[42] Zeidan, F. Y. Application of squeeze film dampers. *Turbomachinery International*, 50–53 (Sept./Oct. 1995).

[43] BT Murphy, B. T., Manifold, S. M., and Kitzmiller, J. R., Compulsator rotordynamics and suspension design. *IEEE Transactions on Magnetics* 33: (1, pt 1) (January 1997).

EXERCISES

5-1. Calculate the stiffness and damping coefficients for a bearing with the following dimensions and parameters using the tables from Someya et al. [8]: bearing diameter, 4 inches; pad length, 3 inches; preload, $m = 0.5$; a 4-pad LBP with a diametral assembled clearance of 0.006". The rotor weight is 4800 lb and assume the CG is in the center between the two bearings. Use the viscosity for ISO-46 oil and assume an operating film temperature of 180°F. The coefficients should be calculated for the following speeds: 3000, 5000, and 7000 rpm.

5-2. Repeat the calculations above but for a preload of 0.6, which will require double interpolation between the two L/D tables and the two preload tables for a 4-pad LBP.

5-3. Assuming that the bearing span is 48 inches, and the CG for the rotor is 20 inches from the left bearing, calculate the load on each bearing and obtain the dynamic coefficients for each bearing, as shown in Problems 5-1 above.

5-4. The pedestal supporting the bearings in Problem 5-1 above was assumed to be infinitely rigid. Using the coefficients obtained in problem 1; provide the equivalent dynamic coefficients accounting for the flexibility of the bearing supports or pedestals. Assume that the supports are symmetric and use 500,000 lb/in. Repeat for an asymmetric support with the vertical being 500,000 lb/in and the horizontal stiffness 300,000 lb/in.

5-5. Calculate the minimum, nominal, and maximum preload for the following shaft journal bearing configuration: shaft journal 2.000″ + 0.000/−0.0005″, pad machined bore diameter 2.005″ + 0.001″/−0.000, pad assembly bore diameter 2.003″ + 0.001″/ −0.000.

Answer: Min preload = 0.182, Nominal preload = 0.348, Max preload = 0.5

5-6. Calculate the minimum, nominal, and maximum preload for the following shaft journal bearing configuration: shaft journal 4.000″ + 0.000/−0.0005″, pad machined bore diameter 4.008″ + 0.001″/−0.000, pad assembly bore diameter 4.006″ + 0.001″/ −0.000.

Answer: Min preload = 0.117, Nominal preload = 0.228, Max preload = 0.333

APPENDIX: SHAFT WITH NO ADDED WEIGHT

Input table of beam and station definitions, more than one beam per station is ok

Station # stnum	Length in length	OD in oda	ID in ida	Weight Density lb/in^3 rhoa	Elastic Modulus psi ea	Shear Modulus psi ga	Added Weight lb awt	Added Ip lb-in^2 aip	Added It lb-in^2 ait
1	3.0000	6.0000	0.0000	0.283	29.0E+6	10.9E+6			
2	3.0000	6.0000	0.0000	0.283	29.0E+6	10.9E+6			
3	3.0000	6.0000	0.0000	0.283	29.0E+6	10.9E+6			
4	3.0000	6.0000	0.0000	0.283	29.0E+6	10.9E+6			
5	3.0000	6.0000	0.0000	0.283	29.0E+6	10.9E+6			
6	3.0000	6.0000	0.0000	0.283	29.0E+6	10.9E+6			
7	3.0000	6.0000	0.0000	0.283	29.0E+6	10.9E+6			
8	3.0000	6.0000	0.0000	0.283	29.0E+6	10.9E+6			
9	3.0000	6.0000	0.0000	0.283	29.0E+6	10.9E+6			
10	3.0000	6.0000	0.0000	0.283	29.0E+6	10.9E+6			

Station # stnum	Length in length	OD in oda	ID in ida	Weight Density lb/in^3 rhoa	Elastic Modulus psi ea	Shear Modulus psi ga	Added Weight lb awt	Added Ip lb-in^2 aip	Added It lb-in^2 ait
11	3.0000	6.0000	0.0000	0.283	29.0E+6	10.9E+6			
12	3.0000	6.0000	0.0000	0.283	29.0E+6	10.9E+6			
13	3.0000	6.0000	0.0000	0.283	29.0E+6	10.9E+6			
14	3.0000	6.0000	0.0000	0.283	29.0E+6	10.9E+6			
15	3.0000	6.0000	0.0000	0.283	29.0E+6	10.9E+6			
16	3.0000	6.0000	0.0000	0.283	29.0E+6	10.9E+6			
17	3.0000	6.0000	0.0000	0.283	29.0E+6	10.9E+6			
18	3.0000	6.0000	0.0000	0.283	29.0E+6	10.9E+6			
19	3.0000	6.0000	0.0000	0.283	29.0E+6	10.9E+6			
20	3.0000	6.0000	0.0000	0.283	29.0E+6	10.9E+6			
21	3.0000	6.0000	0.0000	0.283	29.0E+6	10.9E+6			
22	3.0000	6.0000	0.0000	0.283	29.0E+6	10.9E+6			
23	3.0000	6.0000	0.0000	0.283	29.0E+6	10.9E+6			
24	3.0000	6.0000	0.0000	0.283	29.0E+6	10.9E+6			
25	0.0000								

6

FLUID SEALS AND THEIR EFFECT ON ROTORDYNAMICS

FUNCTION AND CLASSIFICATION OF SEALS

The main function of a noncontacting fluid seal is to reduce the leakage flow rate between a rotating shaft or rotor and a stationary housing. A clearance space around the shaft must exist in high-speed machines to avoid excessive wear and heating from friction. Some leakage is inevitable, and results in axial fluid velocities through the seal in the direction of pressure drop, along with a nonuniform pressure distribution around the seal. Seals can thus produce lateral forces on the shaft (from the fluid pressure) that can have a strong effect on the rotordynamic characteristics of the machine. Although bearing damping can reduce synchronous whirl amplitudes and suppress rotordynamic instability in some applications, the location of the damping at the bearings even when it is large is often not effective for the whirl modes of interest in relatively long flexible shafts that are commonly used in steam turbines and high-pressure compressors. This is because the bearings are usually located at the ends of the shaft where response amplitude (motion) is low and the effective damping cannot be developed. Conversely, seals are often located near antinodes of the whirling modes shapes and thus can be much more effective, either as destabilizers (labyrinth seals) or as vibration dampers (damper seals).

Figure 6-1 shows the working pressures and the designed fluid flow path through a two-stage centrifugal compressor or pump. Since

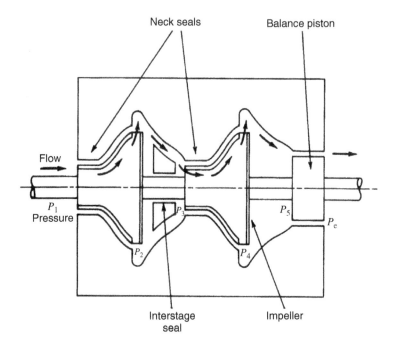

Figure 6-1   Flow path, fluid pressures, and seals in a two-stage compressor.

$P_5 > P_4 > P_3 > P_2 > P_1 = P_e$, the working fluid will tend to leak backward through the impeller eye and interstage seals, and forward past the balance piston seal. The fluid velocity in a fluid seal has both axial and circumferential components, and the nonuniform pressure within the seal generally varies with time due to rotor motion within the seal clearance.

The major types of fluid seals used in turbomachinery are as follows:

1. *Plain seals*. These can be straight, as in Fig. 6-5, or tapered as in Fig. 6-2a, or stepped as in Fig. 6-2b. Plain seals are generally used in pumps. In geometry, they are similar to journal bearings but the clearance/radius ratio is usually larger to avoid rotor/stator contact. The recent advances in engineered polymers allows the use of tighter clearances for improved efficiency.

2. *Floating ring oil seals*. In these seals (Fig. 6-3) lubricating oil is introduced between the two seals at a pressure just above suction pressure. Sweet oil then leaks to the left through the outer ring and to the right where it mixes with the process gas (sour oil). The ring translates (whirls or vibrates) laterally with the rotor, but does not spin or rotate ($U_J = 0$). They are still used in many high-pressure multistage centrifugal compressors, but many have been replaced by

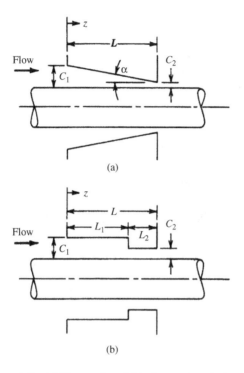

Figure 6-2 (a) Tapered and (b) stepped seals from [1].

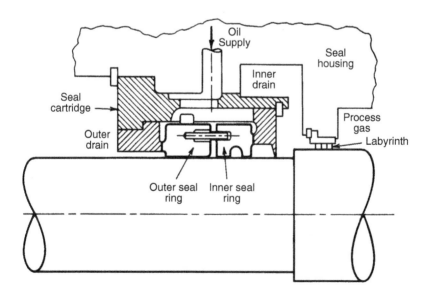

Figure 6-3 Floating ring seal. From [2].

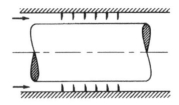

Figure 6-4   Labyrinth seal, teeth on stator.

dry gas seals that have a rotating disk running against a stationary plate with grooves usually spiral in shape. This creates a very small axial gap where a sealing gas is introduced. The use of a dry gas seal basically converts a radial seal in the case of floating oil seals to an effectively axial seal with minimum lateral influence. However, this adds mass to the rotating shaft and whatever damping the oil in the floating seals introduced is now gone.

3. *Labyrinth seals*. The circumferential blades in axial tandem produce a "tortuous passage" (Webster's dictionary) for the leaking fluid (see Fig. 6-4). The blades may be on the stator or on the rotor or on both. These seals are widely used in all types of rotating machinery due to their simplicity and ease of manufacture.

4. *Contact seals*. These are also referred to as mechanical seals where there is no clearance between the rotating and stationary parts. Because of the rubbing, these seals are used mostly in low speed pumps, or where the working fluid can act as a coolant and where the contact element can be rubber, leather, or plastic material. Their effect on rotordynamics is unknown and probably minimal except for some damping from friction.

5. *Damper seals*. These are designed to add damping to the rotor-bearing system. The two main types are pocket damper seals and honeycomb (or hole pattern) seals.

6. *Brush seals*. These can be classified as contact type seals when designed for line to line or with a slight interference that allow a very small amount of leakage. The element that contacts the shaft is made of flexible wire bristles.

Plain Smooth Seals

Because of their similarity to journal bearings, it is tempting to analyze plain seals with Reynolds bearing theory. However, there are some major differences that render the Reynolds theory inapplicable, even if fluid

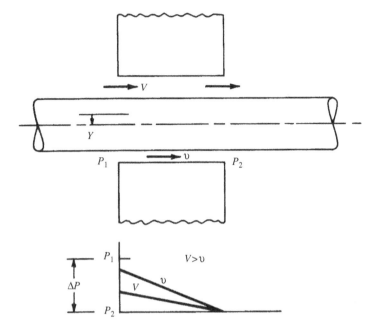

Figure 6-5   The Lomakin effect in a plain seal.

inertia effects are added in. The high axial fluid velocity through the seal and the relatively large radial clearance produce a highly turbulent flow condition in the seal, which violates the Reynolds assumption of laminar flow. If the working fluid is a gas, compressibility effects may also be important. The axial pressure drop across a plain seal causes it to have a radial stiffness, independent of shaft rotation. A radial deflection of the shaft in the seal (e.g., down in Fig. 6-5) produces a smaller clearance in the direction of the displacement. The axial fluid velocity is slower in this region than in the high clearance area at the top. The Bernoulli principle produces higher pressures in the low-velocity region. The total axial pressure drop across the seal is caused by two main factors: an entrance loss, and a friction loss along the length of the seal. As Fig. 6-5 shows, the entrance loss is greater in the high clearance region (at the top), so the average pressure is lower there. The pressure difference produces a restoring force $F_y = -K_s Y$ upward, opposing the displacement Y. This radial stiffness K_s of a plain seal, which occurs even for zero shaft speed, provided the axial pressure difference Δp exists, is known as the Lomakin effect, named after its first investigator [3].

Black and his colleagues [4–6] and Childs [7,8] (later) have formulated and extended Lomakin's theory in terms applicable to the rotordynamic analysis of centrifugal pumps. The Lomakin direct stiffness (K_{XX}) coefficient is given by

$$K_s = 4.7R \left(\frac{\Delta P}{\lambda}\right) \left(\frac{\sigma}{1.5 + 2\sigma}\right)^2 \qquad (6\text{-}1)$$

where
λ = friction coefficient for the axial flow = $0.079/\text{Re}^{.25}$ for turbulent flow, and
$\sigma = \lambda L/C$

Here R, L, and C are the radius, axial length, and radial clearance of the seal, respectively.

If the direct stiffness equation (6-1) were the only effect of the plain seal, then its effect on critical speeds would be easily and accurately predictable. Black, Childs, and others have shown, however, that K_s increases with shaft speed (at constant ΔP) and that the seal also produces cross-coupled stiffness (K_{XY}), direct and cross- coupled damping (C_{XX} and C_{XY}), and direct inertia (D_{XX}) coefficients [6,7]. The effect of these coefficients is discussed in the sections on labyrinth seals and damper seals below. Consider also that the pressure drop ΔP will vary with speed in most turbomachines, and it can be seen that the rotordynamic effects are quite complex. Fleming [1] showed that the direct stiffness of stepped or tapered seals with converging clearance is 2–14 times higher than for a straight seal with clearance equal to the minimum of the converging seal. Experimentally measured stiffness and damping coefficients for several geometries of plain seals were published by Nelson et al. [9].

Floating Ring Seals

The floating rings in an oil seal when properly operating are free to translate laterally. The only force exerted on the rotor is the sliding friction between the floating ring and the mating stationary housing and transmitted through the oil film between the ID of the floating ring and the rotor shaft. This friction force increases with the axial pressure drop across the ring, and when this is coupled with wear between the contact sliding surfaces on the face of the seal and the seal housing, the frictional force can become excessive. Once the friction force becomes excessive, the ring locks in place and acts as a journal bearing. Kirk and Miller [2] have analyzed the transient dynamics of the seal ring and its effect on the imbalance response and whirl stability of a multistage compressor. When they are operating properly, oil ring seals can have a favorable effect in attenuating imbalance response at the critical speeds. If the rings become locked and operate as journal bearings, they can induce the rotordynamic instability known as oil whirl or oil whip. To suppress or eliminate the instability of the locked system, the effective axial hydrodynamic length of the ring can be reduced, but this tends also to reduce the damping effect on imbalance response at the critical speeds. The locked-up condition and how to prevent it has been analyzed and discussed by Semanate and San Andres [10–12].

Conventional Gas Labyrinth Seals

The most common shaft seal for gas applications in high-speed turbo-machinery is the labyrinth seal. The photograph in Fig. 6-6 shows a teeth-on-stator (TOS) labyrinth seal with 19 teeth or "blades." This seal is split horizontally for ease of installation in a horizontally split compressor case, but can be solid when used in vertically split compressors or turbines. In aircraft turbine engines, and some process compressors, the teeth are usually on the rotor (TOR). In such configurations (TOR), the labyrinth teeth are designed to run against an abradable material on the stator. These applications allow for a tighter clearance and therefore improved efficiency as the teeth cut grooves and wear in the material on the stator. Labyrinth seals are designed to be noncontacting, although rubs do sometimes occur, which can cause rapid wear of the seal and heating of the shaft. Rubs can also induce subsynchronous forward whirl that tracks

Figure 6-6   Labyrinth seal half-section.

the running speed at a fixed fraction of the synchronous frequency, or backward whirl with frequency at an eigenvalue of the system. But aside from rubs, labyrinth seals in certain high-pressure applications generally have a deleterious effect on rotordynamic stability.

Any force on the shaft can be modeled mathematically with linearized coefficients, and a destabilizing follower force is modeled with the cross-coupled stiffness coefficients $K_{XY} = -K_{YX}$. The gas pressure distribution around labyrinth seals produces large $K_{XY} = -K_{YX}$ force coefficients. If the seal also has direct damping coefficients $C_{XX} = C_{YY}$, then the total resulting follower force F_ϕ is given by

$$F_\phi = \left(K_{XY} - C_{XX}\dot{\phi} \right) r \qquad (6\text{-}2)$$

where $r = \sqrt{X^2 + Y^2}$ is the whirl radius. Since the whirl velocity is $V_t = r\dot{\phi}$, we can divide Eq. 6-2 by V_t and reverse the sign to get the effective damping coefficient to be

$$C_{\text{eff}} = C_{XX} - K_{XY}/\dot{\phi} \qquad (6\text{-}3)$$

Labyrinth seals generally have negative effective damping C_{eff}, as defined in Eq. 6-3. When comparing two shaft seals at the same speed and operating conditions, the best seal from a rotordynamic viewpoint will have the largest value for C_{eff}. Large values for C_{eff} are obtained by increasing the sum $C_{XX} + C_{YY}$ or by making the algebraic sum of $K_{XY} - K_{YX}$ small or negative. For stability calculations, the whirl speed $d\phi/dt$ in Eq. 6-3 will be a natural frequency of the rotor–bearing system. For unbalance response calculations the whirl speed will be the shaft speed ω. See Chapter 3 for an analysis of the effect of C_{eff} on synchronous response to unbalance. The results are surprising when C_{eff} becomes negative. The response to unbalance can actually be reduced. Consider Eq. 6-2. The first term is destabilizing (provided that $K_{YX} = -K_{XY}$); the second term is stabilizing. When the shaft speed ω is substituted for the whirl frequency, the ratio of the two terms is called the *whirl-frequency ratio f_w*. That is,

$$f_w = \frac{K_{XY}}{C_{XX}\omega} \qquad (6\text{-}4)$$

The numerical value of f_w is sometimes used as a measure of the destabilizing influence of a seal or bearing. To get a useful interpretation of this number, consider a rotor–bearing system with only one seal or bearing and with no other damping or destabilizing components. Unstable whirling, when it occurs, will be at a natural frequency. For marginal stability, with the real part of the eigenvalue $\lambda = 0$, the follower force f_ϕ in Eq. 6-2 is zero, so

$$\frac{K_{XY}}{C_{XX}\dot{\phi}} = 1 \tag{6-5}$$

Simultaneous satisfaction of Eqs. 6-4 and 6-5 requires

$$f_w = \frac{\dot{\phi}}{\omega} \tag{6-6}$$

This is the ratio of whirl speed to shaft speed. For example, seal A with $f_w = 0.5$ would produce instability at a threshold shaft speed of twice the natural frequency, while seal B with $f_w = 0.8$ would have a threshold speed of only 1.25 times the natural frequency. Seal B is more destabilizing because the system becomes unstable at a lower shaft speed.

In the early 1960s, during a development program for the U.S. Navy, an aircraft engine compressor became rotordynamically unstable on a test stand at General Electric Aircraft Engines in Cincinnati. Alford [13] published a paper describing the problem and hypothesizing two possible causes. One of these was negative damping from the labyrinth seals. His mathematical model relied on an assumption that pressure differences could exist across the diameter of a labyrinth seal without circumferential flow (which was neglected in his model). Cross-coupled forces were not considered. The most notable prediction of Alford's analysis was that seals with clearances converging in the direction of flow would have negative damping coefficients ($C_{XX}, C_{YY} < 0$), and seals with diverging clearances would have positive damping coefficients ($C_{XX}, C_{YY} > 0$). The predicted coefficients were quite large for both cases, and were grossly overpredicted.

Alford's 1965 model was limited so as to only predict the direct damping coefficients of labyrinth seals with two blades and choked flow. It modeled only the axial flow and neglected circumferential flow effects. This model was later extended to multiple blades and unchoked flow by Murphy and Vance [14]. They predicted that a ten-bladed diverging labyrinth of 200 mm (7.8 in) diameter and with a pressure ratio of 10 could have a direct damping coefficient equal to 87,560 N-sec/m (500 lb-sec/in), which is about the same as a typical squeeze film bearing damper of similar dimensions. However, more than 10 years of testing labyrinth seals by the first author have shown that the direct damping coefficients of conventional labyrinth seals are, in fact, very small and never approach even a small percentage of the values predicted by Alford's theory. Instead, the dominant rotordynamic coefficient of conventional labyrinth seals is cross-coupled stiffness that reduces the effective damping and is often destabilizing to rotor whirl. The reason for this failure of Alford's theory is that conventional labyrinth seals have continuous and unobstructed annular grooves so that pressure variations across the seal diameter cannot exist without large circumferential flow rates. Unfortunately, it is the circumferential swirl that produces the undesirable cross-coupled stiffness. The eventual understanding of these facts by the first author led to the invention of the pocket damper seal (TAMSEAL) as described later in the chapter.

Zierer and Conway [15] describe measurements made with labyrinth seals at low inlet pressures in a rotordynamic test rig. Straight-through labyrinth seals were installed in the rotordynamic rig instrumented for coastdown measurements of imbalance response (Bode plots). Figure 6-7 shows a cross-section drawing of the test rig.

Both teeth-on-rotor and teeth-on-stator gas seals were tested, each with twelve blades, 173 mm (6.81 in) blade diameter, and 102 mm (4.11 in) total length. The nominal blade tip radial clearance was 0.5 mm (20 mils). The teeth on stator seal was tested with the blade tip clearances diverging (in the direction of flow), uniform, and converging. The teeth-on-rotor seal was tested with uniform clearances. The inlet air pressure to the seals was varied from 1.7 to 14.6 bars (25 to 200 psig) with the last blade exhausting to the atmosphere. There was no induced preswirl other than that produced by the rotating shaft. Coastdown tests of all the seals were performed on a rotordynamic test rig to show their effect on synchronous response to imbalance when passing through a 3700-rpm critical speed. The synchronous response to imbalance was generally increased by all the seals at inlet pressures up to about 11.2 bars (150 psig). The worst

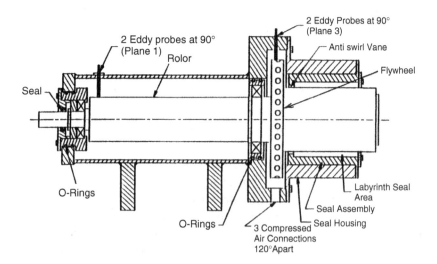

Figure 6-7   Zierer and Conway's test rig.

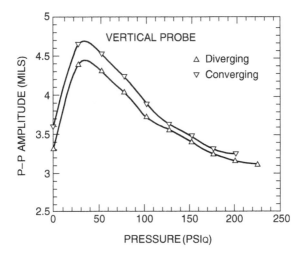

Figure 6-8   Peak unbalance response with diverging and converging labyrinth seals versus pressure.

case was for the teeth-on-rotor seal at about 2.7 bars (35−45 psi) inlet pressure where the rotor whirl amplitude was increased from 0.1 mm (3.75 mils, peak-to-peak) to over 0.13 mm (5 mils). At higher inlet pressures, above 13 bars (176 psig), the rotor whirl amplitude was slightly decreased in most cases. The peak responses (at the critical speed of the test rig) of both seals mostly increased with the upstream pressure as shown in Fig. 6-8, which indicates negative effective damping. Rap testing showed that the seals actually had a small amount of positive damping at zero speed, so it is reasonable to assume that the swirl induced by spin of the rotor decreased the effective damping (with K_{XY}, $-K_{YX}$) and made it negative in most cases.

Experimental research by Benckert and Wachter [16] showed that the destabilizing cross-coupled stiffness, K_{XY}, $-K_{YX}$, is mainly caused by circumferential swirl of gas around the seal. Their measurements indicated that swirl is generally increased by higher shaft speeds from friction at the shaft/gas interface. This effect is usually greater when the seal teeth are on the rotor (TOR). They showed that the swirl can be blocked, or even reversed in direction, by placing angled vanes (*swirl brakes*) at the entrance of the seal, thus improving the effective damping of the seal (less negative or becoming positive). Their measurements is discussed with more detail below in connection with pocket damper seals.

In centrifugal compressors, the level of preswirl at the seal entrance (without swirl brakes) depends on its location in the compressor. The worst situation is where the entrance is next to the back side of an impeller, so that the gas flows radially inward from the impeller discharge to the seal entrance. In this case the ratio of swirl velocity to shaft surface velocity may exceed 1.0, which will produce large cross-coupling in the seal. A common modification is to route the discharge gas directly to the interior of the seal (*shunt injection*), which has often stabilized an otherwise unstable compressor. Probably the main effect is to change the direction of flow on the back side of the impeller from inward to outward, thus reducing the swirl at the seal entrance and perhaps also changing the force coefficients of the impeller. See, for example, Camatti et al. [17].

Laos [18] conducted tests of a six-bladed (six-"teeth") labyrinth seal at low pressures in the Turbomachinery Laboratory at Texas A&M University. The seal dimensions were 4" (shaft) diameter, 1" length, and 0.17" tooth spacing, with 0.008" diametral clearance. The seal was installed in a rotordynamic test rig for measurements of unbalance response while coasting down through a critical speed. Figure 6-9 shows the measured Bode plots for several different inlet pressures to the seal up to 40 psig. The exit pressure was atmospheric. The plots show a small amount of damping. However, at inlet pressures above 40 psig the seal induced a violent instability in the test rig.

Picardo [19] measured the stiffness and damping coefficients of straight-through teeth-on-rotor labyrinth seals at 1000-psi inlet pressures and various pressure ratios. The results are compared with the predictions of the best computer codes available, which generally overpredicted and underpredicted the measurements by large margins. The measurements generally show negative effective damping under most test conditions, but the values are not reliably predicted by the codes.

The negative effective damping indicated in Fig. 6-8 at low pressures and measured by Picardo at high pressures cannot be reliably predicted by contemporary computer codes for labyrinth seals. The flow and the pressure fields in labyrinth seals are much more complex than in fluid film

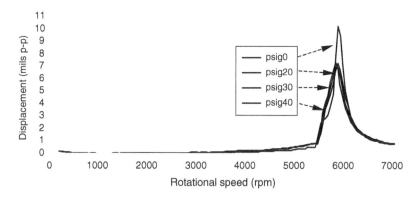

Figure 6-9 Unbalance response with a labyrinth seal.

bearings, where the computer predictions of force coefficients are much more accurate and reliable. Iwatsubo [20] made photographs through a transparent labyrinth seal wall that reveals a very complex flow pattern, containing vortices, around the annulus. To have reliable numbers for the force coefficients of labyrinth seals, tests must be conducted using the exact seal geometry and size of interest. In view of this shortcoming, and because of the limited (or negative) damping offered by labyrinth seals, the authors suggest that turbomachinery engineers should consider using a pocket damper seal or a honeycomb seal (described in following sections) instead of a conventional labyrinth seal for high-pressure applications and when stability is of a concern. These newer types of *damper seals* can be predicted much more reliably as described below.

Pocket Damper Seals

The pocket damper seal (PDS) was invented and first tested in 1991 at Texas A&M University by the first author and Richard Shultz. It is a dual-purpose sealing and vibration-damping device that can be used in any turbomachine with compressible working fluids. It provides both pressure sealing and a remarkable amount of direct damping to attenuate vibration and rotor whirl. It has been used to solve rotordynamic instability problems in a number of multistage, high-pressure centrifugal compressors. (In steam turbines this type of vibration problem is called *steam whirl*.)

The PDS can produce much more effective rotordynamic damping than any other type of seal or bearing damper at the frequencies of interest in multistage compressors. The damping mechanism does not rely on viscous effects. A brief explanation of the theory follows. Figure 6-10 shows a cross section illustrating the original design of the pocket damper seal. A more recent design with improved characteristics is described later, but the damping mechanism is more easily explained with the original design.

The blades, or teeth, of this PDS design are arranged in axial "active" pairs with partition walls equally spaced around the circumference to form pockets. The inlet blade, or upstream tooth, of each pair is identical to a labyrinth seal tooth. The downstream blade, or tooth, of each pair has either (1) a larger clearance to the shaft or (2) slots or notches to allow leakage flow out of the pockets that is not modulated by the shaft vibration. Experiments have shown that the configuration with slots produces more damping. The partition walls around the circumference form the pockets that isolate volumes of dynamically varying pressure. In between each active pair is an inactive annulus with no partition walls so the pressure can reequalize around the circumference of the seal. Since the inactive annulus does not contribute damping forces, it usually is made with a shorter axial pitch. This design allows the journal vibration to modulate the

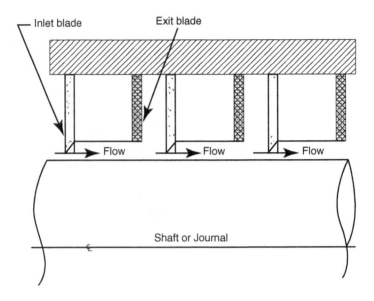

Figure 6-10 A six-bladed pocket damper seal (original design).

flow into the pockets through the inlet blade clearance while minimizing the modulation of the exit flow downstream (through the slots or through the larger clearance). With this design the dynamically varying pressure forces will always oppose the seal journal vibratory velocity, which is the very definition of a damping force. A secondary function of the pocket partition walls is to block the destabilizing circumferential swirl of the fluid around the journal and thus perform the function of swirl brakes.

This configuration makes a seal for which the direct damping predictions of Alford's theory become much more accurate. This is why a similar, but extended, theory for this seal design (based on an assumption that the average pressure in the seal cavities can be calculated based on axial mass flow rates into and out of the cavities) gives fairly accurate predictions and has been verified by laboratory measurements. It is important to understand that the damping produced by the pocket damper seal is not due to viscous effects. It originates from the compressibility of the working fluid and dynamic pressure variation around the circumference of the seal. The dynamic pressures in the pockets are quite large and can produce large damping forces on the shaft.

Optimizing the volume of the pockets is based on the use of the Helmholtz effect to maximize the dynamic pressures [21]. A review of the Helmholtz resonator shown in Fig. 6-11 and its characteristics can help in understanding this effect on the pocket damper seal. There are two primary components in the Helmhotlz resonator: an air cavity, and a port that connects it to the rest of the system (usually a pipe). As the

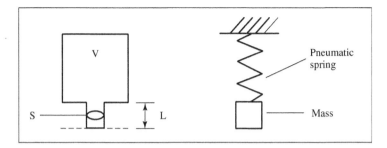

Figure 6-11 Helmholtz resonator model.

flow of air in the pipe pulsates, the air in the cavity acts like a spring and it also has mass. The resulting mass/spring equivalent will resonate at its natural frequency. In the pocket damper seal, the resonant frequency induced by the pocket amplifies the dynamic pressure in the pocket produced by the modulated flow through the seal clearances. The dynamic pressure opposes the vibratory velocity to produce the damping.

Honeycomb Seals

Honeycomb seals have been used in aircraft engines as an abradable seal running against the teeth on the rotor. In process compressors they usually have a smooth rotor and have been used since the early 1960s, initially with the primary objective of reducing leakage. Their stabilizing effects were not realized until much later through the work of Childs et al. [22–24]. Since fluid rotation and circumferential flow tend to increase the cross-coupled stiffness and reduce the stability margin, it seems reasonable to expect better stability when the rotor is smooth and the stator is rough. The smooth rotor with no steps or grooves means that we have a smaller surface area to impart circulation to the gas or fluid, and a rough stator means that the fluid circulation (the mechanism that causes instability) will be slowed and its destabilizing effect reduced. Both of these effects shown in the schematic in Fig. 6-12 reproduced from Childs [25] are helpful in terms of stability. Another factor that the honeycomb cells provide is the ability to build pressure in the pockets and thus provide direct stiffness and damping. The direct stiffness tends to shift the frequency of the forward mode up and thus improves stability,[*] while the direct positive damping will counter the destabilizing effects introduced by the cross-coupled stiffness and thus improve stability.

Zeidan et al. [26] showed how a honeycomb seal can be used to replace a long labyrinth seal for the purpose of eliminating subsynchronous vibrations in the back-to-back compressor shown in Fig. 6-13. Attempts to

---

[*]Not always. See the section below on negative stiffness effects.

Figure 6-12   Schematic of a smooth rotor running against a rough stator (honeycomb) reproduced from Childs [25].

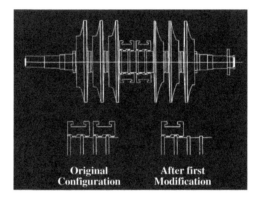

Figure 6-13   High-pressure injection compressor with labyrinth center seal from [26].

stabilize this compressor through the use of swirl brakes at the impeller eye seals plus shunt holes in the center seal failed to produce the desired results. The use of shunt holes is a swirl canceling modification where gas is bled from a higher pressure and introduced into the seal in a direction opposite to shaft rotation, as shown in Fig. 6-14. This has been used successfully in some compressors as reported by Zhou [27], but failed in this high-pressure compressor.

Honeycomb seals used in high-pressure centrifugal compressors are typically made from stainless steel or Hastelloy X sheets that are brazed to the backing material. The bore is then ground or *electric discharge machined* (EDM) to achieve the desired clearance over the shaft. This lengthy process makes the delivery relatively long, and the use of stainless steel or Hastelloy X for the cells necessitates a larger clearance to the

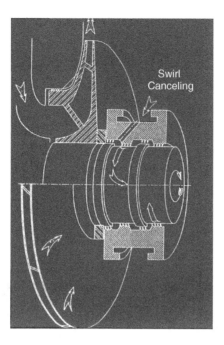

Figure 6-14   Shunt holes or swirl canceling modification.

shaft. Furthermore, the integrity of the brazing is questionable in high-pressure applications, and in some instances has failed, causing a loss of thrust balance and overloading of the thrust bearing. These drawbacks with the honeycomb seals led to the development of hole-pattern seals as an equivalent alternative, and is discussed in the next section.

Hole-pattern Seals

The hole-pattern seal, shown alongside the honeycomb seals in Fig. 6-15 for comparison, was introduced through the work of Childs et al. [28,29] in the mid 1980s to mitigate the shortcomings mentioned above with the honeycomb seals. The hole pattern is generated by precision drilled round bottom holes (these holes are not through but are typically drilled to a depth similar to that in honeycomb seals.) They can also be of any shape if sinker electric discharge machining is used in the manufacturing process. They can be made out of the more "tolerant to rub" materials typically used for seals such as aluminum, bronze, or engineered polymers. This allows the use of a tighter clearance and thus for a more effective seal as well.

The hole-pattern seals were originally used in pump seals and had a relatively lower cavity to total area ratio, as shown in Fig. 6-16. This particular pump seal was designed by Childs and built by Bearings Plus,

Figure 6-15 Side-by-side comparison of honeycomb and hole-pattern seals.

Figure 6-16 Hole-pattern seal used in a pump application.

Inc. for one of the Gulf Coast process plants. The blades or slots in the face of the seal provided a brake (brake the circumferential flow) to straighten the flow (making it predominantly axial) as the process liquid entered the balance piston seal. Childs et al. [28,29] conducted analysis and testing of these seals for use in compressible gas applications. The seals have been used to replace both labyrinth seals and honeycomb seals for their ability to reduce leakage (due to the tighter clearances) and increased damping.

A retrofit was performed on four high-pressure centrifugal compressors based on the analysis provided by Childs. This analysis showed a dramatic reduction in the leakage coupled with improved stability. The outcome was to manufacture the hole-pattern seal shown in the schematic of Fig. 6-17 with the swirl brakes on the flanged face of the seal. The density ratio, defined as the area of the cavities divided by the total area, must be

Figure 6-17   Hole-pattern seal for a balance piston used in a high-pressure
barrel-type centrifugal compressor.

maximized to approach the stability levels of the honeycomb seals. Childs
et al. [28,29] reported a density ratio of 0.69 as the maximum obtained for
the referenced test program, but in recent applications for the hole-pattern
seals values as high as 0.78 were achieved.

Brush Seals

Industrial brush seals have been developed initially for jet engines to provide an effective means of increasing performance through reducing leakage during the initial installation and through the operating cycles that see
significant side loads and shaft excursions. These seals achieve lower leakages by utilizing an array of bristle material (usually a high-temperature
alloy such as Haynes 25, a nickel–cobalt–chromium–molybdenum alloy).
The bristles are oriented at typically a 45-degree angle in the direction
of rotation as shown in Fig. 6-18. This angle provides flexibility when
the shaft contacts the bristles and allows the seal to run with a slight
interference in the case of jet engines.

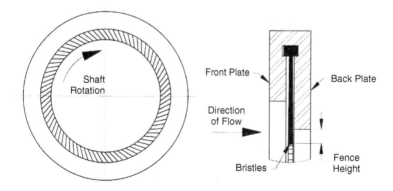

Figure 6-18   Typical brush seal configuration.

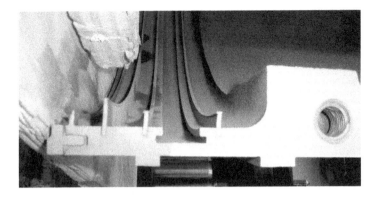

Figure 6-19 Compressor discharge labyrinth damage adjacent to brush seal (brush seal not shown but typically rolls in the hook fits between lady teeth).

The brush seals use and development was then naturally expanded into the heavy-duty industrial gas turbines, where they are used in different sections of the turbine and compressor, and as bearing buffer seals. An example that justifies their use is deduced from the damage to the adjacent labyrinth teeth shown for the compressor discharge section in Fig. 6-19. Differential thermal expansion between the casing and the rotor coupled with excursions and transient upset conditions cause contact between the rotor and the stator labyrinth. The labyrinth teeth yielded and rolled over, resulting in a clearance much larger than the installed clearance. The brush seal offers an excellent solution for these conditions and allows the unit to recover with minor losses in efficiency. Leakage rates compared to labyrinth seals have been reduced by 40–80% with the use of brush seals.

Brush seal rotordynamic effects have been examined by Conner and Childs [30] on a four-stage brush seal. They determined that the cross-coupled stiffness is very low and generally stabilizing. They also found that the brush seal coefficients were independent of the preswirl, which is not the case with labyrinth seals. Labyrinth seals require swirl brakes to enhance their stability. Flow visualization reported by Hendricks [31] show the flow through the bristles to be complex and somewhat similar to flow through porous media. Therefore, the brush seals have a built-in swirl brake mechanism, which explains why they are not sensitive to preswirl.

The use of brush seals has been extended to power generation and process steam turbines. An example of their use in high-pressure steam turbine packing is shown in Fig. 6-20. Their use in steam turbines and process compressors started in the early 2000s and continues to grow as the need for more efficient and robust machines is in higher demand. Their use in these applications must consider the relatively higher pressures in

Figure 6-20   Brush seals installed in high-pressure steam packings.

comparison to jet engines and gas turbines, as well as their effect on the rotordynamics of the machine. Process machines typically have critical speed margins and stability margins that must be satisfied. They are also sensitive to heat generation from contact between the bristles and the shaft. In these applications, the brush seals often have a small clearance to the shaft. This makes the flow and leakage calculations more challenging, since a model for the flow through the bristle pack as well as a model for the flow under the bristle must be made. The other alternatives include the use of *computational fluid dynamics* (CFD), or reliance on testing and empirical data to generate an adequate prediction tool.

In addition to being used as stand-alone sealing elements replacing labyrinth and other seals, brush seals are also used in combination with labyrinth seals, as shown in Fig. 6-20. Brush seals have also been used in other hybrid configurations, for example, in combination with a pocket damper seal. This is referred to as a *brush hybrid seal* (BHS) and results of this configuration compared to labyrinth and conventional pocket damper seals are shown in a later section of this chapter. San Andres [32,33] and other researchers are investigating even more hybrid arrangements for the brush seals.

UNDERSTANDING AND MODELING DAMPER SEAL FORCE COEFFICIENTS

Fluid forces acting on the rotor from bearings, seals, impellers, and so on, generally have radial and tangential (or radial and normal) components relative to the whirl orbit. The force component F_ϕ shown in Fig. 6-21

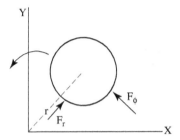

Figure 6-21   Force components on the rotor.

is destabilizing, whereas F_r is not. Direct damping produces negative F_ϕ. Most rotordynamic computer codes require F_r and F_ϕ to be converted into X-Y (Newtonian coordinates). Thus, we have force coefficients K_{XX}, K_{XY}, C_{XX}, C_{XY}, etc. as described in Chapter 3.

The radial force component F_r is modeled in X-Y coordinates with direct stiffness coefficients K_{XX} and K_{YY}. It has a weak effect on whirl stability, but a strong effect on the natural frequency and critical speed. The cross-coupled damping coefficients $C_{XY} = -C_{YX}$ also produce F_r, but these coefficients are small in gas seals.

The most common destabilizing force F_ϕ produced by gas seals is represented in X-Y coordinates by K_{XY} and K_{YX} with equal magnitude and opposite sign. The combination $K_{XY} > 0$, $K_{YX} < 0$ is forward driving, as shown in Fig. 6-21; the combination $K_{XY} < 0$, $K_{YX} > 0$ is backward driving (on the whirl orbit).

ALFORD'S HYPOTHESIS OF LABYRINTH SEAL DAMPING

Although it was incorrect, J. S. Alford's theory described above for labyrinth seal damping was the seed that led to the invention of the pocket damper seal. The self-excited subsynchronous whirl in an axial flow compressor for an aircraft turbojet engine was said by Alford to be caused by negative damping from labyrinth seals. Aircraft engines have ball and roller bearings, so oil whirl could not be a factor. Alford hypothesized two different mechanisms that could drive the whirl: (1) negative damping from the labyrinth seals (of the greatest interest here), and (2) an aerodynamic force due to the tip clearance variation around the circumference of the axial compressor blades. The latter mechanism can be modeled with cross-coupled stiffness coefficients as shown in chapter 3. This type of destabilizing mechanism (any *follower force* from cross-coupled stiffness) later became known as *Alford's force*

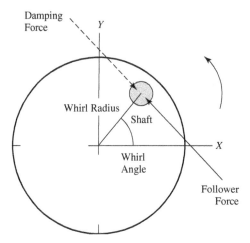

Figure 6-22 Damping and follower forces on a whirling rotor.

through the usage of J. C. Wachel and others. See Fig. 6-22 and note that a negative damping force will add to the follower force F_ϕ in Fig. 6-21, while positive damping will oppose the destabilizing follower force.

Alford illustrated his seal damping theory with a seal model that has two blades and choked flow. Figure 6-23 by Ertas [34] reproduces Fig. 6-7 from Alford's paper [13], and shows the cross section of two teeth in a tooth-on-stator labyrinth seal. The figure shows how the leakage flow into and out of the annular seal cavity will be modulated at different rates by vertical vibration of the rotor when the blade clearances are different. Alford's analysis considered only the axial flow and the compressibility of the gas, but ignored the circumferential flow. For harmonic vibration $\delta(t)$ in the vertical direction, it predicts that the pressure $p(t)$ in the seal cavity above and below the rotor will vary at the same frequency and will have a phase angle relative to the

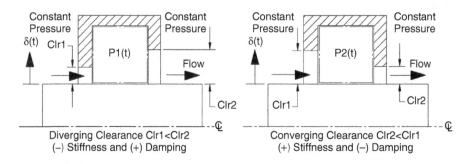

Figure 6-23 Axial flow through a seal cavity. From Ertas [34].

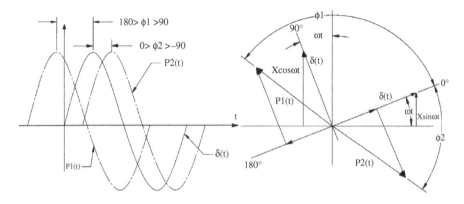

Figure 6-24 Phase of the pressure relative to the vibration $\delta(t)$. From Ertas [34].

displacement. The analysis predicts a leading phase when the clearances are diverging (positive damping) and a lagging phase when the clearances are converging (negative damping). See Fig. 6-23, left and right, and Fig. 6-24.

The phase angle is generally not 90°, so the seal also has a stiffness coefficient. To understand this, note that the vibratory velocity leads displacement by 90°. Damping is force from velocity, and stiffness is force from displacement. Zero phase angle would produce pure positive stiffness and no damping. If this analysis is extended into two dimensions (horizontal as well as vertical vibration), it predicts a tangential force component on a circular orbit that is "damping" (positive damping) for diverging clearances and a follower force (negative damping) for converging clearances.

The measurements described above by Zierer and Conway [15] showed the rotordynamic performance of straight-through labyrinth seals under conditions that are realistic for many turbomachines. Both teeth-on-rotor and teeth-on-stator gas seals were tested, each with twelve blades. Some seals had converging clearances, others diverging. The tests demonstrated conclusively that Alford's damping theory does not apply to conventional labyrinth seals. Taken all together, their measurements indicated: that (1) the labyrinth seals produced both direct damping and cross-coupled stiffness coefficients, (2) some of the damping coefficients may be negative, and (3) the effective damping from the combined coefficients was very small. For all the seals tested by Zierer and Conway, the measured effective damping coefficient was never more than 5 lb-sec/in, and was negative in most cases. Alford's theory predicts large direct damping coefficients for the seal with diverging clearances, which is radically inconsistent with all the measurements on conventional labyrinth seals.

CROSS-COUPLED STIFFNESS MEASUREMENTS

Conventional labyrinth seals produce large values of cross-coupled stiffness. As briefly mentioned above, Benckert and Wachter [16] conducted experiments in the late 1970s showing that the gas swirl in conventional labyrinth seals, and the resulting pressure distribution around the seal when the rotor is displaced radially, can produce a destabilizing follower force that is proportional to the radial deflection. This can be modeled mathematically using cross-coupled stiffness coefficients with opposite signs. They measured direct and cross-coupled stiffness coefficients for a wide range of see-through and interlocking labyrinth geometries. Their test procedure involved displacing the rotor statically in the X direction a distance and measuring the resulting circumferential pressure distribution in the labyrinth annular cavities. Thus, they could not measure damping (with no vibratory velocity). Integrating the circumferential pressures yielded a reaction force component in the direction of displacement X (direct stiffness), and a component perpendicular to the displacement Y (cross-coupled stiffness).

A seal (or a bearing) can have cross-coupled stiffness that is not destabilizing, if the two coefficients K_{XY} and K_{YX} have the same algebraic sign [35]. In fact, the combination is slightly stabilizing if the two coefficients have the same sign. An extended test procedure is required to determine the algebraic signs, which some previous investigators have apparently failed to perform [35]. One way to do the extended testing involves repeating the test with shaft rotation (and swirl) reversed. A questionable example is one of the seal types tested by Benckert and Wachter [16], which used four circumferentially spaced anti-swirl webs in the annular cavities between the labyrinth seal blades. This modification to the straight-through labyrinth seal created five rows of four circumferential cavities, or *pockets*. Their configuration was similar to a pocket damper seal, but differed in that there were no notches in any of the blades. The pocket partition walls are seen to block the swirl of the gas and so would be expected to reduce the follower force. The published result by Benckert and Wachter showed the follower force was larger with the partition walls, but there is no mention of a test with reversed shaft rotation. Experiments conducted at Texas A&M have shown that the gas swirl, and the follower force, is practically eliminated by partition walls in the annular cavities of labyrinth seals.

INVENTION OF THE POCKET DAMPER SEAL

The first author came to the realization around 1990 that labyrinth seals could have large direct damping coefficients, as predicted by Alford, if

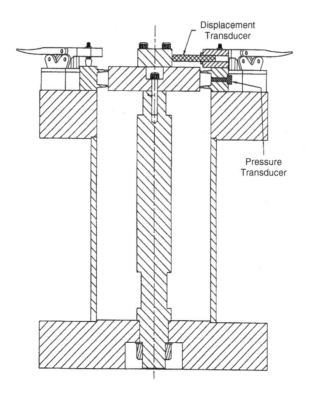

Figure 6-25 Shultz's test rig.

they had partition walls (pockets) and a diverging clearance. In the fall of 1990, Richard Shultz decided to build a test rig to test this idea for his Master's thesis research project [36]. Since shaft rotation has a weak effect on direct damping, it was decided to build a nonrotating test rig. A cross-section drawing of Shultz's design is shown in Fig. 6-25. The vertical cantilevered shaft has the seal journal attached at the top. The shaft and journal are enclosed inside of a cylindrical pressure vessel (a large-diameter pipe). Air is supplied through the side of the pipe and escapes through the seal clearance at the top. The measured pressure inside the pipe is the seal inlet pressure. In Shultz's tests, the natural frequency was 200 Hz for the fundamental mode of the vertical shaft in lateral vibration. Later investigators modified the shaft to produce different frequencies, e.g., Laos [18]. Damping from the seal causes the free vibration to decay more rapidly and can be measured as the logarithmic decrement of the time trace.

Shultz first tested a conventional 4″-diameter labyrinth seal with two blades (teeth), as shown in Fig. 6-25. One blade had twice the clearance, and the seal could be turned around so the clearances could either be converging or diverging in the direction of air flow. All results were very similar to those obtained by Zierer as described above, i.e., very small direct damping from the labyrinth seal at inlet pressures up to 40 psig.

Shultz then added partition walls to form four circumferential pockets. Figure 6-26 shows the free vibration with no air blowing through the seal, and Fig. 6-27 shows the free vibration with 70-psig inlet pressure.

Shultz's seal hardware was fabricated before the analysis was completed, so the seal parameters (pocket depth and clearance ratio) were not optimized. Nevertheless, this first *pocket damper seal* (PDS) as tested produced ten times the damping of the labyrinth seal at 200 Hz. The damping was later found to be much larger with optimized parameters and larger still at lower frequencies.

Shultz verified the theory of operation of the pocket damper seal by measuring the dynamic pressure in opposite pockets during free vibration. Figure 6-28 shows the location of the pressure transducers in pockets on opposite sides of the journal. Figure 6-29 shows the measured pressures. The channel 2 pressure leads the displacement by 105° in this test, which produces mostly damping and a small negative stiffness. (Velocity leads displacement by 90°.) The channel 1 pressure is located 180° from channel 2.

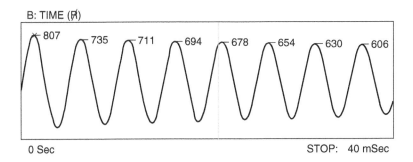

Figure 6-26   Free vibration decay, Shultz's seal, no air flow.

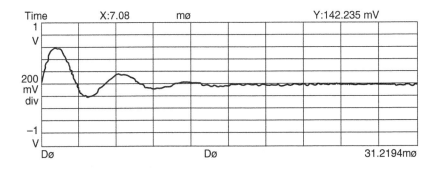

Figure 6-27   Free vibration decay, Shultz's seal, 70-psig inlet pressure.

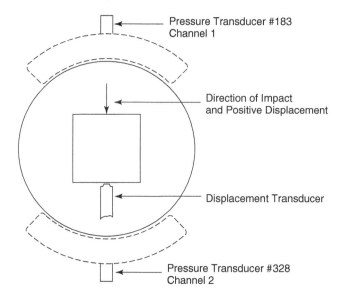

Figure 6-28 Pocket pressure measurement locations.

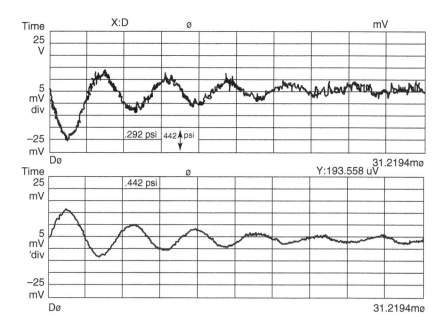

Figure 6-29 The pocket pressures measured by Shultz [36].

POCKET DAMPER SEAL THEORY

As stated earlier, the reason for the failure of Alford's theory to correctly predict large direct damping from conventional labyrinth seals is that they have continuous and unobstructed annular grooves, so that pressure variations across the seal diameter cannot exist without large circumferential gas velocities. Alford's analysis would have been correct if he had incorporated baffle walls between the blades of his labyrinth seal, which is the basis of the pocket damper seal invention by Vance and Shultz. A more rigorous Alford-type analysis based on the assumption that the control volumes are unconnected (i.e., blocked) around the circumference was developed by Sundararajan and Vance [37] with contributions from Dr. San Andres at Texas A&M University. It was applied first to the development of a piston-operated bearing damper, and later to the pocket damper seal (after Shultz's measurements). Thus, the first pocket damper seal analysis used fundamentally the same mathematics as Alford's 1965 analysis on labyrinth seals. The theory is based on the dynamic compressibility of the gas. It ignores the gas viscosity and inertia. It has been shown to be a useful engineering tool for pocketed seals, as it produces predictions for the damping coefficients that are within 10 percent of most direct damping measurements that have since been made on pocket damper seals at inlet pressures up to 200 psi. At higher pressures it underpredicts the damping, but correctly predicts the trends of the coefficients versus frequency and pressure drop. The pocket damper seal design not only makes a seal with very high direct damping, it also blocks the circumferential flow (swirl) and eliminates the destabilizing cross-coupled stiffness.

It is worthwhile to repeat here that the damping produced by the pocket damper seal is not primarily due to viscous effects. Most of the damping comes from the compressibility of the working fluid. The dynamic pressures in the pockets are quite large and can produce large damping forces on the shaft. Analysis of the governing pressure phenomena in these seals is not as complex as that in labyrinth seals. The equations in the code are for a single control volume analysis. No CFD (computational fluid dynamics) analysis with finite elements is required. Viscosity is not a major variable and the Reynolds number is not the dominant Pi group. Boundary layers, friction factors, and turbulence (all are important for labyrinth seals) are much less important in the PDS analysis. It has been verified experimentally that circumferential swirl is negligible in the PDS [38, 39]. All of this means that the analytical prediction of the damping produced by pocket damper seals will always be more accurate than what is possible for conventional labyrinth seals, unless the speed and accuracy of CFD analysis is greatly advanced. Also, the minor role of viscosity suggests that pocket damper seals are an ideal solution to vibration problems in oil-free turbomachinery. The pocket damper seal has been patented. The patent owner is Texas A&M University. It is licensed and marketed under the trade name TAMSEAL by KMC, Inc. and Bearings Plus, Inc.

ROTORDYNAMIC TESTING OF POCKET DAMPER SEALS

One test typically used to evaluate the rotordynamic effect of gas seals is to install the seal of interest into a rotor–bearing system and measure the following results, both with and without gas blowing through the seal: (1) the critical speeds (effect of stiffness), (2) the critical speed amplitudes (effect of damping), and (3) any subsynchronous frequencies (effect on stability). The main advantage of this rotordynamic testing is that these particular results are usually of the greatest interest. The disadvantage is that the force coefficients can only be approximately determined, and the damping and cross-coupled terms cannot be separated. Much of the early testing of pocket damper seals was rotordynamic testing to demonstrate their effectiveness to suppress vibration.

Laos [18] conducted coastdown measurements of synchronous response to unbalance for a pocket damper seals with four blades (teeth). Two versions were tested: one with four and the other with eight circumferential pockets. Two test seals were installed back to back to cancel the thrust on the rotor bearings. A cross section and photograph of the seal cartridge are shown in Figs. 6-30 and 6-31. The partition walls were made from keystock and were bolted to the seals' body. Therefore,

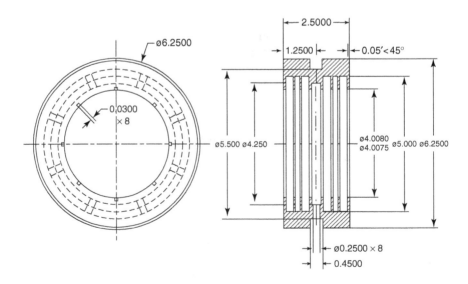

Figure 6-30   Cross section of Laos' pocket damper seal cartridge.

Figure 6-31   The seal cartridge in the tests by Laos.

only one seal body was needed to perform experiments with different numbers of partition walls. The air inlet was through radial holes at the center of the cartridge and from there the leakage flow turned in the axial direction and crossed three cavities in each of the two seals. The second, or middle, cavity was inactive for damping since it had no partition walls around the circumference. It served only to equalize the pressure around the circumference to act as the uniform inlet pressure for the third cavity downstream. This inactive cavity was eliminated in a more recent design to be described later, which greatly increases the damping and allows the designer to choose either positive or negative stiffness.

All the blades had the same radial clearance of 0.127 mm (5 mils). The first and third cavities had divergent axial area flow. In this design, the leakage flow area at the outlet of each cavity is higher than at the inlet because of notches or slots that are machined into the downstream blade. The flow through the notches is not modulated by the shaft vibration, which produces the required phase angle for damping.

Figure 6-32 shows the measured Bode plots at eight different inlet pressures for the seal with eight circumferential pockets. It can be seen that the test rig becomes overdamped by the seal at pressures over 40 psig. Notice also that the critical speeds are a few hundred rpm lower when the seal is active, due to the negative stiffness of the seal.

IMPEDANCE MEASUREMENTS OF POCKET DAMPER
SEAL FORCE COEFFICIENTS (STIFFNESS AND DAMPING)
AND LEAKAGE AT LOW PRESSURES

Laos used impedance measurements from shaker tests to identify the stiffness and damping coefficients of the seal in Figs. 6-30 and 6-31 at

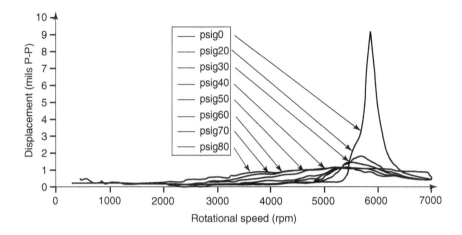

Figure 6-32 Unbalance response with a pocket damper seal at different inlet pressures.

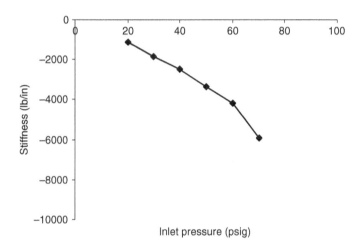

Figure 6-33 Measured direct stiffness of Laos's seal.

low pressures. Figure 6-33 shows the measured direct stiffness coefficient as a function of inlet pressure. The negative values explain why the critical speed peaks in Fig. 6-32 are moved to the left when the seal is operational (pressurized). Figure 6-34 shows the measured direct damping coefficients that suppressed the critical speed amplitudes in Fig. 6-32. Other test rig measurements at low pressures by Li et. al. [38, 39] showed that the cross-coupled stiffness coefficients of a pocket seal are very small (negligible) and have the same sign (which is stabilizing).

One of the early criticisms of the pocket damper seal design was that the different clearances or slots would make the leakage larger than a

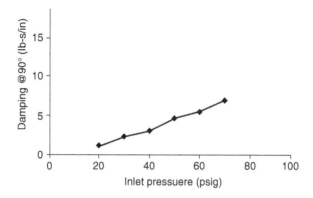

Figure 6-34   Measured direct damping of Laos's seal.

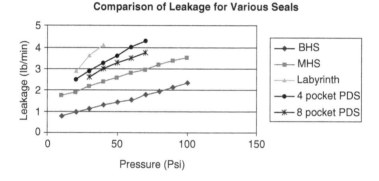

Figure 6-35   Measured leakage at low pressures of several seals.

labyrinth seal. Figure 6-35 shows measured leakages of some of the seals just discussed, all in the Fig. 6-30 cartridge. The top line on the graph (largest leakage) is for the six-bladed labyrinth seal of Fig. 6-9. The line below that is for the four-bladed pocket damper seal of Fig. 6-31 with four circumferential pockets, and the third lowest leakage is from the four-bladed pocket damper seal of Fig. 6-31 with eight circumferential pockets. These three seals had the same diameter, overall length, and radial clearance. The two lowest-leaking seals are hybrid designs using brush elements and wire mesh elements contacting the journal [40].

   An explanation of why the four-bladed pocket damper seals leak less than the six-bladed labyrinth seals is based on (1) restriction of circumferential flow in the pocket damper seals, and (2) the longer blade spacing (pitch) of the pocket damper seal. When the seal journal is off center, which is always the case, leakage flow will be biased toward the larger clearance, but this is prevented by the pocket partition walls. Experiments described at the end of this chapter show how increased pitch of the teeth reduces leakage. These effects should be considered when comparing leakage against traditional labyrinth seals. Much more damping is obtainable from pocket damper seals as the number of blades is reduced within a given axial length.

THE FULLY PARTITIONED PDS DESIGN

As described up to this point, the original pocket damper seal design had an inactive circumferential cavity (a labyrinth annulus with no pockets/separation walls) between each pair of active cavities. It will be referred to henceforth as the *conventional* design. Figure 6-36 by Ertas [34] shows an example with six blades (teeth) and three pairs of teeth or blades with active pockets. The downstream blades of each active cavity are shown with slots or notches instead of a larger clearance. The design with slots has been shown experimentally to produce more damping. This conventional pocket damper design usually has negative stiffness.

The *fully partitioned* design, as illustrated in Fig. 6-37, produces more damping than the conventional design along with the capability to produce positive stiffness (or negative stiffness under some operating conditions, if desired). This important design variation was first tested by Li, Kushner, and DeChoudhury at Elliott [41]. Figure 6-38 shows the seal hardware. Notice that the partition walls extend unbroken along the full length of the seal so that all the circumferential grooves are divided into pockets. Reference [41] calls this the "slotted" design, but it is also fully partitioned in addition to being slotted.

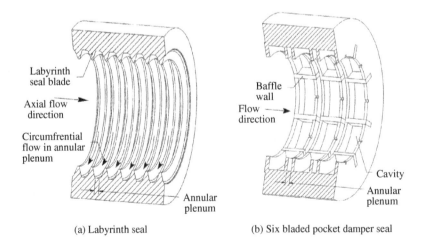

(a) Labyrinth seal   (b) Six bladed pocket damper seal

Figure 6-36 The conventional pocket damper seal design with six blades.

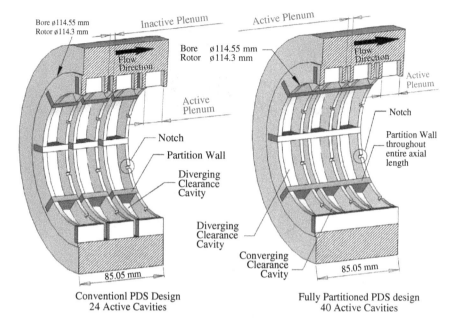

Figure 6-37 The conventional PDS (left) and the fully partitioned PDS (right). From [34].

The rotordynamic test rig used to test this seal at Elliott is shown in Fig. 6-39. As usual in this type of rig, two identical seals were tested back to back in order to cancel the axial thrust on the rotor. Figure 6-40 shows the measured baseline Bode plots (with the seals unpressurized). Notice the critical speed of 6400 rpm at the outboard horizontal location.

Figure 6-41 shows the Bode plots with the seals pressurized at two different inlet pressures between the two seals. The critical speed amplitude is cut in half when the seals are pressurized at 14.5 bars. The change in bandwidth also indicates a large amount of damping. Notice also that the critical speeds have been raised to 6500 and 6700 rpm, respectively, which indicates a positive stiffness contribution from the fully partitioned seal.

Figure 6-42 shows that preswirl of the gas at the seal inlet (43 m/sec tangential velocity) has a small effect on the effective damping from the seal. The conventional pocket damper seal was tested in the same test rig. Figure 6-43 shows the measured critical speeds for the two cases, which indicate positive stiffness for the fully partitioned pocket damper seal and negative stiffness for the conventional one. Results also showed more damping for the fully partitioned design even though the design was not optimized since a code was not yet available.

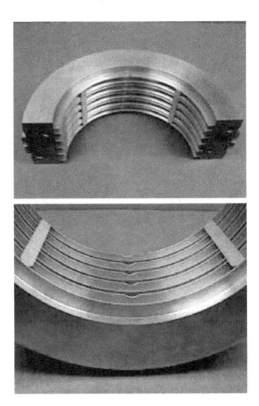

Figure 6-38   A fully partitioned PDS tested by Elliott [41].

Computer codes for both types of pocket damper seals have now been written to support the research in the Turbomachinery Laboratory at Texas A&M, and for the patent licensee. The code for the fully partitioned version was much more challenging. The inlet and exit pressures for each active cavity in the conventional design are constant (in the inactive cavities), whereas the pressures in all the cavities of the fully partitioned design are dynamic. The quoted text immediately following is from Ertas [34], with the figure number changed to be compatible with the present document.

Figure 6-44 shows a cross-sectional view of the two types of damper seals where the upper portion of the figure represents a conventional pocket damper seal composed of active plenums separated by inactive plenums or annulus. An active plenum for a PDS is defined as an annular plenum that has partition walls or barriers inserted in eight equally spaced angular locations creating eight pressurized pockets around the circumference of the seal. These pockets of pressurized gas can be modeled as individual control volumes that contain a dynamic pressure that varies with time. An inactive

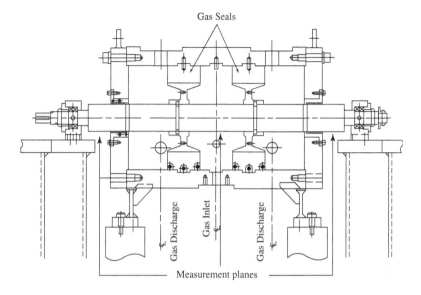

Figure 6-39 The rotordynamic test rig from [41].

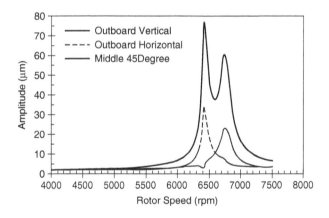

Figure 6-40 Measured baseline Bode plots on the test rig of Fig. 6-39

plenum or annulus does not have partition walls, therefore possessing small dynamic pressure modulations for the applicable frequency ranges of rotating equipment. For the conventional PDS analysis each control volume is separated from adjacent control volumes by a constant pressure, therefore allowing one to solve the conservation of mass equations for each control volume separately. The configuration shown in the bottom portion of Fig. 6-44 must utilize a coupled analysis between the conservation of mass equations.

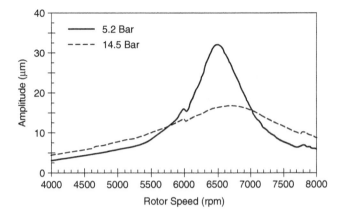

Figure 6-41   Outboard horizontal Bode plots with the fully partitioned seals under two inlet pressure conditions.

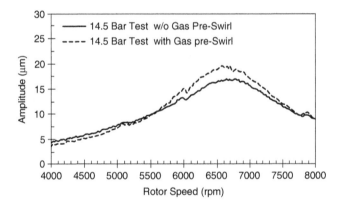

Figure 6-42   The effect of preswirl at the inlet.

Implementing walls throughout the seal length requires all the internal pressures in the seal to be time dependent (dynamic), resulting in the mass flow equations for the internal blades to depend on two dynamic pressures, therefore coupling the analysis between each control volume. This ultimately requires the simultaneous solution of $2n$ (n equals the number of control volumes) equations to determine stiffness and damping, rather than the single-equation solution for the conventional PDS analysis.

Predictions from the code for the fully partitioned design reveal that it is capable of wide variations in stiffness and damping coefficients. The design variables are variable pitch, pocket depth, and slot or notch size. For example, varying the depth of the last cavity downstream can produce either positive or negative stiffness in some applications.

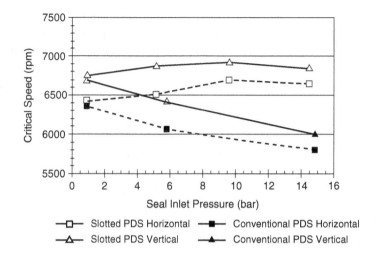

Figure 6-43   Critical speed comparisons for the two types of damper seals.

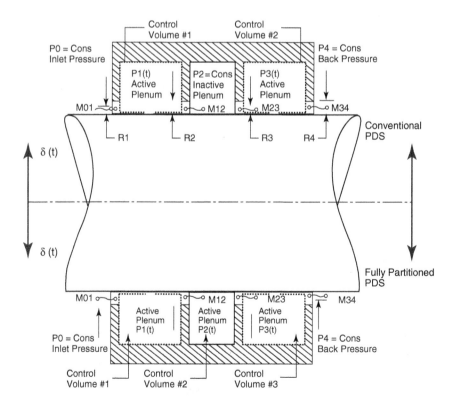

Figure 6-44   Stand-alone cavities versus sequential cavities [34].

EFFECTS OF NEGATIVE STIFFNESS

Contrary to common intuition, negative stiffness from a seal is not necessarily destabilizing, and in some cases is desirable. For example, if a machine is operating just above a critical speed then negative stiffness can bring that critical speed down a few hundred rpm to create more separation margin. Also, if a damper seal has maximum damping at low frequencies, then negative stiffness can lower the natural frequency into a region with more damping available and make the eigenvalue more stable. In this case, the low stiffness of a squeeze film damper accomplishes the same thing as the negative stiffness in the seal. Most surprising of all is that the logarithmic decrement of a machine with no cross-coupling can be *raised* by installing a seal with negative stiffness. These statements are illustrated and explained below with examples, using an eight-stage centrifugal compressor as a model.

Figure 6-45 shows the computer model of the compressor rotor generated using (XLRotor) with 35 stations and 140 degrees of freedom. This model is for a back-to-back high-pressure centrifugal compressor design with a center labyrinth seal located at midspan between stages 4 and 8. It has a well-documented history [42, 43] of rotordynamic instability that took about a year to solve after its delivery on a platform in the North Sea in the early 1970s. During that time, the first author worked with J. C. Wachel at Southwest Research to model the instability. Lund's landmark paper on eigenvalue computations [43] used this compressor as its example (see Chapter 7).

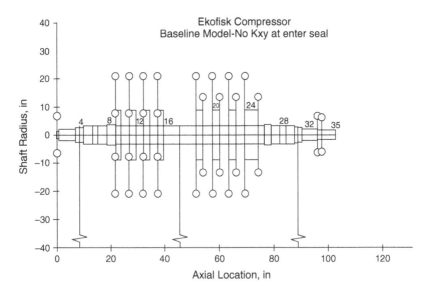

Figure 6-45 Computer model of the Ekofisk 8-stage centrifugal compressor.

Damped Eigenvalue Mode Shape Plot

Ekofisk Compresssor
Baseline Model-No Kxy at center seal

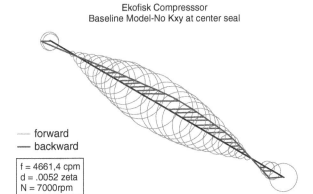

— forward
— backward

f = 4661,4 cpm
d = .0052 zeta
N = 7000rpm

Figure 6-46 Whirl mode of the Ekofisk compressor with no center seal.

With tilting-pad bearings (5 pads, load on pad) and no force coefficients from the center labyrinth seal, the eigenvalue of lowest forward frequency is computed to be 4661 cpm, which is close to the measured frequency of the instability. The real part of the eigenvalue gives a damping ratio $\zeta = 0.0052$, which is barely stable. The corresponding *log dec* (logarithmic decrement) is 0.033. Figure 6-46 shows the computed mode shape.

The "bearing" to ground at station 17 (midspan) in Fig. 6-45 represents the force coefficients of a center labyrinth seal. For a seal (imaginary seal) that has no damping or cross-coupling, but only negative stiffness $K_{XX} = K_{YY} = -50,000$ lb/in, the computed whirl frequency is reduced by 306 cpm to a frequency of 4355 cpm from 4661 cpm−4355 cpm and the damping ratio is slightly *raised* to $\zeta = 0.0057$, which corresponds to a log dec $\delta = 0.036$. The mode shape remains identical to Fig. 6-46. This can be explained as follows: If there are no destabilizing forces ($K_{XY} = 0$), the critical damping ratio of the mode is given by

$$Z = \frac{C}{2m\omega_n} \tag{6-7}$$

where C is the modal damping coefficient and ω_n is the natural frequency (close to the whirl frequency). Reducing the whirl frequency with negative stiffness will make the denominator smaller and raise the damping ratio. The log dec is raised as well as it is a function of the damping ratio:

$$\delta = \frac{2\pi\xi}{\sqrt{1-\xi^2}} \tag{6-8}$$

But past experience has shown that *raising* the whirl frequency of a centrifugal compressor by increasing the stiffness improves stability. This

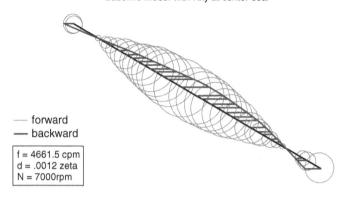

Damped Eigenvalue Mode Shape Plot

Ekofisk Compresssor
Baseline Model-with Kxy at center seal

—— forward
—— backward

| f = 4661.5 cpm |
| d = .0012 zeta |
| N = 7000rpm |

Figure 6-47   Whirl mode of the Ekofisk compressor with a destabilizing center seal.

is due to the presence of cross-coupled stiffness. To illustrate, consider the Ekofisk compressor of Fig. 6-45 with a center seal that has only cross-coupled stiffness $K_{XY} = 5000$ lb/in, $K_{YX} = -5000$ lb/in (similar to some labyrinth seals). The computed whirl frequency is 4662 cpm and the log dec becomes negative, $\delta = -0.0075$. Figure 6-47 shows the computed whirl mode and the eigenvalue.

If the seal can somehow be modified to produce positive direct stiffness $K_{XX} = K_{YY} = 50,000$ lb/in (still retaining the $K_{XY} = -K_{YX}$), then the whirl frequency is raised to 4947 cpm and the log dec becomes slightly less negative at $\delta = -0.0054$. Here we see that introducing positive stiffness alone is slightly stabilizing to the system. In the actual case of the Ekofisk compressor the whirl frequency was raised and the instability suppressed by shortening the bearing span and increasing the shaft diameter.

Cross-coupled stiffness from the seals changes the situation because it reduces the effective damping. Labyrinth seals have cross-coupled stiffness that does not vary much with frequency. With cross-coupled stiffness K_{XY}, $-K_{YX}$ acting, the effective damping coefficient C_{eff} of a seal for subsynchronous whirling at frequency ω_d is given by

$$C_{\text{eff}} = C_{XX} - K_{XY}/\omega_d \qquad (6\text{-}9)$$

where C_{eff} is the direct damping coefficient of the seal. The effective damping C_{eff} goes up as the whirl frequency ω_d goes up, and the whirl frequency can be raised by stiffening the system. This can be understood physically by studying Fig. 6-48 from [45].

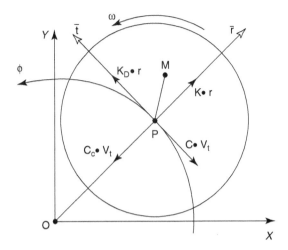

Figure 6-48 Rotor forces from damping and cross-coupled stiffness. From [45].

Figure 6-48 shows forces acting on a rotor that is whirling in a circular orbit. The damping force CV_t increases with whirl frequency, but the cross-coupled stiffness force K_D does not. Thus, raising the natural frequency will improve stability, *assuming that cross-coupled stiffness and damping coefficients remain constant with frequency*, which is almost true for labyrinth seals. But the development of damper seals, with frequency-dependent force coefficients, has changed the rules and made this conclusion no longer universally applicable except for labyrinth-type seals.

Even without frequency dependence it can be shown that the effective damping of a seal has a much stronger influence on stability than the stiffness. For example, consider installing a pocket damper seal in the Ekofisk compressor modeled in Fig. 6-45. The design parameters (pocket depth, pitch, notch size) of the 5" L × 7.7" D seal can be varied to produce direct damping $C_{XX} = 1000$ lb-sec/in with either positive or negative stiffness. The resulting damping ratio with zero seal stiffness is $\zeta = 0.639$ with log dec $\delta = 4.01$. This huge damping lowers the first eigenvalue to 3700 cpm (see Fig. 6-49).

FREQUENCY DEPENDENCE OF DAMPER SEALS

Different types of damper seals can have different frequency dependence of the force coefficients. Figure 6-50 shows a honeycomb seal [17], which has frequency dependence opposite to that of pocket damper seals. For

Damped Eigenvalue Mode Shape Plot

Ekofisk Compressor
TAMSEAL, 1000 lb-sec/in and zero stiffness

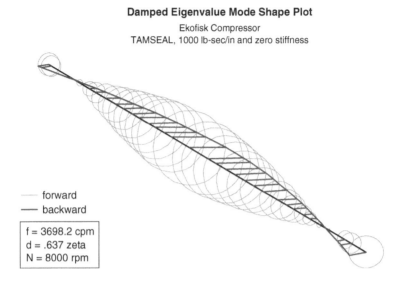

--- forward
— backward

f = 3698.2 cpm
d = .637 zeta
N = 8000 rpm

Figure 6-49 Whirl mode of the Ekofisk compressor with a pocket damper seal.

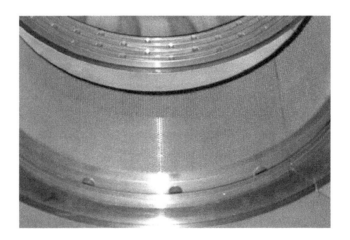

Figure 6-50 Honeycomb seal from [17].

example, Figs. 6-51 and 6-52 show how the stiffness and damping coefficients of a particular pocket damper seal vary with frequency. Compare this with the frequency dependence of the honeycomb seal, as shown in Figs. 6-53 and 6-54.

The pocket damper seal of Figs. 6-51 and 6-52 was designed for a multistage compressor that had a whirl frequency of 70 Hz. Notice in

Direct Stiffness

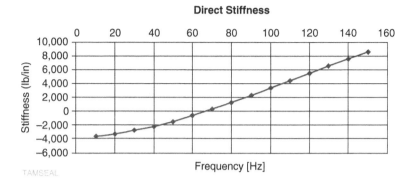

Figure 6-51 Frequency dependence of the stiffness for a pocket damper seal.

Direct Damping

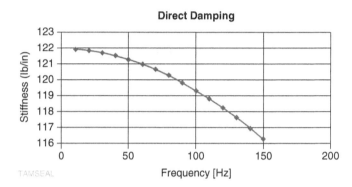

Figure 6-52 Frequency dependence of the damping for a pocket damper seal.

Fig. 6-51 that the stiffness just becomes positive at this frequency. Some engineers engaged in the specification and use of gas seals have a bias (often based on a false intuition) against negative stiffness. This bias may become a design constraint for the seal designs that generate negative stiffness. This particular (conventional) PDS could have been redesigned to have even more damping if negative stiffness was allowed at 70 Hz. Notice in Fig. 6-52 that the damping becomes larger at lower frequencies, so that negative stiffness in this case could lower the whirl frequency and produce more damping. Adding a squeeze film damper, with its low stiffness, could produce the same favorable effect.

Camatti et al. [17] describe an opposite situation where the effective damping of a honeycomb seal decreases and becomes negative as the whirl frequency is lowered. Figure 6-53 shows how the effective damping (Eq. 6-9) of this seal varies with frequency, with and without shunt holes. Figure 6-54 shows the frequency dependence of the stiffness for different tapered clearances (axial variation of the clearance). Notice that this seal

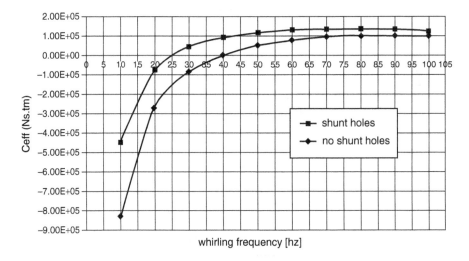

Figure 6-53 Effective damping of a honeycomb seal versus frequency [17].

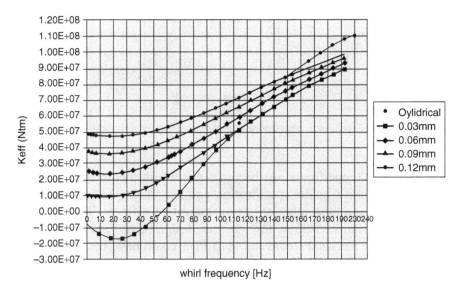

Figure 6-54 Effective stiffness of a honeycomb seal versus frequency, for different tapers [17].

would probably be incompatible with a squeeze film damper. The feature of squeeze film dampers that makes them effective is their low stiffness, and this could put the damper seal into a region of negative effective damping.

Moore et al. [46] describe another case for a compressor with a honeycomb damper seal. In their case the seal is distorted by the gas pressures, producing negative stiffness that drops the whirl frequency into a range of negative effective damping. The solution was to machine an offsetting convergent profile and set the mean clearance high enough so that tolerances could not cause a critical speed reduction into the negative damping region.

A compressor manufacturer used a honeycomb seal to eliminate a subsynchronous instability problem. Although the honeycomb damper seal achieved its initial objective of eliminating the instability, the increase in direct stiffness of this seal pushed the critical speed too close to the minimum operating speed, violating the API acceptable separation margin requirements. The manufacturer replaced the honeycomb damper seal with a pocket damper seal, which also eliminated the subsynchronous vibrations, but it introduced negative stiffness that kept the critical farther away from the operating speed range. Therefore, there are advantages to the negative stiffness with this type of seals, in addition to the fact that this characteristic can be varied without affecting the direct damping very much.

LABORATORY MEASUREMENTS OF STIFFNESS AND DAMPING FROM POCKET DAMPER SEALS AT HIGH PRESSURES

The Conventional Design

The conventional (original) design layout of the PDS had circumferential grooves with no partition walls (laby annulus), between each pair of active circumferential pockets, as illustrated in Fig. 6-37. Reference [45] presents measured frequency dependent stiffness and damping coefficients for 12-bladed and 8-bladed conventional pocket damper seals (PDS). Figure 6-55 shows the 8-bladed seal.

The test program was subdivided into four different seal configurations. Rotating experimental tests were carried out with inlet pressures at 1000 psig (69 bars), a frequency excitation range of 20–300 Hz, and rotor speeds up to 20,200 rpm. The downstream (exit) pressure was varied to produce different pressure drops across the seal. The testing method used to determine direct and cross-coupled force coefficients was the mechanical impedance method, which required the measurement of external shaker forces, seal housing accelerations, and dynamic displacement of the seal (relative to the shaft) in two orthogonal directions. Figure 6-56 shows the test apparatus. It was originally designed and constructed for bearing measurements by Childs and Hale [47], and had been later modified for gas seal measurements (e.g., Dawson [48]). The housing for the pocket damper seals was designed and constructed by Ertas for these experiments. In addition to the impedance measurements, dynamic

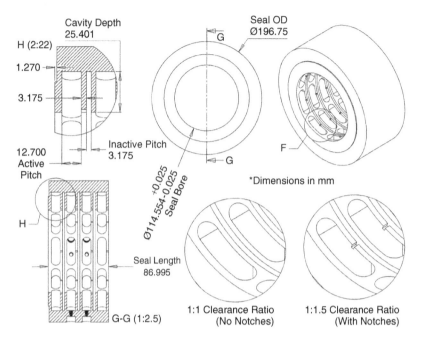

Figure 6-55 The 8-bladed PDS. Tests described in [45].

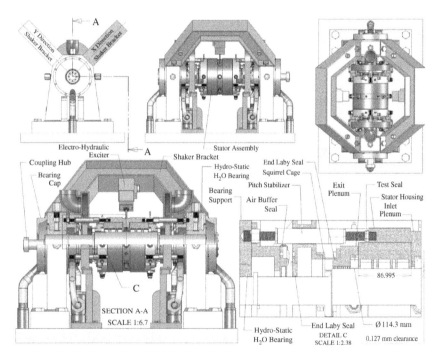

Figure 6-56 The test apparatus used in the tests of [45].

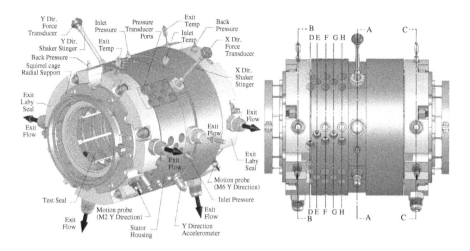

Figure 6-57   The seal housing used in the tests of [45].

pressure was measured in the individual seal cavities of the 8-bladed PDS. Figure 6-57 shows the new seal housing with its instrumentation.

The results of the tests show that the 8-bladed PDS possessed significantly more positive direct damping and negative direct stiffness than the 12-bladed seal. It also allowed less leakage. The results from the dynamic pressure response tests show that the diverging clearance of this PDS design strongly influences the dynamic pressure phase and force density of the seal cavities. The impedance tests also revealed the measurement of *same-sign* cross-coupled (cross-axis) stiffness coefficients for all the seal configurations, which indicate that the seals do not produce a destabilizing influence on rotor–bearing systems. This was confirmed later by pressure measurements in the seal cavities for a 6-bladed version [34].

Some representative measurement results from [45] are shown here. Figures 6-58 and 6-59 show the direct stiffness and damping coefficients for the straight-through and diverging clearance 8-bladed PDS configurations at 10,200 and 20,200 rpm. The effect of pressure ratio in the straight-through configuration seems to be most predominant in the lower frequency ranges of 20–80 Hz, but for higher frequencies the effect of pressure ratio on the direct damping is small.

The Fully Partitioned Design

Reference [34] describes tests carried out by Ertas to compare the performance of the fully partitioned PDS (FPDS) versus the conventional design. Figure 6-60 shows the two test seals, 4.51" diameter by 3.35"

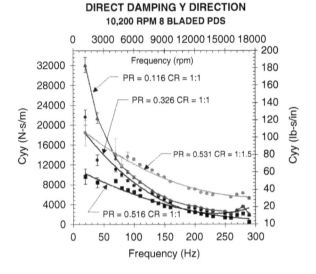

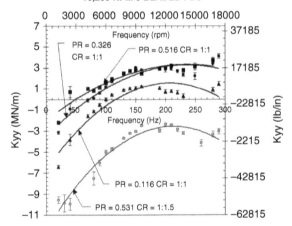

Figure 6-58   Measured damping and stiffness coefficients at 10,200 rpm for the 8-bladed PDS [45].

long, both of which had a 0.005-inch (0.127-mm) radial blade clearance from the rotor. The conventional design has partitioned walls inserted for three pairs of blades, creating three axial rows of eight equally spaced angular cavities (active plenum) separated by two annular plenums (annulus) with no cavities (inactive plenum). The notch on the downstream side of each cavity creates a diverging clearance pocket, which has been shown to significantly increase the direct damping by shifting dynamic pressure phase relationship to system excitation. As

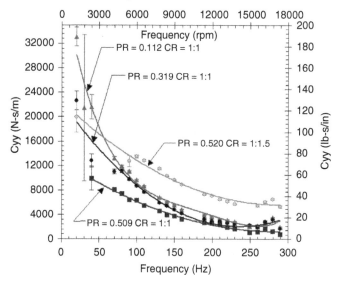

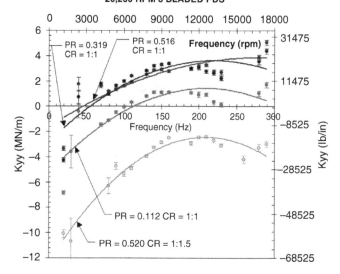

Figure 6-59 Measured damping and stiffness coefficients at 20,200 rpm for the 8-bladed PDS [45].

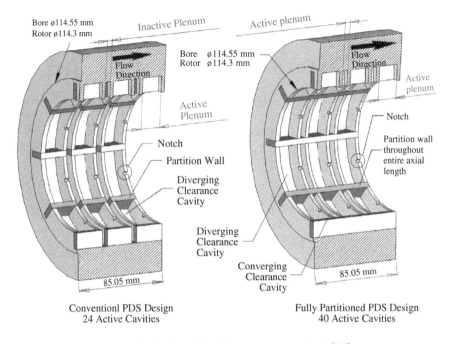

Bore ø114.55 mm
Rotor ø114.3 mm

Inactive Plenum

Active plenum

Bore ø114.55 mm
Rotor ø114.3 mm

Flow
Direction

Active
plenum

Flow
Direction

Active
Plenum

Notch

Notch

Partition wall
throughout
entire axial
length

Partition Wall

Diverging
Clearance
Cavity

Diverging
Clearance
Cavity

Converging
Clearance
Cavity

85.05 mm

85.05 mm

Conventionl PDS Design
24 Active Cavities

Fully Partitioned PDS Design
40 Active Cavities

Figure 6-60   Details of the two test seals in [34].

described earlier, the FPDS design incorporates baffle walls throughout the entire axial length of the seal creating sequentially placed cavities that have diverging and converging clearance geometries. This design has no continuous annular plenums and is a fully active seal. Unlike the conventional PDS design, the FPDS design has 24 diverging cavities (3 pairs times 8 pockets in each pair) and 16 (2 pairs times 8 per pair) converging clearance cavities, yielding a total of 40 active cavities in the seal.

The test apparatus was the same as described in the previous section for the conventional PDS (Figs. 6-56 and 6-57), and the same two test methods were employed. The first method was the mechanical impedance method, which required the measurement of the external shaker forces, relative stator–rotor motion, and stator acceleration. The second method was the dynamic pressure response method, which required the measurement of the dynamic cavity pressures in combination with stator motion.

Figure 6-61 shows the measured damping and stiffness measurements versus vibration frequency. The measured cross-coupled stiffness coefficients were very small and of the same-sign (stabilizing). Figure 6-62 shows a comparison of the direct and cross-coupled coefficients from the conventional PDS with those from the FPDS. The force coefficients are almost independent of shaft speed for both seals.

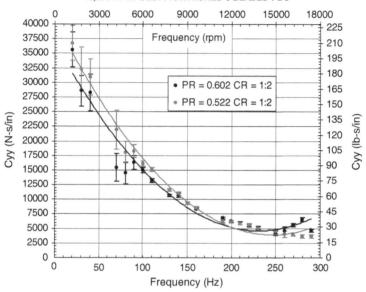

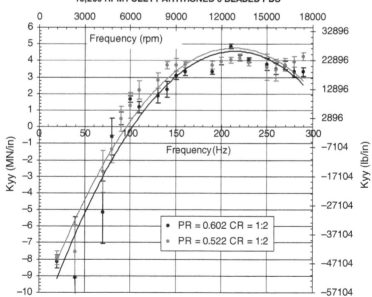

Figure 6-61   Measured damping and stiffness coefficients at 10,200 rpm for the 6-bladed FPDS [34].

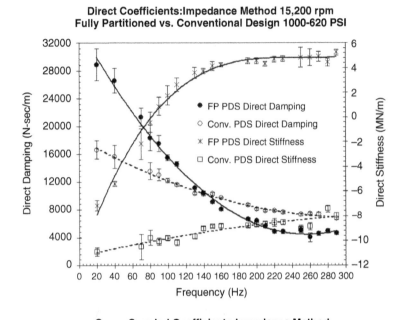

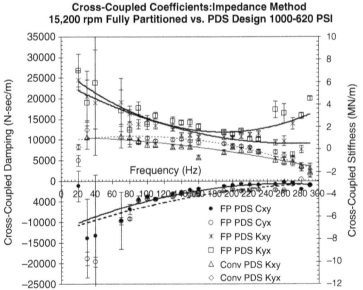

Figure 6-62 FPDS coefficients compared with PDS coefficients. From [34].

FIELD EXPERIENCE WITH POCKET DAMPER SEALS

Two Back-to-Back Compressor Applications

The first two field applications of pocket damper seals took place in the early 1990s, and were reported by Richards et al. [49] in 1995. The first application was by KMC, Inc. for two high-pressure injection compressors that were undergoing full-speed, full-load testing at the compressor manufacturer test facility. The latter application was suggested by John Platt of BP (formerly Amoco) after seeing the successful laboratory tests of pocket damper seals in the Turbomachinery Laboratory at Texas A&M University. In both cases, the subsynchronous vibration was eliminated through the use of the (then) new pocket damper seal configuration. Most of the following figures and much of the text are adapted from [49].

Case 1:   The first case involved a compressor train with a low-pressure compressor (LPC) and a high-pressure compressor (HPC). Here the high subsynchronous vibration showed up on both compressors during the full-load, full-pressure, full speed (FL/FP/FS) testing. The compressor train was rated at a maximum continuous speed (MCS) of 11,493 rpm. The stability threshold speed was reached at about 7500 rpm on the LPC and at 10,600 rpm on the HPC. Both compressors in the train were of the back-to-back configuration with a relatively long center seal. The subsynchronous vibration was believed to be the result of high aerodynamic cross-coupling introduced by the conventional grooved rotor and stationary labyrinth center seal. Furthermore, the location of the seal at an anti-node for the first forward mode (Fig. 6-63) increased the modal influence of the center seal on the stability of both compressors.

Due to the instability, it became necessary to reexamine the rotordynamic characteristics of both compressors in much greater detail. Special considerations were given to the seals and bearings, making use of the data obtained in the tests to help zero in on the source of the discrepancy

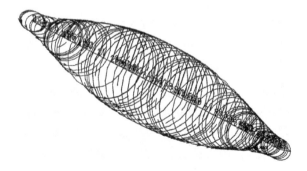

Figure 6-63   First forward mode the LP compressor from [49].

between the earlier predictions and the shop test results. The threshold speed of the LPC was much lower than the HPC compressor, thus making it the more difficult case to stabilize. A very thorough rotordynamic analysis was performed by the OEM, the bearings and seals supplier, and by the end user. The rotor model was verified by impact testing and measurement of the free–free modes. These measurements were used to benchmark the computer-generated rotor model and, thus, account for the stiffening effects of the shrunk-on wheels and shaft sleeves. This process verified that the mass-elastic properties of the rotor were accurately accounted for. Next, the fluid film bearings were modeled. The existing spherical pivot tilt pad bearings were analyzed at the nominal, minimum, and maximum extremes of the manufacturing tolerances to obtain their stiffness and damping coefficients.

The next step in the analysis was to simulate and benchmark the unstable operation observed during testing. The maximum continuous speed of the LPC was 11,493 rpm. During the FL/FP/FS test, subsynchronous vibration was observed at an operating speed of 7500 rpm. The subsynchronous vibration frequency was about 3675 rpm (61.25 Hz). This corresponded to the first forward mode shown in Fig. 6-63, which has an anti-node at the center seal. This suggested that the cross-coupling in the center seal was a significant modal influence in comparison to other elements in the dynamic system, and might be the major source of instability. A stability analysis was performed using the calculated bearing coefficients along with cross-coupled stiffness at the center labyrinth seal from Childs and Scharrer [50], and at the impellers using Wachel's formula [43]. The analysis predicted stable operation when the theoretical values obtained from the seal programs were used. However, the shop testing indicated otherwise. This suggested, as other cases have in the past, the inability of existing seal programs to adequately model the cross-coupled stiffness in labyrinth seals. To match the test results, the cross-coupling at the center seal was increased to approximately five times the theoretically predicted value. This constitutes a very significant shortcoming with the present analysis tools for labyrinth seals.

Since the location of the center labyrinth seal would have the most influence on the stability of the machine (due to its size and location), it was concluded that modifying this seal would constitute the most direct approach to solving the problem. Previous to this case, honeycomb seals had replaced labyrinth seals where they had been shown to provide better damping and eliminate subsynchronous whirl. However, the delivery schedule for these compressors was not compatible with the relatively long lead time required for the manufacture of a replacement honeycomb

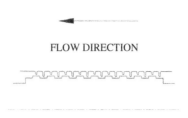

Figure 6-64   Center seal configuration for the LPC [49].

seal. Furthermore, the rotordynamically desirable honeycomb configuration consists of a smooth rotor and a honeycomb stator. This would require machining the steps off of the rotating shaft (Fig. 6-64) to provide a smooth rotor. This in turn would mean a repeat of the high-speed balancing operation on a total of eight rotors (three trains with six rotors plus two spare rotors). These limitations, which are not uncommon, dictated an alternative solution that could be accomplished in a relatively short time, and without any change to the rotating components on the shaft.

The optimum pocket damper seal configuration generally consists of a smooth rotor and a bladed (toothed) stator. However, due to the limitations imposed in this case that precluded any modifications to the rotating components, a less than optimum configuration was used. This configuration consisted of a pocket damper seal facing the grooved shaft as shown in Fig. 6-65. In this configuration, the active pockets were positioned over the relatively narrow high step on the shaft sleeve, leaving the low step for the inactive annulus.

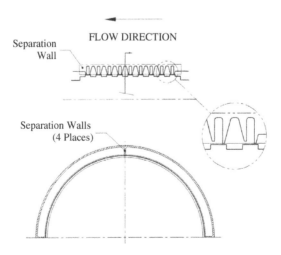

Figure 6-65   Pocket damper seal for the LPC [49].

The pocket damper seal proved to be a very rugged and reliable source of damping for these three compressors. Its ability to withstand high pressures and to serve as a quick drop-in replacement for the labyrinth seals made it especially advantageous in comparison to honeycomb seals.

Case 2:   The second case involved a set of four compressor trains. The compressors had six impellers in a back-to-back configuration. They were installed on a platform in the North Sea and were employed in a high-pressure injection process. Figure 6-66 shows the rotor with six impellers. Surge caused the labyrinth clearances to open up and this resulted in high subsynchronous vibrations, as shown in Fig. 6-67.

Considerable effort was required to reduce the vibration to acceptable levels. In the mid-1980s, one of these machines was converted to a dry

Figure 6-66   Rotor with six impellers and center seal location.

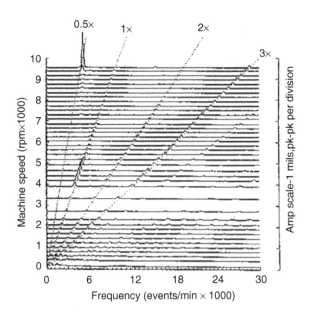

Figure 6-67   Compressor vibration spectrum with conventional labyrinth seals.

gas lubricated face seal. After the conversion, that machine was inoperable. Subsynchronous vibration levels of as high as 5 mils peak to peak were exhibited and, on at least one occasion, the center balance piston labyrinth clearance was wiped open to 0.065 (65 mils) from 10 mils original clearance.

Extensive rotordynamic analyses were conducted by both the manufacturer and the user. These analyses were done using a conventional rotor model that incorporated the shaft for stiffness and the sleeves and impellers as added masses and inertias. The modeling of forces on the rotor was limited at first to the bearing coefficients. To calibrate the rotor model, the suspended rotating assembly was subjected to an experimental modal analysis. The results of this analysis were compared to the predicted free–free modes of the rotor model. Good agreement was found.

However, the predicted eigenvalues on bearings did not agree well with the observed behavior of the machine. In search of better agreement, attempts were made to calculate the stiffness and damping coefficients of the labyrinths. The codes existing at that time provided wildly differing values for these coefficients. Reasonably good agreement was finally achieved with the inclusion of 50,000 lb/in cross-coupled stiffness at the center balance piston location in the computer model.

Several modifications were attempted, including several variations of shunt injection against rotation at the center balance piston labyrinth location, anti-swirl devices at the impeller eyes, and optimized bearings with two degrees of freedom pads. The combination of all of these modifications rendered the machine operable as long as the labyrinth clearances were maintained at blueprint values. However, due to the installation and problems with the control system, these machines were subjected to occasional surge conditions. When this occurred, the labyrinth clearances would be increased and the machine then became unstable. At this point, squeeze-film damper supported bearings were installed and the dry gas seals were removed and replaced with the original oil film floating seal rings. These modifications rendered the machines marginally operable. However, as the process continued to evolve, the suction pressure continued to reduce, requiring the machines to operate at higher speeds to achieve the increasing differential pressure requirement. Eventually, the machines again became unstable. In 1991, a honeycomb seal was installed at the center balance piston. This modification on three of the machines, together with the squeeze film dampers and the anti-swirl devices, completely stabilized them. The predicted logarithmic decrement increased from 0.2 to approximately 2.0, and the machines became operable again.

Due to a decrease in gas volume available, one of the machines remained unmodified and was idled as an emergency spare. It was decided to install a pocket damper seal constructed of a polyimide-amide

Figure 6-68 The pocket damper seal made of Torlon.

polymer (PAI) material commonly known by the trade name Torlon at the center balance piston location. The PAI material is a high-strength, high-temperature engineering polymer. Figure 6-68 shows the seal made of this material. Note that the seal has only four blades (teeth). This application called for maximum damping, even at the sacrifice of some leakage. Figures 6-69 and 6-70 show how the computed damping coefficient and the leakage vary with the number of blades.

This modification, by itself, produced acceptable stability in the machine. Figure 6-71 shows the computed whirl mode along with its very stable logarithmic decrement. Figure 6-72 shows the measured vibration spectrum. Figure 6-73 shows the synchronous response in a coastdown test, where the critical speed is barely discernable.

The use of the PAI material in the manufacture of the pocket damper seal, the robust design of these seals, and careful engineering of tolerances and thermal properties, enabled the design of a seal that runs

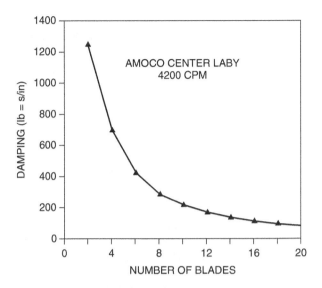

Figure 6-69   Damping from the pocket damper seal versus number of teeth [49].

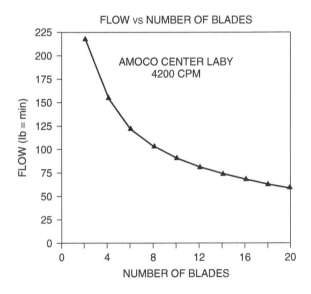

Figure 6-70   Pocket damper seal leakage versus number of teeth [49].

at closer clearances than conventional labyrinth seals. This machine had several full-pressure surge events associated with emergency shutdowns after installation of the pocket damper seal. There was no observable degradation in the mechanical dynamic behavior of the machine due to those events.

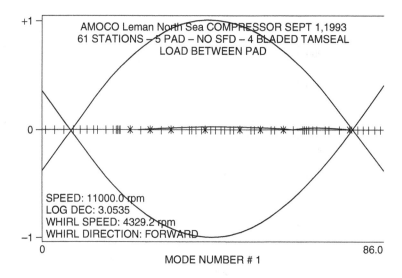

Figure 6-71   Computed stable whirl mode with the pocket damper seal.

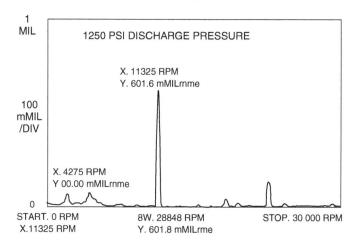

Figure 6-72   Measured vibration spectrum with only the pocket damper seal.

This conventional pocket damper seal was designed before the fully partitioned design had been developed. Today, the same damping could be obtained using more blades, thus allowing less leakage. More discussion of leakage issues is addressed in a later section.

A Fully Partitioned Application

Unlike the application just described, this case led to a pocket damper seal of the fully partitioned design with a large number of blades

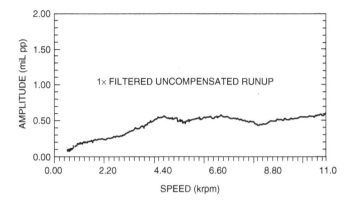

Figure 6-73 Synchronous imbalance response in a coastdown test.

(teeth). Most of the following text and figures are adapted from Li et al. [51].

In 1998, two identical equipment strings of high-pressure centrifugal compressors were delivered to an offshore platform located in the South Atlantic Ocean for natural gas re-injection service. The two compression strings are referred to as TCA and TCB, respectively. Each string consisted of two centrifugal compressors driven by a gas turbine through a gear increaser. Both centrifugal compressors were of the back-to-back configuration. The compressor train was rated at a design speed of 11,316 rpm and the design power was 8790 kW (11,787 hp). The equipment trains successfully passed the mechanical test at the manufacturer's facility per API 617 requirements. However, during commissioning in 1999, the high-pressure compressor of train TCA exhibited subsynchronous vibration at 4800 cpm, its first damped natural frequency. The onset instability rotating speed was about 11,300 rpm, at which the subsynchronous vibration first appeared. As shown in Fig. 6-74, the subsynchronous vibration decreased or disappeared when the speed was reduced slightly.

The rotordynamics model of the high-pressure compressor rotor was reexamined. In the rotor stability analysis the impeller aerodynamic cross-coupling stiffness had been accounted for, using Wachel's formula and the manufacturer's empirical equation, respectively. The center seal in a back-to-back compressor is located at an axial position where the destabilizing force is more effective to excite unstable rotor whirl. The analysis also showed that the rotor was sensitive to the bearing clearance of the original spherical seat, tilt pad journal bearing. Figure 6-75 illustrates the high-pressure centrifugal compressor rotor, having a total of nine stages arranged in two back-to-back sections. Compared to other impeller eye

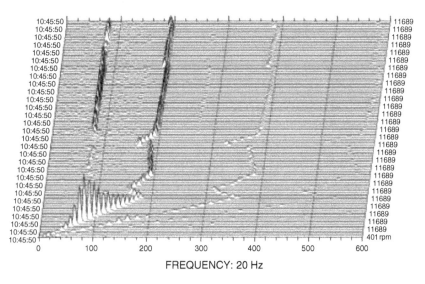

FREQUENCY: 20 Hz

Figure 6-74 Waterfall plot from the high-pressure TCA compressor [39].

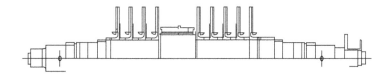

Figure 6-75 The nine-stage HP compressor rotor [51].

seals and shaft seals, the center seal was exposed to the largest differential pressure (64 bars).

For a larger stability margin, two types of gas damper seals have been used in the field to solve instability problems in centrifugal compressors. These include the honeycomb seal and the pocket damper seal. The very conservative approach that users are always constrained with dictates that solutions for equipment in the field do not make major hardware alterations that are not easily reversible. This case requires that any change in the center seal configuration should be reversible in case the modification does not eliminate the problem or makes it worse. Other factors, such as machining the rotating components on the shaft (necessary for a honeycomb conversion) and redoing high-speed balance, add cost and time constraints to the conversion. A carefully designed pocket damper seal addresses these concerns, since it could match with the existing rotating part on the shaft so that no change is required on the rotor. A pocket gas damper seal was examined for field retrofit in this case. The proposed pocket seal was specifically designed to ride against the existing stepped (grooved) rotating sleeve surface. The diameter of the pocket damper seal is 146.05 mm (5.75 in) with an *L/D* ratio of 0.94 and the depth of the damper cavity is 6.35 mm. The diametral clearance of the pocket damper seal is 0.874 mm, which is same as the clearance designed for the original center labyrinth seal. The upstream and downstream pressures for the damper seal are 176.5 and 112.2 bars, respectively. As shown in Fig. 6-76, the designed pocket damper seal has 22 teeth. Meanwhile, the original shunt injection holes remained on the right end of the seal. Slight modification was made on the shunt hole to induce high-pressure gas flow into the seal in a tangential direction against the rotation. In the previous seal the shunt holes introduced the gas radially on the shaft. A computer analysis predicted that using the pocket damper seal would result in a positive aero-log decrement even in the worst case with the minimum bearing clearance. Both a Flexure Pivot bearing and a pocket damper seal were ordered and shipped to the offshore platform. The high-pressure compressor was upgraded with the new bearing and center damper seal in September 2000 without any modifications on the rotor or the casing. The compressor train then worked at its design operating condition with a low vibration and the vibration level was stable. The subsynchronous vibration component was completely eliminated from the vibration waterfall plots, as shown in Fig. 6-77.

The following comments and information are from the first author of this book (Vance, not [41]), based on research and analysis performed after this case was published in 2002.

Due to an oversight in the hurried design and manufacturing process, the seal was actually produced as the fully partitioned design, even though all the published testing and analysis up to that time was for the conventional design. Figure 6-78 is a photograph of the seal hardware, which verifies the fully partitioned construction. It should also be pointed out that the theory and computer code developed to date for fully partitioned pocket damper seals would not apply rigorously to an application with a grooved shaft. The groove in the shaft will surely allow circumferential leakage from pocket to pocket.

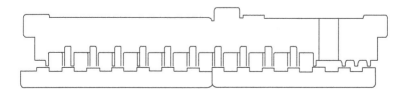

Figure 6-76 Schematic of the pocket damper seal installed in the HP compressor.

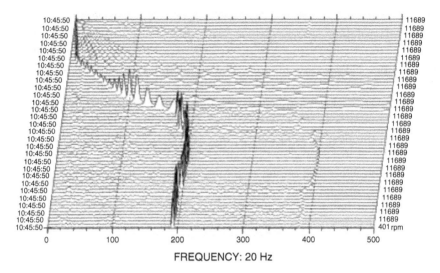

Figure 6-77 Waterfall plot from the HP compressor with the pocket damper seal.

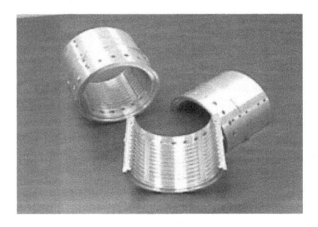

Figure 6-78 The pocket damper seal for the Petrobas HP compressor.

DESIGNING FOR DESIRED FORCE COEFFICIENT CHARACTERISTICS

Computer simulations and a limited amount of experimental verification [34,45] have shown that the conventional and fully partitioned pocket damper seal designs offer a wide range of force coefficient values that can be varied with a convenient set of design variables. This is illustrated here with an example, using the seal dimensions and operating pressures from the Amoco compressor case presented above. The shaft diameter is 7.8", the seal length is 8.44", and the allowable minimum seal clearance is 0.01". The working gas is methane, with seal inlet pressure 1200 psia and discharge pressure 600 psia. The natural frequency to be stabilized was 4200 cpm. The computer codes used were written by Ahmed Gamal in the Turbomachinery Laboratory, sponsored by a consortium of users.

The Conventional PDS Design

Figure 6-69 is repeated here as Fig. 6-79 to show how the damping of the conventional design (that was actually used) varies with the number of blades (teeth). Four blades were chosen. The cavity depth was optimized for maximum damping at each point on the graph. In some cases, optimum depth for the conventional design is too large to be practical for a retrofitted seal. At that time it was believed that there should be a 2:1 ratio of active cavity length to inactive cavity length (active pitch/dead pitch).

It is now known that the inactive cavity length (dead pitch) can be held constant as the number of blades is varied. Figure 6-80 shows how this change affects the damping when the inactive cavity length is held to 0.25". The cavity depth is held to 0.5". Figure 6-81 shows the negative stiffness associated with this design. Figure 6-82 shows the leakage. Figures 6-83 and 6-84 show how the damping and stiffness of this design (with four blades) vary with whirl frequency.

Notice that if the second critical speed of this compressor is 12,000 rpm (200 Hz), this seal (with four blades) would raise it slightly. If it were

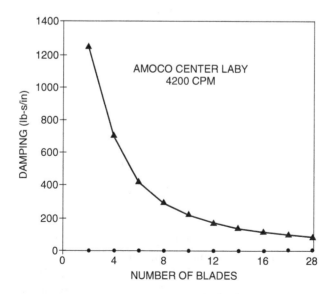

Figure 6-79 Damping of the original Amoco PDS design.

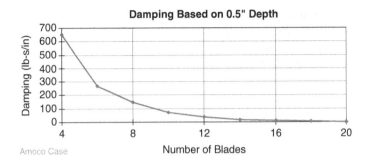

Figure 6-80   Damping of the conventional PDS with constant dead pitch.

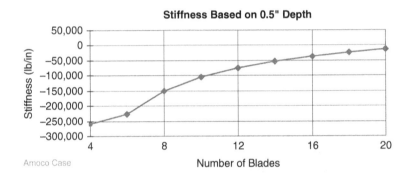

Figure 6-81   Stiffness of the conventional PDS with constant dead pitch.

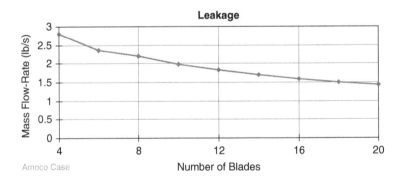

Figure 6-82   Leakage of the conventional PDS with constant dead pitch.

desired to lower the second critical speed, reducing the cavity depth to 0.25" can accomplish this. Figures 6-85 and 6-86 show the result of this modification to the conventional PDS. Notice that the stiffness is negative at 12,000 cpm and the damping at 4200 cpm is only slightly reduced. Notice also that the damping at 12,000 rpm is increased.

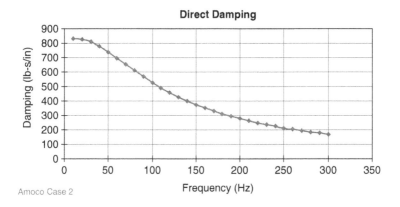

Figure 6-83   Damping versus frequency for the conventional PDS.

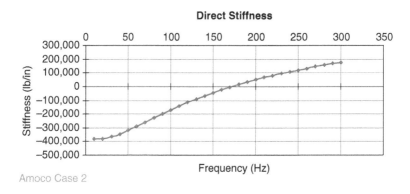

Figure 6-84   Stiffness versus frequency for the conventional PDS.

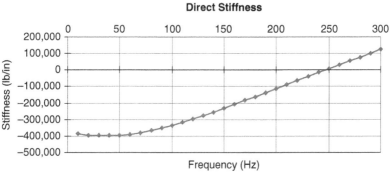

Figure 6-85   Stiffness versus frequency for the conventional PDS with reduced pocket depth.

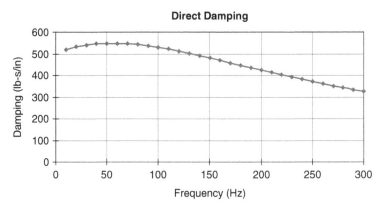

Amoco Case 2 w/ .25" cavity depth

Figure 6-86 Damping versus frequency for the conventional PDS with reduced pocket depth.

The only clear advantage of the conventional design is its ability to produce negative stiffness, if it is desired to lower the critical speeds. The two major disadvantages of the conventional design are that (1) To get large amounts of damping, a relatively small number of blades are required, thus increasing leakage, and (2) positive stiffness cannot usually be obtained at frequencies below about 200 Hz. The fully partitioned design completely eliminates these constraints, and allows the designer to obtain a wide range of force coefficients with a much larger number of blades (teeth) to reduce the leakage.

The Fully Partitioned Pocket Damper Seal

If the PDS for the Amoco case could have been designed fully partitioned, the current computer code predicts the following characteristics, with 0.125" pocket depth, uniform cavity lengths, and uniform clearances. Twelve blades can be used to achieve the same amount of damping at 4200 cpm as the four-bladed conventional PDS, which reduces the leakage from 2.8 lb/sec to less than 1.5 lb/sec. Figures 6-87 to 6-89 show the damping, stiffness, and leakage versus the number of blades (teeth), to be compared with Figs. 6-80 to 6-82. Figures 6-90 and 6-91 show the damping and stiffness versus frequency (with 12 blades), to be compared with Figs. 6-83 and 6-84.

Computed results also predict that the fully partitioned PDS would not be as sensitive to rubs or thermal distortion producing a diverging clearance, when compared to a honeycomb or hole pattern seal (recall Figs. 6-53 and 6-54). Figure 6-92 shows the new stiffness versus frequency

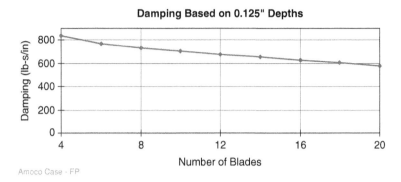

Figure 6-87   Damping versus number of teeth, for the fully partitioned PDS for the Amoco compressor.

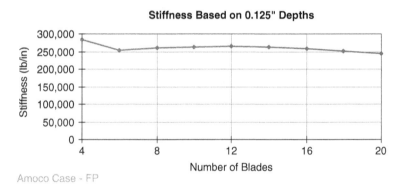

Figure 6-88   Stiffness versus number of teeth, for the fully partitioned PDS for the Amoco compressor.

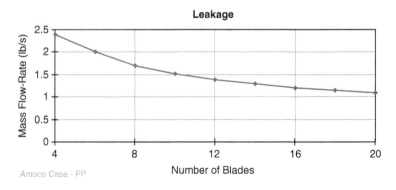

Figure 6-89   Leakage versus number of teeth, the fully partitioned PDS for the Amoco compressor.

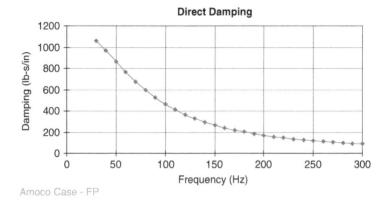

Amoco Case - FP

Figure 6-90 Damping versus frequency of the fully partitioned PDS with 12 blades, for the Amoco compressor.

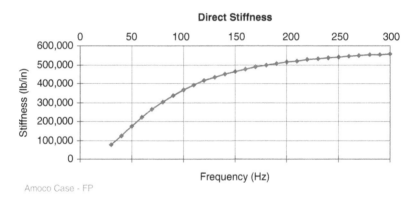

Amoco Case - FP

Figure 6-91 Stiffness versus frequency of the fully partitioned PDS with 12 blades, for the Amoco compressor.

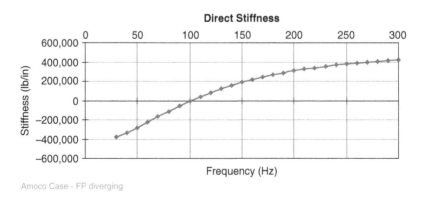

Amoco Case - FP diverging

Figure 6-92 Stiffness of the fully partitioned PDS with twelve blades and diverging clearances, for the Amoco compressor.

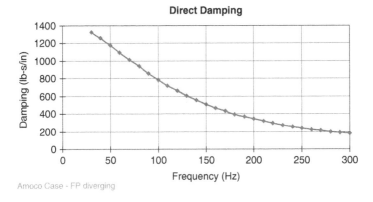

Figure 6-93 Damping of the fully partitioned PDS with 12 blades and diverging clearances, for the Amoco compressor.

for the FP PDS of Fig. 6-91 after the clearance has been made diverging from 10 mils at the entrance to 21 mils at the exit. The new stiffness is only slightly negative at 4800 cpm (80 Hz). Figure 6-93 shows how the damping has been greatly increased (from 600 to 900 lb-sec/in).

Leakage Considerations

Pocket damper seals generally have more damping with fewer blades (teeth), which leads to a trade-off between damping and leakage, if a labyrinth seal is to be replaced with a PDS. The trade-off may not be as severe as it seems. Gamal et al. [52] have measured and analyzed the leakage of pocket damper seals at high pressures. The two most interesting results of that study are that (1) seals with teeth of rectangular cross section leaked less than those with beveled blades, and (2) a seal with twelve blades (teeth) leaked more than a seal with eight blades. The seal with twelve blades had beveled teeth and closer blade spacing (shorter pitch) since both seals were of the same axial length.

Gamal and Vance [53] carried out leakage experiments on labyrinth seals and pocket damper seals at both high and low pressures. Some of the following description of results is a condensed version of the conference presentation of reference [53] by Gamal in Montreal. The effects on pocket damper seal (PDF) leakage of two main seal design parameters—cavity depth and blade spacing (pitch)—were examined through a series of nonrotating tests in the Turbomachinery Laboratory. The PDS tests were sponsored by a consortium of PDS users organized by Bearings Plus, Inc.

A reconfigurable seal design was used in a modified version of the test rig in [53], which enabled the testing of three-, four-, five-, and six-bladed seals of different geometries with supply pressures of up to 100 psia. Leakage measurements were made for each of thirteen seal configurations. The tests results show that changing the cavity depths of a four-bladed seal from 500 to 40 mils and then to 20 mils reduced leakage by up to 4 and 7 percent, respectively. The same changes to the cavity depths of a five-bladed seal resulted in up to a 4 percent reduction and a 2 percent increase in leakage, respectively. These results point to the preferred use of the fully partitioned design, since it generally has much smaller cavity depths than the conventional design.

Gamal's tests on the effect of blade spacing produced even more significant results, showing a drop in leakage of up to 16 percent at low pressures and up to 8 percent at high pressures for one six-bladed seal that was tested. This conventional pocket damper seal had an initial constant blade spacing of 0.125 inch., which was then increased to 0.5 inch. To test more realistic PDS designs, seals with nonconstant pitch lengths were also used. These seals had inactive cavities (or secondary cavities in the case of a fully partitioned seal) that were shorter than the active (or primary cavities). The configurations of three pairs of these seals are summarized in the table. The drop in leakage resulting from an increase in pitch is less dramatic than that observed for the constant pitch seal (and for the labyrinth seals tested) because fewer cavities were affected by the changes in pitch. In the six seal configurations tested, only the two inactive (secondary) cavity lengths were changed; a change in the length of every cavity would have had a more significant effect.

Gamal's tests on a three-bladed and a four-bladed seal with the same overall lengths, but with different pitch lengths, showed (see the graphs below) that the two seals had almost identical leakage rates, indicating that blade spacing should be considered as important a factor as the number of blades under certain conditions (Figs. 6-94 to 6-96). Recall that tests by Laos (previously described) showed that a six-bladed labyrinth seal had higher leakage than a four-bladed pocket damper seal with the same clearance. While blade pitch and profile were partially responsible for this result, it can be assumed that the partition walls in the PDS (which virtually eliminates circumferential flow) were also a contributing factor to the PDS's superior leakage performance. This assumption is supported by the results of the tests by Gamal that compare the leakage through a six-bladed conventional and fully partitioned pocket damper seals. Both the conventional and the FP-PDS had partition walls in the three active (or primary) cavities, but the FP-PDS also had partition walls in the two secondary cavities. The FP-PDS exhibits superior leakage reduction and the difference between the performances of the two seals increases at higher pressures (Fig. 6-97).

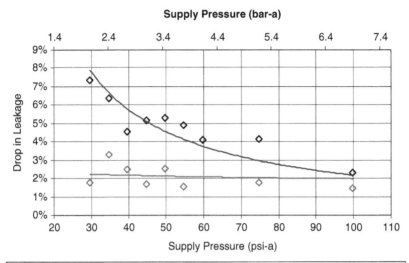

Figure 6-94 Drop in leakage resulting from increasing six-bladed conventional PDS blade spacing upper curve seal lengths: 1.625 to 2.125 inches (30 percent increase of length). Lower curve seal lengths: 2.125 to 2.5 inches (17 percent increase of length)

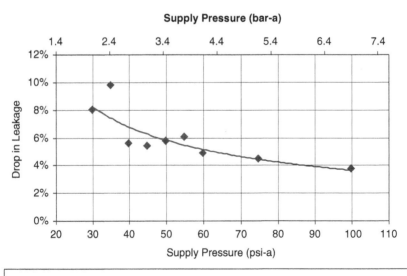

Figure 6-95 Drop in leakage resulting from increasing six-bladed FP-PDS blade spacing. Seal lengths: 2.125 to 2.5 inches (17 percent increase of length).

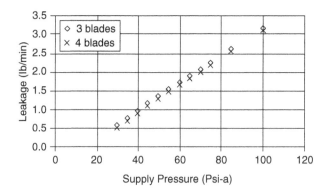

Figure 6-96 Leakage through conventional pocket damper seals with three and four blades. Seal lengths: Both 0.875 inch; three blades with 0.25-inch spacing and four blades with 0.125-inch spacing (plus blade thicknesses).

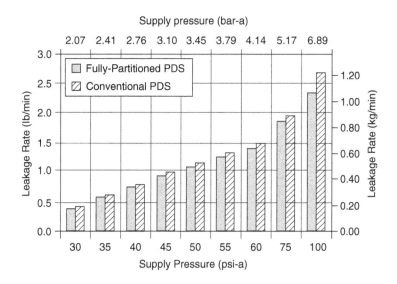

Figure 6-97 Leakage through conventional and fully partitioned pocket damper seals with six blades.

All of the graphs of leakage versus number of blades presented in previous sections were computed under a traditional assumption that blade spacing (pitch) has no effect, and that shaft eccentricity has no effect. If Gamal's findings above are taken into account, the leakage reduction achieved with adding blades (within a constrained length) is less, and greater consideration can be given to reducing the number of blades to achieve more damping. Gamal also found that the increase in leakage that results from offsetting the seal journal (off-centered in the clearance) is less for pocket damper seals when compared to straight-through labyrinth seals. This can be attributed to the blockage of circumferential flow in the PDS. This effect was also ignored in the previous leakage graphs, and in all leakage computations known to the authors.

SOME COMPARISONS OF DIFFERENT TYPES OF ANNULAR GAS SEALS

Ahmed Gamal has generated graphs comparing the damping performance of different types of seals on a normalized basis. The data are from his experiments and from Dr. Dara Childs in the Turbomachinery Laboratory at Texas A&M University. The normalization scheme was developed for hole pattern seals and honeycomb seals by Childs. It is not dimensionless. Figure 6-98 shows that the normalization scheme works as long as the seals are of the same type. Figure 6-99 shows a comparison of normalized damping for several different types of seals.

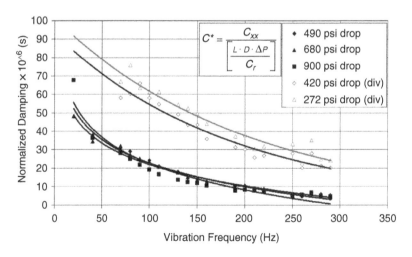

Figure 6-98 Normalized damping of eight-bladed pocket damper seals (straight-through and diverging).

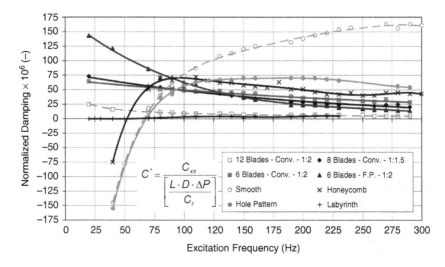

Figure 6-99 Annular gas seal normalized effective damping comparison. Conv. and F.P. designate conventional and fully partitioned pocket damper seals.

Much of the material in this chapter is adapted from papers published by the ASME and by the Turbomachinery Symposium. They are listed in these references at the end of the chapter. Many were written or co-written by the authors of this book. The latter part of this chapter is taken mainly from [54].

REFERENCES

[1] Fleming, D. P. High stiffness seals for rotor critical speed control. ASME Paper No. 77-DET-lO (1977).

[2] Kirk, R. G., and Miller, W. H. The influence of high pressure oil seals on turbo-rotor stability. *ASLE Transactions* 22:14−24 (1979).

[3] Lomakin, A. A. Calculation of critical speed and securing of dynamic stability of the rotor of hydraulic high pressure machines with reference to forces arising in the seal gaps [in Russian]. *Energomashinostroenie* 4: 1−5 (April 1958).

[4] Black, H. F. Effects of hydraulic forces in annular pressure seals on the vibrations of centrifugal pump rotors. *Journal of Mechanical Engineering Science* 11: 206−213 (1969).

[5] Black, H. F., and Murray, J. L. The hydrostatic and hybrid bearing properties of annular pressure seals in centrifugal pumps. The British Hydromechanics Research Association, Paper No. RRI026 (October 1969).

[6] Black, H. F., and Jenssen, D. N. Dynamic hybrid properties of annular pressure seals. Paper 9, Advanced Class Boiler Feed Pumps, Fluid Plant

and Machinery Group, Institution of Mechanical Engineers, September 1970; also *Proceedings of the Institution of Mechanical Engineers (London)*, 184:92–100 (1970).

[7] Childs, D. W. Dynamic analysis of turbulent annular seals based on Hirs lubrication equations. *Journal of Lubrication Technology* 105:429–436 (July 1983).

[8] Childs, D. W. Finite-length solutions for rotordynamic coefficients of turbulent annular seals. *Journal of Lubrication Technology* 105:437–444 (July 1983).

[9] Nelson, C., Childs, D., Nicks, C., and Elrod, D. Theory versus experiment for the rotordynamic coefficients of annular gas seals, part 2: constant-clearance and convergent-tapered geometry. *Journal of Tribology*, 433–438 (1986).

[10] Semanate, J., and San Andrés, L. Thermal analysis of locked multi–ring oil seals. *Tribology International* 27:3 197–206 (1994).

[11] Semanate, J. and San Andrés, L. Analysis of multi-land high pressure oil seals. *STLE Tribology Transactions* 36: 661–669 (1993).

[12] Semanate J., and San Andrés, L. A quasi-static method for the calculation of lock-up conditions in floating ring oil seals. Proceedings of the IV Congreso Latinoamericano de Turbomaquinaria, Queretaro, Mexico, December, 1993, pp. 55–64.

[13] Alford, J. Protecting turbomachinery from self-excited rotor whirl. *Journal of Engineering for Power*, 333–344 (1965).

[14] Murphy, B. T., and Vance, J. M. Labyrinth seal effects on rotor whirl stability. Proceedings of the Second International Conference on Vibrations in Rotating Machinery, Cambridge, England, September 2–4, 1980.

[15] Vance, J. M., Zierer, J. J., and Conway, E. M. Effect of straight-through labyrinth seals on rotordynamics. Proceedings of the ASME Vibration and Noise Conference, Albuquerque, New Mexico, 1993.

[16] Benckert, H., and Wachter, J. Flow induced spring constants of labyrinth seals. ImechE, Proceedings of the Second International Conference, Vibrations in Rotating Machinery, Cambridge, England, 1980, pp. 53–63.

[17] Camatti, M., et. al. Instability of a high pressure compressor equipped with honeycomb seals. Proceedings of the 32nd Turbomachinery Symposium, Houston, Texas, September 8–11, 2003.

[18] Laos, H. *Experimental Investigations of Damper Seal Performance*.Thesis, Texas A&M University, 1999, pp. 17 and 58.

[19] Picardo, A. M. *High Pressure Testing of See Through Labyrinth Seals*. Thesis, in Texas A&M University, 2003.

[20] Iwatsubo, T. Evaluation of the instability forces of labyrinth seals in turbines or compressors. NASA CP2133, Workshop on Instability Problems in High performance Turbomachinery, Texas A&M University, College Station, Texas, May 1980.

[21] Vance, J. M., Sharma, A., and Jayakar, N. Effect of frequency and design parameters on pocket damper seal performance. IMECE 2002-32561, Proceedings of Fluid Structure Interactions, ASME Winter Annual Meeting, New Orleans, Louisiana, November 17–22, 2002.

[22] Childs, D., Nelson, C., Nicks, C., Scharrer, J., Elrod, D., and Hale, K. Theory versus experiment for the rotordynamic coefficients of annular gas seals, part I: test facility and apparatus. *Journal of Tribology*, 108, 426–432 (1986).

[23] Childs, D., Elrod, D., and Hale, K. Annular honeycomb seals: test results for leakage and rotordynamic coefficients: comparisons to labyrinth and smooth configurations. *Journal of Tribology*, 111, 293–301 (1989).

[24] Childs, D., Elrod, D., and Ramsey, C. Annular honeycomb seals: additional test results for leakage and rotordynamic coefficients. IFToMM, Proceedings of the Third International Conference on Rotordynamics, Lyon, France, 1990, pp. 303–312.

[25] Childs, D. W. *Turbomachinery Rotordynamics*. New York: Wiley, 1993, page 293.

[26] Zeidan, F., Perez, R., and Stephenson, E. The use of honeycomb seals in stabilizing two centrifugal compressors. Proceedings of the 22nd Turbomachinery Symposium, 1993, pp. 3–16.

[27] Zhou, R. Instability of multistage compressor K 1501. Rotordynamic Instability Problems in High-Performance Turbomachinery, Texas A&M University, NASA Conference Publication 2443 (1986) pp. 62–75.

[28] Child, D., and Kim, C-H. Test results for round-hole pattern damper seals: optimum configurations and dimensions for maximum net damping. *ASME Transactions, Journal of Tribology* 108: 605–611 (1986).

[29] Yu, Z., and Childs, D. A comparison of experimental rotordynamic coefficients and leakage characteristics between hole-pattern gas damper seals and a honeycomb seal. *ASME Transactions, Journal of Engineering for Gas Turbines and Power* 120: 778–783 (1998).

[30] Conner, K., and Childs, D. Rotordynamic coefficient test results for a four-stage brush seal. *Journal of Propulsion and Power* 9, No. 3 (May–June 1993).

[31] Hendricks, R., Flower, R., and Howe, H. A brush seals program modeling and developments. NASA Technical Memorandum 107158, Ninth International Symposium of Transport Phenomena in Thermal-Fluids Engineering, Pacific Centre of Thermal Fluids Engineering, Singapore, Republic of Singapore, June 25–28, 1996.

[32] Andrés, L., Baker, J., and Delgado, A. Measurements of leakage and power loss in a hybrid brush seal. Proceedings of ASME Turbo Expo 2008: Power for Land, Sea and Air, GT2008, Berlin, Germany, June 9–13, 2008.

[33] Andrés, L., Delgado, A., and Baker, J. Rotordynamic force coefficients of a hybrid brush seal: measurements and predictions. Proceedings

of ASME Turbo Expo 2009: Power for Land, Sea and Air, GT2009, Orlando, Florida, USA, June 8–12, 2009.

[34] Ertas, B. H., Rotordynamic Force Coefficients of Pocket Damper Seals, Ph.D. Dissertation in Mechanical Engineering, Texas A&M University, August 2005.

[35] Ertas, B., and Vance, J. M. The influence of same-sign cross-coupled stiffness on rotordynamics. DETC2005-84873, Proceedings of the ASME Design Engineering and Technical Conference, Long Beach, California, September 24–28, 2005, and *Journal of Vibrations and Acoustics*, 2007.

[36] Shultz, R. M. *Analytical and Experimental Investigations of a Labyrinth Seal Test Rig and Damper Seals for Turbomachinery*. Thesis, Texas A&M University, 1997.

[37] Sundararajan, P., and Vance, J. M. A theoretical and experimental investigation of a gas operated bearing damper for turbomachinery, part I: theoretical model and predictions. Proceedings of the 14th Vibration and Noise Conference, Albuquerque, New Mexico, ASME DE-Vol. 60, pp. 67–83.

[38] Li, J., Ransom, D., San Andres, L., and Vance, J. Comparison of predictions with test results for rotordynamic coefficients of a four-pocket gas damper seal. ASME *Journal of Tribology* 121: 363–369 (1999).

[39] Li, J., San Andrés, Aguilar, R., and Vance, J. M. Dynamic force coefficients of a multiple blade, multiple pocket gas damper seal: test results and analytical validation. *ASME Journal of Tribology* (99 TRIB 35) 122: 317–322 (2000).

[40] Laos, H. E., Vance, J. M., and Buchanan, S. E. Hybrid brush pocket damper seals for turbomachinery. ASME *Journal for Gas Turbines and Power* (99-GT-16) 112: 330–336 (2000).

[41] Li, J., Kushner, F., and De Choudhury, P. Experimental evaluation of slotted gas pocket damper seals on a rotating test rig. GT-2002-30634, Proceedings of ASME Turbo Expo 2002, Amsterdam, The Netherlands.

[42] Doyle, H. E. Field experiences with rotordynamic instability in high-performance turbomachinery. Rotordynamic Instability Problems in High-Performance Turbomachinery, NASA CP 2133, Texas A&M University Workshop, May 12–14, 1980, pp. 3–4.

[43] Wachel, J. C. Rotordynamic instability field problems. Rotordynamic Instability Problems in High-Performance Turbomachinery, NASA CP 2250, Texas A&M University Workshop, May 10–12, 1982, pp. 3–6.

[44] Lund, J. W. Stability and damped critical speeds of a flexible rotor in fluid-film bearings. *Journal of Engineering for Industry* 96:682 (1974).

[45] Ertas, B. H., Gamal, A., and Vance, J. M. Rotordynamic force coefficients of pocket damper seals. ASME *Journal of Turbomachinery*, 725–737 (2006).

[46] Moore, J. J., Camatti, M., Smalley, A. J., Vannini, G., and Vermin, L. L. Investigation of a rotordynamic instability in a high pressure centrifugal compressor due to damper seal clearance divergence. 7th IFToMM Conference on Rotor Dynamics, Vienna, Austria, September 25–28, 2006.

[47] Childs, D. W., and Hale, K. A test apparatus and facility to identify the rotordynamic coefficients of high speed hydrostatic bearings. *ASME Journal of Tribology* 116:337–334 (1994).

[48] Dawson, M. A comparison of the static and dynamic characteristics of straight-bore and convergent tapered-bore honeycomb annular gas seals. Thesis, Texas A&M University, 2000.

[49] Richards, B., Vance, J. M., Paquette, D. J., and Zeidan, F. Y. Using a damper seal to eliminate sub-synchronous vibrations in three back to back compressors. Proceedings of the 24th Turbomachinery Symposium, Houston, Texas, September 26–28, 1995, pp. 59–71.

[50] Childs, D., and Scharrer, J. An Iwatsubo based solution for labyrinth seals: comparison to experimental results. *Journal of Engineering for Gas Turbines and Power* 108: 325–331 (1986).

[51] Li, J., Choudhury, P., and Tacques, R. Seal And bearing upgrade For eliminating rotor instability vibration in a high pressure natural gas compressor. ASME Paper No. GT-2002-30635, Proceedings of ASME Turbo Expo 2002, Amsterdam, The Netherlands, June 3–6, 2002.

[52] Gamal, A. M., Ertas, B. H., and Vance, J. M. High-pressure pocket damper seals: leakage rates and cavity pressures. Proceedings of GT2006 ASME Turbo Expo 2006: Power for Land, Sea and Air, May 8–11, 2006, Barcelona, Spain.

[53] Gamal, A. M., and Vance, J. M. Labyrinth seal leakage tests: tooth profile, tooth thickness, and eccentricity effects. Proceedings of GT2007 ASME Turbo Expo 2007: Power for Land, Sea and Air, Montreal, Canada, May 14–17, 2007.

[54] Childs, D. W., and Vance, J. M. Short course on "Annular Gas Seals and Rotordynamics." 36th Turbomachinery Symposium, Texas A&M University, September 10–13, 2007.

7

HISTORY OF MACHINERY ROTORDYNAMICS

Knowledge of the history of rotordynamics is valuable because it reveals the basic nature of the various types of problems encountered when designing and developing rotor–bearing systems for various applications. The history also illustrates, in an interesting way, the surprising rotordynamic phenomena that were encountered and that are contrary to intuition. It is a common assumption of the uninitiated that making a rotor assembly to spin smoothly and stably on bearings should be a simple task. Machinery pioneers in the early days of the industrial revolution quickly discovered a contrary reality. Since then, the problems encountered have become more challenging and serious as rotating machines have become faster and more powerful through the years.

History is subjective and depends greatly on the viewpoint of the writer, even in a technical field. Most rotordynamic investigations have been motivated by machine problems or failures. In the early days, the problems and failures were not seen so much as a weakness of the manufacturer, but more as a general lack of the requisite knowledge. In that atmosphere the reports of analysis and testing could be quite honest and complete, and were publicized as a service to the engineering community. This situation has gradually disappeared as industry became more competitive and society more litigious. A good example can be seen in the publication of articles in *General Electric Review*, which provided much of the early knowledge about machinery rotordynamics in the early 1900s and then disappeared in the 1950s.

Several authors have already written histories of rotordynamics and their works will be drawn on heavily here. Notably, E. J. Gunter [1], R.E.D. Bishop [2], C. B. Meher-Homji [3], A. Dimarogonas [4], and

D. W. Childs [5] have written accounts that are fairly comprehensive when taken together. These articles and most other references available to the present authors were written in the English language. There can be no claim here for completeness, as other important reports may exist in other languages without proper recognition. In addition, there have been many contributors to the rotordynamic literature that are not noted here due to constraints of space. Many of those omitted would add to the clarity or applicability of the original contributions cited here, if space allowed.

THE FOUNDATION YEARS, 1869–1941

Rotating machines began to be manufactured in significant numbers concurrently with the development of waterwheels for hydraulic power in the early 1800s and steam turbines in the late 1800s. The first dynamics problem encountered was the critical speed, where the vibration produced by rotor unbalance is magnified by resonance with the natural frequency of the system. Rankine's first analysis of the critical speed phenomenon in 1869 was only partially correct, and the resulting dominant design philosophy was to keep running speeds below the first critical speed. This philosophy was challenged by the experiments of DeLaval in the 1890s that showed the possibility of successful supercritical operation. Early investigators did not recognize the importance of the bearing properties, so the first analytical models had simple supports at the bearing locations (infinite stiffness, no moment restraint). After the experiments of Beauchamp Tower on oil film bearings and the associated theoretical development by Osborne Reynolds in 1886, the importance of the bearings was only gradually recognized. Consequently, most of the published articles in this period can be categorized as either shaft dynamics investigations (without a realistic bearing model) or as bearing investigations (without a realistic shaft model). By the early 1900s, both Delaval in France and Parsons in England were successfully running steam turbines at supercritical speeds. In 1919, Jeffcott published a correct analysis of the critical speed inversion with damping included. As a result, the predominant design philosophy was changed so that supercritical operation of turbomachinery became an accepted practice. Whirl instability then appeared as the next major problem in rotordynamics. It was first addressed by Newkirk and Kimball in the 1920s by a series of experiments, and theoretically analyzed by D. M. Smith. In the 1930s, an influence coefficient method for two-plane balancing of rotors was published by Thearle. During and after WWII there was rapid progress in increasing the size and power density of turbomachines, along with more powerful methods for analysis and testing.

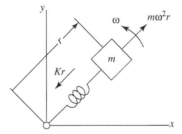

Figure 7-1   Rankine's analytical model and free-body diagram.

Shaft Dynamics

In 1869 the first rotordynamic investigation published in the English language was by Rankine [6], an engineer whose name is well known for his work on the thermodynamic cycle in steam turbines. Rankine's analytical model is shown in Fig. 7-1. The rigid mass m in Fig. 7-1 represents the center of mass of a rotor whirling at shaft speed ω. The spring represents the bending stiffness of the shaft. Under the D'Alembert concept of dynamic equilibrium, the restoring force kr of the spring is supposed to be opposed by the so-called *centrifugal force* $m\omega^2 r$. The requirement for the two forces to be equal is $\omega^2 = k/m$. Using this model, Rankine correctly predicted the critical speed in rad/sec to be $(k/m)^{1/2}$, where the system is in a state of indifferent equilibrium, i.e., the rotor can whirl at any amplitude r when running at the critical speed. This concept is useful for eigenvalue analysis of undamped linear systems where the natural frequency always turns out to be independent of the whirl amplitude.

However, Rankine incorrectly predicted that speeds higher than the critical speed would be unstable. Gunter [1] and others state that the reason for this incorrect result was a failure to include the Coriolis force in the model. A more basic mistake is that Rankine considered the centrifugal force to be a real force. Rankine reasoned that this "force" grows by the square of the speed while the restoring force kr of the shaft does not; therefore, the amplitude r would have to become unbounded at supercritical speeds. But real forces on a body are always from (1) contact with another body, (2) fluid pressure acting on a surface area, or (3) fields such as gravity or electromagnetics. None of these real forces depend on motion of the body. The centrifugal force does not fit properly into any of the three categories of real forces. A correct interpretation of Rankine's model at the critical speed is that the spring force kr acts on the mass m to produce a centripetal acceleration $\omega^2 r$ toward the center of rotation. There is no other force acting. Even today, students of rotordynamics often get into trouble when they treat the centrifugal force or the Coriolis force as a real force in their analytical models. See Meriam and Kraige [7] for additional discussion of the shortcomings of D'Alembert's principle.

Rankine's incorrect conclusion that supercritical operation (above the first critical speed) is impossible undoubtedly influenced some engineers afterward to design rotating shafts for maximum bending stiffness, i.e., short bearing span and large diameter. But others discovered experimentally that supercritical operation is not only possible but in some cases can be desirable. See the discussion of the work by DeLaval, Jeffcott, and Parsons below.

In 1894 Dunkerley [8] published an article in which he showed that the critical speeds of a uniform rotating shaft on simple supports are the same as its natural frequencies of transverse vibration. This is remarkable, since the two motions are quite different; one is a circular orbit while the other is a planar oscillatory motion. Dunkerley suggested that any unbalance of the rotating shaft would excite the natural frequencies to produce critical speeds with large, synchronous whirling amplitudes. Dunkerley also gave a simple formula for calculating critical speeds of shafts with several wheels or disks spaced at different locations along the shaft. The formula neglects the effect of gyroscopic moments, but was still being used in some machine design textbooks as late as the 1960s. Dunkerley's formula (for two wheels) is

$$\frac{1}{\omega_{cr}^2} = \frac{1}{\omega_1^2} + \frac{1}{\omega_2^2} \tag{7-1}$$

where ω_1 and ω_2 are the critical speeds of the shaft with wheels 1 and 2 alone, respectively, and ω_{cr} is the critical speed with all wheels installed.

At this point it is interesting to consider the major contribution of Lord Rayleigh to vibration theory, although his work was not specifically focused on rotor dynamics. First published in 1877, Rayleigh's book on the *Theory of Sound* [9] gave a method for calculating natural frequencies of undamped vibration by equating the maximum kinetic energy to the maximum strain energy during the vibration cycles, where the total energy was noted to be conserved at all times. Consider a straight line harmonic motion $r\sin(\omega t)$ of the mass in Fig. 7-1, where ω is the natural frequency. Note that the kinetic energy is maximum when the spring is unstrained and zero when the strain energy in the spring is maximum. The maximum displacement is r and the maximum velocity is $r\omega$. Thus, we have an equation $\frac{1}{2}kr^2 = \frac{1}{2}m(r\omega)^2$ from which the natural frequency $\omega = \sqrt{k/m}$. But in Rankine's problem, where the motion is a circular orbit, there is no transfer of energy from strain to kinetic. And yet the method still yields the correct answer, even for more complex rotor models (thus supporting the conclusion of Dunkerley, but not explaining it).

During the decade of 1880–1889 Delaval was building experimental steam turbines, which needed to run at high speeds to match blade velocities with steam nozzle velocities. Delaval became convinced that rotors could run supercritically and finally demonstrated this for the German patent office with a disk mounted on a slender flexible shaft around 1895. The experimental model ran through the first critical speed with an unbalanced disk and ran smoothly at higher speeds. This demonstration of the critical speed inversion, contrary to Rankine's conclusions, was explained by analytical models developed by Foppl [4] and DeLaval [4]. These models had an unbalanced disk on a flexible shaft with simple supports, but had no damping, so they predicted a sudden phase change when running through the critical speed and predicted unbounded whirl amplitudes at the critical speed. The complete theory of smooth critical speed transition with bounded whirl amplitudes remained unexplained until two decades later with the work of Jeffcott, who included damping in the model.

The classic book on steam turbines published in 1905 by Stodola [10] describes the various successful turbines of his time and contains a number of sections [11] describing the rotordynamics investigations of Delaval, Dunkerley, and himself. He apparently was the first to correctly state the two-plane balancing theorem for rigid rotors, to describe the determinant method and the matrix iteration method for calculating critical speeds of shafts with multiple wheels, and to describe the possibility of dynamic interactions between the rotor and the supporting foundation. Stodola also analyzed the stability of supercritical operation, but missed the most important results because he ignored the effect of internal friction in the rotor assembly as well as the cross-coupled stiffness from oil-film bearings (which had not yet been discovered).

By the year 1919 many turbomachines were running successfully at supercritical speeds, but vibration-related failures were occurring often enough to promote more advanced studies of rotordynamics. The Royal Society of London commissioned H. H. Jeffcott to undertake an analysis of rotor whirling and critical speeds. His 1919 publication [12] was the first clear mathematical description of rotordynamic response to unbalance with the effect of damping included. His simple model consisted of a flexible massless shaft on simple supports with an unbalanced disk mounted at midspan with its translational velocity resisted by a viscous damping force. Modified in various ways to analyze the effects of different parameters, it has been used by hundreds of subsequent investigators and is still used today as the simplest descriptor of many rotordynamic phenomena. The model (and its variations) is used extensively in Chapter 3 of this book.

Jeffcott restricted his study of the model to the effect of rotor unbalance on rotor whirling, with damping included. He neglected gravity and showed that a circular whirl orbit satisfies the equations of motion. The analysis showed that the unbalance leads the whirl vector by an increasing phase angle as speed is increased, and comes around to the inside of the disk center as the critical speed is traversed (the critical speed inversion). It showed that the critical speed amplitude is restricted to a finite value that can be made smaller by adding damping. It also showed that the inversion occurs more smoothly over a wider speed range as damping is increased (not a sudden jump as DeLaval had predicted without damping in the model). For designers and troubleshooters, Jeffcott's analysis showed that synchronous[1] whirling amplitudes could be reduced by (1) balancing the rotor, (2) adding damping (if near the critical speed), and (3) operating at speeds farther away (either higher or lower) from the critical speed. This continues to be the basis for controlling unbalance response up to the present day.

Jeffcott's analysis, along with the experiments and analysis of DeLaval, encouraged engineers to design rotating machinery with flexible shafts for supercritical operation. Quoting from Jeffcott: "Consider the design of a shaft for a given duty and operating at a given working speed. Then the preceding formulae indicate that it is better from the vibration point of view to design the shaft with its critical speed below the working speed rather than to have a critical speed the same proportion above the working speed." Unfortunately, it was yet to be learned through hard experience with violent subsynchronous whirling and some catastrophic failures that this design philosophy can easily lead to rotordynamic instability at supercritical speeds. These problems have nothing to do with response to unbalance.

In the early 1920s some General Electric (GE) blast furnace compressors that had been designed under this philosophy experienced violent subsynchronous whirling. The whirling frequency was so low that it could be tracked with the human eye, so the problem was named *shaft whipping*. At the GE Research Laboratories in New York, B. L. Newkirk conducted experiments designed to investigate the cause. In his first series of experiments [13] he found that shaft whipping (1) was not affected by rotor balancing, (2) occurred only at speeds above the first critical speed, (3) was always at a frequency approximately equal to the first critical speed, and (4) occurred only with built-up rotors. Subsequent research, described below, would show that this phenomenon was self-excited (i.e., due to internal characteristics of the rotor–bearing system) from internal friction in the rotor assembly or from the oil-film bearing force coefficients.

Working with Newkirk on this problem was A. L. Kimball. In a marvelous demonstration of the power of human intellect to overcome intuitive paradigms, Kimball postulated that internal friction, or damping in the rotor, could produce internal moments at supercritical speeds that would drive the whirling [14]. Even today this idea is met with strong resistance when first presented to students. How can damping produce

---

[1] At the same frequency as shaft speed.

unstable vibration, when it is well known that damping suppresses vibration? Newkirk showed that the internal friction is much greater in built-up rotors with interfaces between the fitted parts, and Kimball later proved and demonstrated his postulate with experiments [15]. He showed that a long horizontal sagging shaft with internal friction would deflect horizontally when rotated. He also wrapped a flexible shaft with copper wire to produce an assembly with a lot of internal friction, and showed that it whirled at supercritical speeds with a frequency of the first critical speed.

In 1933, D. M. Smith [16] published a paper containing correct models of the rotor–bearing system that included shaft flexibility (both symmetric and orthotropic) and bearing support flexibility (both symmetric and orthotropic), and with internal friction modeled as a "cross-coupled" force (horizontal deflections induce vertical forces, and vice versa). Although correct values for many of the parameters were not known, Smith's paper included mathematical solutions for all of the fundamental rotordynamic problems involving internal rotor friction (which he called "rotary damping") and stiffness asymmetry. Among his many useful conclusions that went much farther than Jeffcott, were the following (quoting from [16]):

1. Hence a system which includes rotary damping but no stationary damping is unstable at all speeds above the critical speed; while a system which contains both stationary and rotary damping is unstable at all speeds above a certain transition speed which is higher than the critical speed.

2. If the rotating parts are unsymmetrical, the critical speeds occur in pairs which enclose ranges of violently unstable speed, and in addition, instability may be produced by shaft damping at speeds above the lowest range of unstable speeds.

3. Damping in the flexible bearing supports always favours stability, but damping in the rotating parts should be avoided when the machine has to run above a critical speed.

4. The effect of flexibility of the bearings in producing stability at speeds well above the critical speeds in a rotor with rotary damping was discovered experimentally by Newkirk, but it appears doubtful whether he realized that the unsymmetrical flexibility of the bearings is the essential feature inducing stability even in the absence of damping in the bearings.

5. In practice frictional forces arise in various ways and follow various laws; a severe type of rotary friction sometimes experienced is due to working of the shaft in a cramping fit of the rotor and the effect of this in bringing about instability at super-critical speeds shows itself erratically.

In 1934, Thearle [17] applied Stodola's two-plane balancing theorem and published the method for influence coefficient balancing of rigid rotors (no significant bending, or for balancing at a single speed) that is still predominant in the United States today.

Torsional vibration is another type of rotordynamic problem that caused early failures, especially in drive trains with reciprocating engines. These problems were most common in drive trains with several machine components coupled together, such as an engine driving a blower through a gearbox. The most obvious method of mathematical analysis is to solve sets of coupled second-order differential equations, one for each rotary inertia in the train. But this method results in large matrices that cannot be practically solved without a computer. In 1921 Holzer [18] published a clever interactive method of solving for the natural frequencies of torsional vibration without employing the large matrices.

Bearings

During this period almost all bearings in industrial machinery were oil-lubricated journal bearings with simple circular geometry. The common view throughout most of the 1800s (and held even today by the uneducated) was that the oil simply served to reduce the sliding friction coefficient between the shaft and the bearing. Experiments by Beauchamp Tower on railroad car bearings and the supporting mathematical analysis by Reynolds [19] eventually brought about an understanding of the pressures generated in the oil film by rotation of the bearing journal. Reynolds' work produced a partial differential equation for the oil film pressure around a circular bearing that was impossible to solve in the two dimensions of the bearing surface (without a computer). But Sommerfeld and Harrison [20] attacked the problem using simplifying assumptions for special practical cases and were able to solve for the pressures that produce the force of the bearing on the shaft. The most remarkable and useful result from the standpoint of bearing life was that this force could be large enough to support the load and prevent metal-to-metal contact between the bearing and the shaft. From the standpoint of rotordynamics, the most important result was that the supporting oil film force is not in line with the radial deflection of the shaft from the bearing centerline. There is a force component normal to the radial deflection, now known as the cross-coupled force, and this is what produces the subsynchronous instability at supercritical speeds now known as oil whip. In fact, before cavitation of the oil was recognized, integration of the calculated pressure field by Robertson [21] produced a force that was *entirely* normal to the radial deflection of the shaft. This was the cause of his inaccurate prediction that the rotor would be unstable in asynchronous whirl at all speeds.

Robertson's analysis was probably motivated by some earlier experiments of Newkirk and Taylor [22]. They had conducted experiments with rotor–bearing systems that exhibited shaft whirling and whipping, but that could not possibly contain internal friction in the rotor assembly. They were able to relate the instability to the oil film bearings supporting the rotor. They found that the system became unstable in subsynchronous whirl when the shaft speed reached twice the critical speed, but were not able to successfully explain this phenomenon.

At about the same time as DeLaval (1895), Parsons built his first successful steam turbine in England [23]. As described above, Delaval concluded that the turbine shaft should be flexible to allow the critical speed inversion and smooth operation at supercritical speeds. This design solved the unbalanced response problem, but caused problems of rotor-dynamic instability. Parsons was the first turbine builder to realize (even before D.M. Smith's analysis was published) that the bearing supports in a high-speed machine should be flexible and should have damping, and this eventually proved to be the better design concept for stability at high speeds. Figure 7-2 shows a flexible "squirrel cage" bearing support tested in his first turbine (now standard practice in almost all aircraft turbine engines). It was not incorporated as the preferred support, probably because it did not have a damper (most squirrel cage bearing supports today are combined with squeeze film dampers).

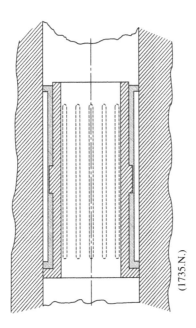

(1735.N.)

Figure 7-2 Parsons' flexible bearing support. From Richardson [23].

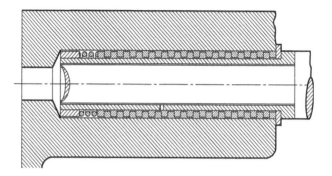

Figure 7-3 Parsons' first bearing damper. From Richardson [23].

Figure 7-3 shows the design adopted for use in Parsons' first turbine. Quoting from Richardson [23], "A light bush fitted on the journal formed the actual bearing. On this bush a series of rings or washers were strung, each alternate ring fitting the bush but not the casing, while those between fitted the casing but not the bush. Near the end there was on both the bush and casing a wide ring for centering the shaft, and a spiral ring, tightened up by a nut and washer, pressed all the rings together." It can be surmised that the idea behind this design was to create a lot of lubricated Coulomb friction at the axial interfaces between the rings. Since purely Coulomb friction cannot limit critical speed amplitudes, it can also be surmised that the design actually produced some viscous damping.

Figure 7-4 shows the bearing damper later incorporated into all the Parsons turbines running over 2000 rpm. This design is now known as a concentric multiland squeeze film damper. Quoting again from Richardson: "Over the bush there were three concentric tubes, held by a collar on the bush at one end, and by a nut at the other. A thin film of oil, fed through the holes, kept the bushes apart, and allowed such play of the shaft as might be caused by vibration, while the viscosity of the oil resisted powerfully any undue motion. This form of bearing was the outcome of

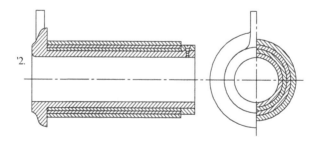

Figure 7-4 Parsons' squeeze film damper. From Richardson [23].

experience with preceding machines, where a series of washers had been adopted. It was found that these washers cut into the bearing bush and into the casing around the bearing. The concentric tubes were found much more satisfactory, and the system is still adopted practically in all Parsons turbines running at speeds over 2000 revolutions per minute."

Parsons steam turbines became the preferred propellant in marine vessels, both commercial and military (navy), first in England, and then in France, Germany, the United States, and other countries around the world. Their horsepower ratings grew from the original 6 hp design (driving an electric generator) to 42,000 hp in each of four Russian battleships built in 1911.

REFINING AND EXPANDING THE ROTORDYNAMIC MODEL, 1942–1963

Along with the development of gas turbines for aircraft propulsion, turbo-electric generators for utilities, and centrifugal machines for petrochemical gas compression, turbomachinery was being designed in the 1940s through the 1960s for many more applications at a variety of speeds and power/weight ratios. The theory that explains most fundamental rotordynamics problems had been published by this time, but the new applications brought new challenges in using the theory to produce practical design analysis.

In 1943 Foote, Poritsky, and Slade [24] published an analysis that expanded on Smith's theory [16] for the effect of stiffness asymmetry in the shaft (rotating) and in the bearing supports (nonrotating). The motivation was to explain the vibrations observed in two-pole turbogenerators, which have orthotropic shaft stiffness asymmetry. The analysis shows the possibility of whirling instability over a narrow speed range unless the bearings have sufficient damping.

In 1945 Prohl [25] published a new method for calculating critical speeds of flexible rotors with many "stations" (wheels or lumped masses along the rotor). The method later became widely known among vibration analysts as the transfer matrix method. Prohl's application was steam turbine rotors. The mathematical method was an extension of Holzer's method (for torsionals) to lateral vibration. The method had been published a year earlier by Myklestad [26], who developed it to calculate natural frequencies of airplane wings.

Up until the 1940s most analytical models of critical speeds and whirling eigenvalues had neglected gyroscopic effects of the spinning wheels. With low-speed machines, the errors in results were generally small. As operating speeds of turbomachines got higher, however, the gyroscopic effects could not be reasonably neglected. In 1948

Greene [27] published a thorough mathematical treatment of gyroscopic effects on critical speeds.

Recall that Robertson's oil film bearing analysis [21] had predicted that turbomachines on oil film journal bearings would be unstable in oil whip at all speeds, but Newkirk's and Taylor's experiments [22] had demonstrated that such machines could be operated at speeds up to twice the critical speed without whirl instability. In the early 1950s, Poritsky [28] and Simons [29] finally revealed the reason for this discrepancy. Poritsky modified Robertson's analysis to include an oil film force component in line with the off-center journal displacement (radial to the whirl orbit). This was justified by Simon's analysis of the oil film bearing, which showed that the force resultant of the oil film pressure distribution can be normal to the radial displacement only if the off-center displacement is small. The film becomes cavitated and vaporized for larger displacements, and this produces the radial force component that allows stable operation of the rotor at speeds up to twice the critical speed.

Problems with oil whip at high speeds provided the incentive to develop new types of oil film bearings with smaller cross-coupling. Figure 7-5, from McHugh [30], shows the various designs in this evolution of the radial bearings. All of these improved fixed-geometry designs are more resistant to oil whirl and whip than a circular journal bearing. The tilt pad bearing was the culmination of this progression, having almost none of the destabilizing characteristics.

The tilt pad bearing was first invented as a thrust bearing in the early 1900s by Kingsbury in the United States and Michell in Australia (separately and almost simultaneously). It was later developed as a radial bearing with high stiffness for machine tools and precision machinery. Experiments at Columbia University, as described by Fuller [31], showed that it was highly resistant to oil whirl and whip. Since then it has become almost universally employed in high-speed centrifugal compressors and other machines susceptible to subsynchronous whirl instability.

In the 1950s it became evident that subsynchronous whirling in steam turbines was somehow related to the steam forces on the turbine wheels. An investigation and analysis by Thomas [32] in Germany showed that the variation of tip clearance around the blades created a resultant follower force on the whirl orbit. This type of subsynchronous instability in steam turbines became known as steam whirl.

The 1950s and 1960s was also a time of concentrated development on squeeze film bearing dampers (SFD). The following discussion has been edited from reference [33] by the first author. A report for the U.S Army [34] shows that design analysis of squeeze film dampers for the Chinook helicopter drive shaft was in progress before 1963. The designs considered were very similar to Parson's designs from the early 1900s, consisting of multiple concentric metallic rings around the ball bearing

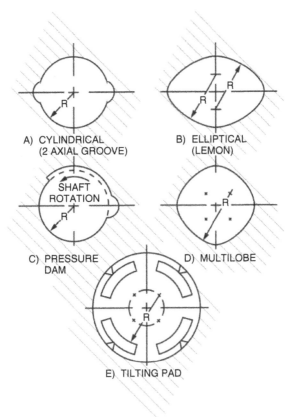

A) CYLINDRICAL
(2 AXIAL GROOVE)

B) ELLIPTICAL
(LEMON)

SHAFT
ROTATION

C) PRESSURE
DAM

D) MULTILOBE

E) TILTING PAD

Figure 7-5   Bearing types to suppress oil whip. From [30].

outer race with viscous lubricant supplied to the clearances between the rings. It was found experimentally that the rings had to be pinned against rotation to avoid nonsynchronous oil whirl. The Chinook drive shaft is 28 feet long and so operates through several flexural critical speeds if bearings are not provided at a number of locations along the shaft. The desire to eliminate the weight of these inboard bearings was the incentive for designing squeeze film dampers with frequency-dependent damping so that each critical speed of the supercritical shaft on two bearings would have its optimum damping coefficient. It was shown by analysis that the frequency dependence could be obtained by making the damper rings massive. Unfortunately, the required mass of the damper rings added up to four or five pounds, so the frequency-dependent characteristic was abandoned. It was found by later research on dampers as described in the next historical section that cavitation and air entrainment in the oil would probably have prevented such precise control of the damping coefficients unless a very high lubricant supply pressure was maintained.

A patent application describing a squeeze film damper was filed on November 23, 1966 in the London patent office. The invention was declared by Rolls Royce, Ltd. and the inventor was Arthur Bill. The specification stated: "An advantageous means of damping vibrations in bearings is to provide a hydrodynamic fluid film, sometimes referred to as a 'squeeze film,' between the bearing and its supporting structure." A U.K. patent number (1,104,478) was assigned and the complete specification was published on November 23, 1968. One of the specific embodiments of the invention was described as an application to axial flow compressor bearings as used in aircraft engines. A significant portion of the specification was devoted to describing means for supplying lubricant to the squeeze film.

A few years earlier in 1963 a research engineer at Rolls Royce had published a technical paper in which vibration control experiments on a rotor test rig were described. The results of these experiments by Cooper [35] are almost certainly the seed of the patent described above, and the reason why squeeze film dampers have been used in most turbojet and turboshaft aircraft engines designed since that time. Squeeze film dampers came to be widely regarded as an essential component of these engines by the 1970s. A turboshaft engineer in the United States explained to the first author in the early 1970s that, even though his company's engines did not have squeeze film dampers designed a priori, the roller bearings actually had a small clearance around the outer race, which collected oil and acted as a squeeze film to attenuate vibration.

Cooper's experiments at Rolls Royce were on a 36″-long, 2″-diameter shaft, with two 6″-diameter wheels spaced 12.5″ apart. The two bearings were mounted outboard of the wheels and were suspended on flexible springs to allow the critical speed inversion to take place at low speeds (around 750 rpm). The flexible supports had enough stiffness asymmetry to split the critical speed into two, spaced about 200 rpm apart. Cooper started out using gas bearings, which have very little damping. He found that, although the rotor ran very smoothly at supercritical speeds, the synchronous whirl amplitude when passing through the critical speed was unacceptably large.

Cooper decided to install hard mechanical stops (eight bumpers) around the shaft in an attempt to limit the critical speed amplitude to 3 or 4 mils. He found that the bumpers prevented inversion and caused violent pounding, which threatened the life of the gas bearings. The bumpers were replaced with a concentric metal ring to restrain the shaft and the same pounding and hammering occurred, with delayed (or no) passage through resonance into the smooth "self-balanced" mode of operation. The life of the gas bearings was threatened by the violent vibration so they were replaced with ball bearings to accommodate the high loads and abusive testing. The same behavior was observed, except that high-frequency vibration was now contributed by the ball dynamics.

After noticing that a hand on the shaft could provide enough damping to smoothly pass through the critical speed, Cooper decided to feed some oil to the clearance between the bumper ring and the shaft and found that the rotor ran smoothly in the inverted mode at very low speeds with the first critical speed not evident. The second critical speed, however, developed into oil whirl. After driving through the oil whirl with severe pounding of the limit ring, inversion was finally obtained with smooth running at high speeds. With regard to this particular test, Cooper states in his paper: "The results were considered worthwhile and they confirmed the usefulness of proceeding with the less simple scheme which followed." In this following scheme "the hydrodynamic control film was established between non-rotating cylindrical members, a bearing fixed in the structure and a journal which followed the translatory motion of the rotor but was prevented from rotating." Figure 7-6 is reproduced from Cooper's paper and shows the tested configuration. One can see that the "bearing" is a large cylinder supported by a stiff central plate. The inside diameter has a very wide central oil supply groove. The squeeze film is formed by two lands on each side of the supply groove. The "journal" is the outside diameter of a sleeve, which is fitted to the outer race of a ball bearing, and which is softly supported by coil springs. This is called a *parallel*

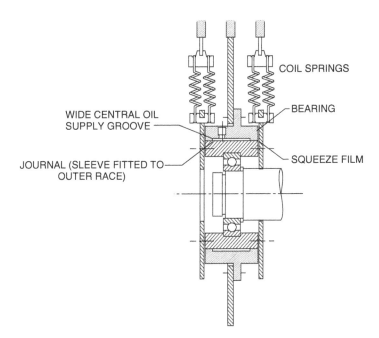

Figure 7-6   Cooper's test apparatus for squeeze film dampers.

arrangement, that is, the "journal" and roller or ball bearing are spring-supported with the "bearing" attached to the engine frame as shown in the next section. In this design, the squeeze film and the spring support are subject to the same deflection, and the dynamic loads are divided unequally between them.

Cooper reported test results for three squeeze film dampers of the Fig. 7-6 configuration, which he referred to as bearings A, B, and C. They had different clearances and land widths. Bearing C, which had the smallest clearance (0.0015″) had the poorest performance, not allowing inversion at any speed with 0.5 oz-in imbalance. Bearings A and B with 0.003″ clearance performed much better. Bearing A, with the narrowest land (0.1″), performed best by allowing inversion to a 0.002″ orbit with 1.0 oz-in imbalance. The implications of these results have since been qualitatively confirmed many times over in tests and computer simulations by other researchers. That is, successful SFD designs must have larger clearances than intuition would suggest, and dampers in series with mechanical stiffness (i.e. shafts) have an optimum value of damping that should not be exceeded by making the lands too large. Because of the low speeds of Cooper's tests, he did not discover the various forms of cavitation which have a profound effect on SFD performance at higher speeds.

Cooper also performed a force balance analysis of the whirling bearing, which confirmed that inversion cannot occur if the squeeze film land is too wide. The results of his analysis also suggested that if inversion does occur, three different whirling orbits could exist with force equilibrium: (1) a small orbit with the imbalance inverted to the inside, (2) a large orbit without inversion, and (3) an orbit of intermediate size. Predictions 1 and 2 have since been verified by experiments by D. C. White [36] that showed the jump phenomenon, where the whirl amplitude changes instantaneously during a runup or a coastdown. The third whirl orbit was shown to be unstable, i.e., jumping to the first or second.

The present authors know of no evidence that suggests Cooper was aware of the earlier SFD developments by Parsons and others. Cooper's work brought SFD technology into aircraft turbine engines, but the development might have progressed more quickly into advanced SFD designs if he had been aware of existing literature at the time.

MULTISTAGE COMPRESSORS AND TURBINES, ROCKET ENGINE TURBOPUMPS, AND DAMPER SEALS, 1964–PRESENT

The design requirements for speed and power of turbomachinery have been rapidly increasing since the 1960s. Three examples are (1) multistage axial compressors and turbines for aircraft turbine engines, with turbine shaft power increasing past 35,000 hp, (2) multistage centrifugal compressors for gas reinjection in oil fields required to produce more than 600 bars (9000 psi) pressure at high discharge rates requiring 20,000 hp (15,000 kW), and (3) rocket engine turbopumps running at speeds over 30,000 rpm driven by 70,000-hp turbines about the size of a frisbee. These kinds of performance numbers result in machines that are likely to be unstable in subsynchronous whirl. Designing multiple stages on one shaft results in long shafts with bending modes, which makes instability difficult to suppress. If the machine must also be lightweight, then the rotors are even more flexible, and this also makes balancing difficult. Most of the significant development over the last three decades has been to develop analysis tools, measurement tools, and vibration damping devices (mostly bearings and seals) to solve these problems.

The long flexible rotors in multistage machines produce multiple critical speeds. In theory, to achieve a perfect state of balance, the number of balance planes should equal the number of critical speeds traversed. Since a large number of balance planes is not always available, methods were developed to achieve optimal results with a limited number of balance planes. The most popular of these methods in the United States is called least-squares balancing, published by Goodman [37] at General Electric in 1964. This method makes use of vibration measurements made at several different speeds. It is a special case of a more general analysis published later by Lund and Tonnesen [38], which allows determination of balance weights in N or more planes to successfully balance a flexible rotor through N critical speeds.

The vibration measurements required by Goodman's method would not be possible without the development of a reliable shaft proximity probe. This development, which allowed noncontacting vibration measurements on a rotating shaft, was effectively done by Don Bently [39]. In fact, none of the progress since the 1960s as described here could have been made without the ability to make these measurements. Noncontacting probes were invented as early as 1923 [40], but they relied on bulky and unreliable vacuum tubes and galvanometers. Bently utilized the transistor and other electronic innovations developed after 1950 to make the eddy–current proximity probe an industry standard for measuring rotating machinery vibration.

J. S. Alford published a paper in 1965 [41] describing problems with self-excited whirl in an axial flow compressor for an aircraft turbojet engine. These engines have ball and roller bearings, so oil whirl could not be a factor. Alford hypothesized two different mechanisms that could drive the whirl: (1) negative damping from labyrinth seals (discussed in Chapter 5), and (2) an aerodynamic follower force due to the tip clearance variation around the circumference of the compressor blades. This latter mechanism became known as Alford's force. It is similar to the steam-whirl mechanism investigated earlier in turbines by Thomas [32].

Subsequent arguments have ensued about whether the same mechanism can excite forward whirl in turbines as well as compressors. Vance and Laudadio [42] demonstrated experimentally that Alford's force does exist and can drive forward whirl in a high-speed blower. Later experiments at General Electric Aircraft Engines [43] showed that the excited whirl direction in turbines depends on the operating speed range.

The Ph.D. dissertation of J. W. Lund [44] in 1966 defined linearized force coefficients in Newtonian coordinates for oil film bearings that are still used today in most computer simulations of rotordynamics. Lund's definition of stiffness and damping coefficients was general enough to include cross-coupling effects, that is, forces produced by motions normal to the force. They are now widely used to model the rotordynamic effects of seals and process wheels as well as bearings. Lomakin [45] had already published a paper in 1958 describing the rotordynamic effect of plain smooth seals in terms of their effective positive stiffness, which is now widely known as the Lomakin effect. Black [46] extended this concept in 1969 and made valuable contributions to the analysis of seal forces in centrifugal pumps.

Stability Problems with Multistage Centrifugal Compressors

Post WWII, along with the increased demands for oil production came the need to reinject large volumes of natural gas back underground at very high pressures. Multistage centrifugal compressors were found ideal for this task, but beginning in the 1970s some became violently unstable in subsynchronous whirl. Two of these cases became somewhat famous in the community of rotating machinery engineers. Due to the huge financial losses incurred by these failures and the probability of repeated occurrences, they became the incentive for the development of improved tools for stability analysis and measurements. Detailed descriptions of these cases now follow.

Kaybob, 1971–72

These compressors were at a large natural gas plant operated by Chevron Standard Limited in Canada. The natural gas was rich in condensate and was to be reinjected into the formation with three large centrifugal compressors operating in parallel. The following account of the Kaybob case is taken in quotes from an ASME Paper by Fowlie and Miles [47], project engineers for Chevron Standard. Words in italics and not in quotes are by the present first author.

"When started in the Kaybob plant, the nine-stage low-pressure compressors vibrated violently. The predominant frequency was between 40 and 60 percent of the speed the machine was running. Amplitude of this fractional frequency vibration reached 10 mils (250 micrometers) peak-to-peak at the bearings. The compressors have oil-film seals. The radial bearings have five tilting pads. Thrust bearings have six pads in the normal load direction and 12 pads in the reverse direction.

"We selected these compressors in part because they were based on successful methanol plant and ammonia synthesis gas designs. Units of the nine-stage design had operated at similar speeds and pressures, but had not been operated in reinjection service. An independent consultant was hired to check critical speeds and rotor response as part of the purchase order. He confirmed the vendor's criticals. He also pointed out that the operating speed would be 2.85 times the low-pressure casing's first critical speed and that some machines operating 2.5 times or higher had experienced resonant whirl under less severe operating conditions. (Resonant whirl is a form of fractional frequency instability where the major vibrational frequency is near the first critical speed frequency even though the machine is running at higher speeds.) We recognized that full-load testing was desirable to find out if vibration instability would really be a problem with these machines. The manufacturer, however, had neither the horsepower nor the pressure capability to do this in his factory. Furthermore, he was confident that instability probably would not occur in these machines. He had done extensive lab test work with a full-size rig. Instability problems in the prototype design had been completely eliminated by design refinement. Also, he had a number of machines in service that were completely free of fractional frequency instability. Thus, we agreed to forego the full-load factory test, but the schedule was set to allow one month earlier completion of the compressors, to permit full-load field testing before plant start-up. One month was not enough. We did test the machines as full as the factory facilities would permit. All of the compressors easily met the API vibration specifications. Five of the six compressors were completely free of unusual vibrations. One unit had shown 0.26 mils (6.6 micrometers) vibration at a frequency of about one-half running speed. This fractional frequency amplitude was about 42 percent of the total vibration amplitude at the upper operating speeds. The manufacturer attributed the fractional frequency vibration to the oil film seals used during the test.

"As planned, the first compressor train was started early for testing. On November 9, 1971, we discovered we had a major problem in the nine-stage low-pressure casing. Vibration was moderate at low speeds. As the speed was raised, the vibration jumped as high as 10 mils (250 micrometers) peak-to-peak, resulting in damage to the bearings, seals, and labyrinths.

"Fig. [4 shown here as Fig. 7-7] shows vibration spectra at several speeds and that the sudden increase was in the fractional frequency component. It can be seen that there is an instability threshold speed, below which the machine is stable, but above which it is violently unstable.

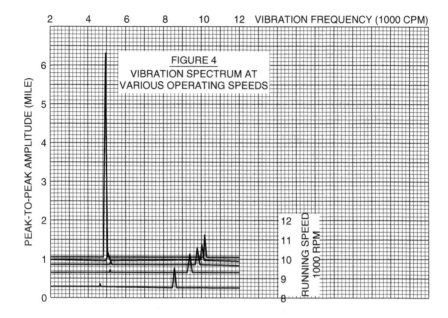

Figure 7-7 Figure 4 from Fowlie and Miles [47].

"Within about ten days, we had eliminated several of the simplest possible causes—opened up clearances of the oil control rings in the tilting pad radial bearings, machined grooves in the oil film seal rings to reduce their ability) to act as bearings, increased the oil film seal ring clearances so that they were more than the bearing clearances. We eliminated the suspected oil whirl generators in the bearings and oil film seals. (Oil whirl is another form of fractional frequency vibration instability. The predominant frequency is a percentage of the running speed, usually about 43 to 47 percent.) None of these changes helped.

"Having checked out the simpler causes, we were left with the more difficult ones. Some we attacked were:

1. Foundation and support resonance
2. Acoustic pulsations
3. Excitation from drive train
4. Coupling lockup
5. Shaft nodal points too close to bearings, hence insufficient bearing damping
6. Bearing instability
7. Oil film seal ring radial lockup
8. Rotor shrink-fits and material hysteresis
9. Bearing effects of inter stage labyrinths
10. Aerodynamic drag."

More than three decades later we have learned that the most probable cause of the Kaybob instability is not on the list; it is the midspan labyrinth seal. Also, we know that number 5 on the list is a design flaw of all these types of compressors that persists even today: it is caused by high bearing support stiffness, which prevents damping from acting at that location. This was originally due to ignorance of rotordynamic theory, but has been perpetuated by API specifications that limit bearing motion and promote shaft bending.

"After trying a 6-in. (152-mm) shorter rotor (by combining the lube and seal oil flows), we could see a significant rise in the instability threshold speed. The machine was still unstable, but could reach higher speed before vibration became excessive. At last, we had a machine we could operate. The next compressor was rebuilt with the short shaft also, and for this one, the impellers were bored out as large as possible for a bigger shaft to gain more stiffness. The second unit was also unstable. but could run at even higher speed than the first. On January 9, two months after first trial, we had two compressors that could be run at part load. The remaining step to full stability was more difficult. Further enlargement of the shaft would require new impellers. . . . This would require boring out the existing hubs and welding in new ones. . . .

"By March 9, we had one compressor that could run at full speed and two that could run at part speed. Now the first two compressors were rebuilt, one at a time, with the welded hub rotors. On May 31, the third one was successfully run, and we had a plant that could make its rating. It had taken 29 weeks. . . .

"Not even the leading rotordynamics consultants could simulate these rotors. . . . We believe full-pressure, full-speed mechanical testing in the factory is essential for new machines near or beyond the boundaries of verified field experience. Research and development in this area are required by the manufacturers and the consultants to develop reliable prediction methods of the instability phenomena."

Ekofisk, 1974–75

The first author of this book was directly involved in this case while working for a consulting group headed up by J. C. Wachel. The following account in quotes is taken from Doyle [48], a project engineer for Phillips Petroleum. Words in italics and not in quotes are by the present author.

"The Ekofisk oil field in the Norwegian Sector of the North Sea was developed by the Phillips Petroleum Norway Group on the premise that early delivery of some of the crude oil would be possible by temporarily reinjecting all of the gas produced. This type of installation would allow production of a part of the oil prior to completion of the gas pipeline to shore. Accordingly, process and compression facilities were installed offshore to separate oil from gas and to compress the gas to approximately 625-bar pressure for injection into the formation. This arrangement would permit producing oil to the equivalent gas capacity of the injection compressors. Although this represented only a portion of the ultimate capacity of the field, nevertheless it was very important to the principals involved since it was the beginning of a return on a very large investment.

"The reinjection compressors receive gas from the separator area at 68-bar pressure and boost it through two parallel trains to 625-bar pressure. Each train consists of two 15,000-kW units in series, with the first unit discharging at 240-bar pressure and the second unit at 625 bars. Each casing contains 8-stage rotors with back-to-back impeller construction. The flow was from the suction through the first four impellers in series out the center of the casing to an interstage cooler and then return to the opposite end of the compressor with flow through the final four impellers and discharge from the center of casing. This arrangement necessitated a long labyrinth in the center to break down discharge pressure from 625 bars to 440 bars." *Years later, after measurements of labyrinth seal force coefficients and more computer simulations of the rotor–bearing system, the center labyrinth was found to be a major contributor to instability in this type of compressor.*

"An attempt was made in March 1974 to commission the reinjection compressors. By June it was evident that we were not going to be able to operate them in their existing state and further that we had a full-grown rotordynamic instability problem with no immediate solution in sight. What followed was a long period of testing which, for the most part, yielded negative results and which has become characteristic of similar situations. First the manufacturer tried the relatively simple changes such as adjusting seal clearances, seal lockup, adding seal grooves, adjusting lube oil temperatures, and numerous bearing configurations. All these many changes required that we operate the compressors to determine their effectiveness. At the conclusion of this period, several months had elapsed with no solution in sight. At this time it was decided by the manufacturer to design a squeeze film bearing that in a subsequent test proved to be successful and was adopted as an interim solution." *All of these hardware modifications and operating tests should normally be part of the manufacturer's research and development program. The rumor among consultants at the time was that they had recently laid off the engineers who would be competent with rotordynamics and that the machines had been designed with a Xerox copier, i.e., scaled up from drawings of smaller machines with a successful history.*

"On Christmas Eve 1974 we started gas injection into the formation, and shortly thereafter we commenced producing crude oil close to the anticipated design rate for that period. Concurrently with the design and manufacture of the new squeeze film bearings the compressor manufacturer started work on a new rotor design incorporating a larger diameter shaft and a slightly shorter bearing span. At about the same time we decided to hedge our position and secure new compressors for the final stage of each string which incorporated a change in design by using two compressor bodies rather than one. Each set of compressors, which were manufactured by different firms, was designed, built, and full-load tested at actual operating conditions in approximately 1 year, a very laudable accomplishment. It is interesting to note that the calculated payout of these new compressors was somewhat less than 1 week in terms of lost crude oil production. We were able to operate the compressors successfully with the squeeze film bearings and continued injecting gas and producing crude oil until the summer of 1975, when the new design rotors were installed. The machines have operated successfully since. They operate at reduced head now since the formation pressure is much lower. They are actually only needed when the gas pipeline is out of service for some reason or when it is operating at restricted capacity."

The following quotes regarding the Ekofisk case are from J. C. Wachel [49], who headed up the rotordynamics investigation done by a consulting company where the first author was employed at the time. Wachel's group made measurements on the compressors in the field (a North Sea platform) and conducted computer simulations to suggest possible fixes.

"This case deals with a much discussed reinjection compressor which experienced excessive nonsynchronous vibrations on startup. Field vibration data will be presented which shows the influence of oil ring seals, aerodynamic cross coupling, and speed on instability frequencies and amplitudes. These areas are of major concern to rotordynamists; however, very little experimental data is available in the open technical literature. The 22,000 horsepower, eight stage compressor with back-to-back impellers (fig. 8) was rated at 8500 rpm, had a design suction pressure of 24.1 MPa (3500 psi), and discharge pressure of 63.4 MPa (9200 psi). The calculated first critical speed of the rotor was 3800 cpm for a bearing span of 206 cm (81 inches). Floating oil seals were located a few inches inboard of the bearings. The compressor originally could not be brought to design speed and pressure without tripping out on high vibrations (Fig. 9) [Fig. 7-8 here].

"The units were monitored by shaft vibration probes which automatically shut down the unit whenever the vibrations exceeded 64 m (2.5 mils); however, due to the monitor's finite response time and suddenness of the instability trip-outs, vibration amplitudes equaling total bearing clearance were experienced. The frequency of the non-synchronous instability was 4400 cpm which was higher than the calculated rigid bearing critical speed of 4200 cpm. This can occur if the floating oil seals

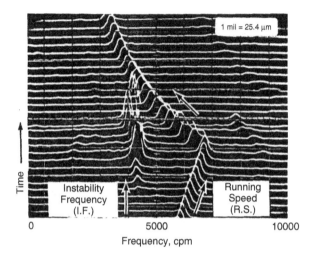

Figure 7-8   Fig. 9 from Wachel [47].

lock up and carry some load, thereby effectively reducing the bearing span. In the computer simulation of this shaft, an effective oil seal stiffness of 286,000 N/cm (500,000 1b/in) was required to calculate an instability frequency of 4400 cpm. Using this stiffness for the oil seals the calculated log decrement reduced to 0.08 compared to 0.3 calculated for the original rotor, neglecting the effect of the seals. Therefore, the calculations indicated that the seals significantly reduced the stability of the unit. To improve the rotor instability, two circumferential grooves were cut into the sealing surface of the seals, the pressure balance of the rings was improved, and the coefficient of friction of the sliding surfaces was reduced. The compressor was still unstable, as can be seen in Figure 10 [Fig. 7-9]." *Many of the tests described by Wachel in this paper involved modifications of the oil seals, whereas it is now known that the labyrinth gas seals, especially the center seal, are much more important for rotor stability.*

"A non-synchronous instability occurred at 4700 cpm; however, instabilities above running speed at 9500 and 10,500 cpm were also excited. As the unit speed reduced, the instability component at 10,000 cpm remained. The rotor was found to be sensitive to the rate of acceleration; therefore, by slowing down the startup procedure, it was possible to operate in the normal speed range.

"To more fully define the stability limits, data was obtained throughout the entire performance map. For a constant speed of 7600 rpm, figure 11 [Fig. 7-10] shows how the aerodynamic loading affects the amplitude of the instability component at 5160 cpm (a forward precessional mode).

"As the suction pressure increased, the amplitude of the instability increased but remained within bounds until a limiting pressure was

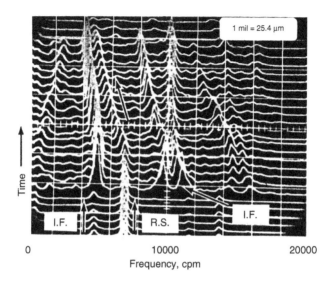

Figure 7-9   Fig. 10 from Wachel [49].

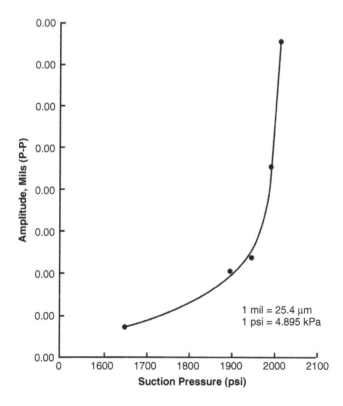

Figure 7-10   Fig. 11 from Wachel [49].

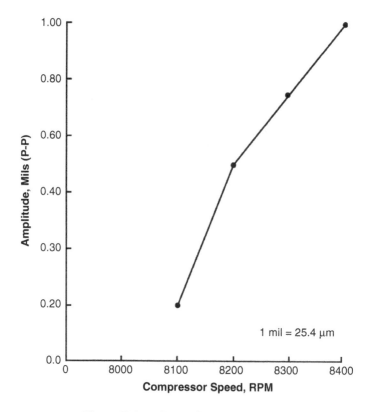

Figure 7-11   Fig. 12 from Wachel [49].

reached. The frequency of the instability component moved from 4400 to 5160 cpm as pressure was increased. To show the effect of speed on the instability the suction pressure was held constant at 10.3 MPa (1500 psi) and the speed increased. The instability amplitude increased almost linearly with speed (fig. 12) [Fig. 7-11].

"Some specifications require that the first critical speed be greater than 0.6 times the running speed to help prevent instabilities. Figure 15 [Fig. 7-12] shows that an instability component at 0.8 times running speed occurred when the running speed was 4000 rpm or slightly above the first critical speed. In this data, the instability occurred when the ratio of running speed to first critical was 1.25, showing that a ratio of running speed to first critical of less than 2:1 does not necessarily ensure that a rotor will be stable. The majority of the instability trip-outs were at speeds where the ratio of running speed to first critical speed was less than 2:1. The data presented shows that the stability frequency characteristics were dramatically changed by changing only the oil seals; however, no change in seal design made this system stable. This indicates that the seals were not the predominant destabilizing factor.

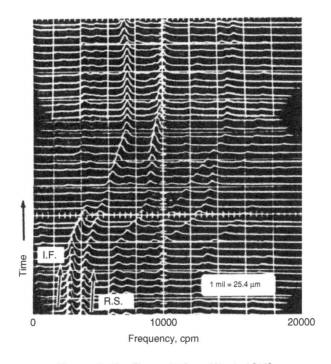

Figure 7-12   Figure 15 from Wachel [49].

"Major efforts were then expended to reduce other destabilizing factors. These changes included aerodynamic changes to the impellers and diffusers, seal modifications, shortening the bearing span, and changing the bearings to 5 shoe, nonpreloaded tilted pad bearings. Log decrement calculations indicated that these changes represented a significant improvement in the rotor stability.

"Even after these modifications were installed, instabilities still occurred. On startup, an instability component (5200 cpm) was present with a fluctuating amplitude. The instability frequency then shifted up to 5800 cpm, and the amplitude jumped to greater than 152 μm (6 mils) in approximately one second. The instability component was particularly sensitive to pressure ratio across the machine, which confirmed that the aerodynamic destabilizing effects were of major importance and over-shadowed other improvements that were made.

"The stability of the unit was markedly improved when a damper bearing was installed in series with the inboard bearing. Squeeze film damper bearings employ an oil film in the space between the outside of the bearing and the case, to which oil is continuously supplied. Stiffness of the damper bearing is usually supplied by a mechanical support such as a squirrel cage cylinder with ribs, welded rod support, corrugated metal ring, or o-rings. Figure 16 [*omitted here*] gives the frequency analysis of the shaft vibrations and two probes monitoring the damper bearing for 8450 rpm with a suction pressure of 23.1 MPa (3350 psi) and a discharge pressure of 56.9 MPa (8250 psi). Instability frequencies were still present, but the added damping from the damper bearing prevented the amplitudes from becoming unbounded. The improvement in the stability characteristics of this rotor illustrates the potential of damper bearings in high pressure applications." *This has become a popular retrofit used by compressor manufacturers to suppress rotordynamic instability. Note that the two damping sources in series result in less damping than either one, so it is not correct to say that damping has been added. Instead, it is the reduced stiffness of the bearing support that allows the reduced damping to be effective.*

"At the time of this study [*1982, the time of Wachel's writing*], the mathematical techniques for predicting instabilities were not as developed as today's procedures and the application of a damper to an industrial compressor involved some tuning to obtain the optimum stiffness and damping for an individual rotor. This can be amply illustrated by the fact that a damper bearing installed in a second identical unit was not successful in eliminating instability trip-outs. After some tuning of the damper bearing, the stability was significantly improved.

"In the process of evaluating the performance of the damper bearing, the oil temperature was varied to determine if it had a significant effect. At an oil temperature of 51°C (124°F), the frequency analysis of shaft vibrations showed instability components at 2800 and 4800 cpm along with the running speed component. Some surprising results were noted as the temperature was lowered. At a temperature of approximately 48.9°C (120°F), the 6 μm (0.25 mil) component at 2800 cpm disappeared [*fig. 18, omitted here*]. Again this points out that very small changes can be significant to the stability of a rotor.

"This case history illustrates that many factors influence the onset, frequency and amplitude of instability vibrations, and that sophisticated mathematical models are required to simulate the instability phenomena measured in these machines. This instability problem was controlled primarily by increasing the shaft diameter to raise the first critical speed, thus significantly increasing the ability of the shaft to withstand the large aerodynamic loading effects. Aerodynamic loading effects are the most predominant destabilizing components in many high pressure systems. In the design stage the designer needs to estimate the level of equivalent aerodynamic loading so that the rotor will have an adequate stability margin. The author has consulted on several instability problems and has developed an empirical formula for estimating the level of aerodynamic loading.

$$K_{XY} = \frac{B\,(hp)(Mol\ Wt)}{Dhf}\,\frac{\rho_d}{\rho_S} \tag{7-2}$$

where

K_{XY} = cross-coupled stiffness coefficient, N/m or lb/in
B = cross-coupling constant, 16 or 105
hp = power, kW or hp
Mol Wt = molecular weight
D = impeller diameter, m or in
h = restrictive dimension in flow path, m or in
f = speed, Hz
ρ_d = density of fluid at discharge conditions, kg/m^3 or lb/ft^3
ρ_S = density of fluid at suction conditions, kg/m^3 or lb/ft^3"

This entirely empirical formula (7-2) is now known as Wachel's equation and is widely used to compute cross-coupled stiffness coefficients for centrifugal compressors. It is now part of the latest API specifications for these types of machines, where K_{XY} is referred to as the Alford force, even though Alford never applied it to centrifugal machines.

Subsequent Developments

The precise source of the rotordynamic instabilities in both of the compressors described above was never determined. Both machines were "fixed" by redesigning the rotor to be stiffer with a shorter bearing span so they became less sensitive to the destabilizing forces, but performance was degraded by the reduced inlet flow path to the impellers. With the test data now available for labyrinth seals, and with the improved computer simulation capability now available, it is fairly evident that the midspan labyrinth seal was a major destabilizing factor in both machines.

The expensive problems experienced with the two multistage compressors described above were the incentive for development of improved computer simulations of rotordynamic instability. The most significant development was an algorithm published by Lund in 1974 [50]. It was an extension of the Myklestad-Prohl method (which only calculated critical speeds) that allowed damping and cross-coupled force coefficients to be included in the model and computed the complex damped eigenvalues of the rotor–bearing system.

When the first author saw Fig. 7-10 (Figure 11 from Wachel), it became the incentive for his development of his torque whirl theory [51]. When it finally became known that the midspan labyrinth seal was the most likely cause of these problems, that became the incentive for development of damper seals for compressors, described below.

NEW FRONTIERS OF SPEED AND POWER DENSITY WITH ROCKET ENGINE TURBOPUMPS

With the advent of the "space race" in the 1970s, turbopumps were developed to feed the liquid-fueled rocket engines. These pumps typically ran at speeds well above 30,000 rpm and were driven by turbines on the same rotor with tens of thousands of horsepower. These machines were (and are) of the same size as the engine in an automobile. The cryogenic fuel being pumped is too cold to allow conventional squeeze film dampers, or elastomeric dampers, to work. There was no previous history or experience with these types of machines, and unfortunate design errors were made that resulted in severe hardware damage and some spectacular explosions. Some of the errors were associated with rotordynamics. During the period of development of turbopumps for the cryogenic fuels used in rocket engines, approximately 50 percent of these machines suffered rotordynamic instability problems at some stage during their development [52]. Probably the most expensive, and certainly the most publicized, of these cases was the hydrogen fuel pump for the space shuttle main engine, described below.

The Space Shuttle Main Engine (SSME) High-pressure Fuel Turbopump (HPFTP) Rotordynamic Instability Problem

Detailed accounts of this case can be found in the papers by Ek [53], Biggs [54], and Childs [55]. Matthew Ek was vice president and chief engineer at Rocketdyne, and Robert Biggs was the project engineer for SSME Systems Analysis during the time of the pumps' development. Dara Childs worked at Rocketdyne and carried out NASA contracts for the turbopump analysis at the University of Louisville. The following discussion is taken mainly from their papers, with some editorial opinions of the present first author added in retrospect with hindsight based on subsequent turbopump developments.

The Space Shuttle Main Engine (SSME) fuel pump consisted of three centrifugal pump impellers (3 stages) driven by two axial flow reaction turbines (2 stages), all on one rotor. See Fig. 7-13 from the ASME paper by Childs [55]. The rotor weighed approximately 130 lb. The pump assembly produced 77,000 hp and weighed 760 lb. The power/weight ratio was an order of magnitude higher than previous turbopump designs.

The rotor assembly is described by Biggs [54] as follows: "The 130 pound rotating assembly is designed around a central drawbolt threaded into the second-stage turbine disc. The turbine disc and the three impellers are rotationally piloted by splined sleeves, which also perform the functions of interstage seals and journal bearings. This stack-up is drawn

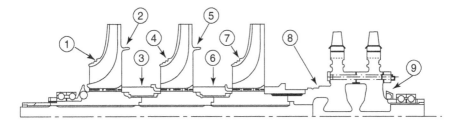

Figure 7-13   Figure 1 from Childs [55].

together by a nut on the first-stage impeller end of the drawbolt. The first-stage turbine disc is bolted to the second-stage turbine disc with a curvic coupling. Radial positioning is accomplished by two sets of angular contact, duplex spring-loaded 45 millimeter ball bearings spaced 23.3 inches apart. The bearings are not lubricated but are cooled with LH2 during operation. Axial positioning is maintained by an LH2 pressure balance with the back side of the third-stage impeller acting as a double-acting, self-compensating balance piston. Balance piston pressure is supplied through an axially sensitive overlapping orifice at the impeller."

Subsynchronous shaft whirl limited operation of the SSME for several months in early 1976. As Biggs describes it [54]: "The HPFTP sub-synchronous whirl was a violent instability which caused a gyration of the rotor in the direction of normal rotation at a frequency of about half of the pump speed. This caused a forward precession of the rotor, which was actually an orbiting of the normal rotating axis. Being a true instability, the whirl was self-initiating and would usually start when the pump speed exceeded twice the first critical speed of the rotating assembly, with an inception frequency equal to the first critical speed (originally about 8,500 rpm). The amplitude would increase rapidly; and within half a dozen cycles, with bending of the rather flexible rotor, the nominal clearances would be breached and internal rubbing would occur at many locations. With clearances closed and bearing supports bottomed out, the system stiffness increased significantly, preventing further increase in amplitude (limit cycle) and raising the first critical speed and, therefore, the whirl frequency. Bearing loads in the limit cycle condition were higher on the turbine end than the pump end by a factor of three, and a significant number of turbine bearing failures were experienced."

Childs had correctly predicted the instability by analysis the year before it occurred [56]. He correctly identified the seals as a major influence, using mathematical models for the seals previously developed by Henry Black [57]. The seals became an even stronger influence after the bearing supports were stiffened by modifications to the original design. This modification, and the fact that the pump was able to run with it, created an unfortunate paradigm in U.S. turbopump design that persists even until today, which is that cryogenic turbopumps must have stiff bearing supports. Ek [53] stated that "system damping and stiffness were additionally augmented by utilizing the interstage seals as axial flow hydrodynamic bearings." In fact, the *only* way to introduce favorable damping into a rotor with stiff bearing supports is through the seals.

The main sources of the instability were stated by Ek to be "turbine cross-coupling and interstage seal effects in combination with low-system stiffness and damping." The whirl inception threshold speed and the first-shaft critical speed were ultimately raised by stiffening the shaft *and* the bearing supports. It appears to the present first author that there never was a good understanding in this program of how shaft stiffness and bearing support stiffness have opposite effects on rotordynamic stability. High bearing support stiffness makes bearing support damping ineffective. Perhaps it was the opinion of program engineers at the time that bearing dampers could not be made to work at cryogenic temperatures. If this were true, then soft bearing supports would not be advantageous. One of the hardware modifications that was tried (unsuccessfully) before stiffening the supports was the addition of Coulomb (sliding friction) dampers, "based on differential motion between the bearing carriers and the turbopump casing" [55]. But this type of bearing vibration damper is notoriously unpredictable and unreliable, and never has been successful in a rotating machine.

As described by Childs [55], computer modeling of the following three design configurations was carried out: (1) the original softly mounted rotor, (2) a stiffly mounted rotor, (3) and an asymmetrically supported rotor with wire mesh bearing dampers. Stiffness asymmetry of bearing supports was known to be stabilizing. The wire mesh dampers were tested and found to possess four times more damping than the Coulomb dampers. The computer simulations showed design (3) to have "better stability and synchronous response characteristics than the stiffly supported design." Nevertheless, the stiffly mounted rotor with modified seals was able to demonstrate acceptable stability in tests on a quick timetable, and it was ultimately used.

The large number of pilot diameters and splined connections in the rotor assembly created a strong susceptibility to excitation of whirl instability by internal friction. Realizing this, NASA funded research on internal friction effects in rotors, but not until ten years later [58]. The results of those studies, as well as later internal friction research by the present first author [59], suggest that internal friction may well have been a significant source of the turbopump instability.

The most positive contribution to knowledge about machinery rotordynamics that came out of the SSME turbopump instability problem was the demonstration that shaft seals can play a major part in suppressing rotordynamic instability and unbalance response. The primary contributor of this knowledge was Dara Childs.

Ten years later, expensive recurring problems with the Rocketdyne turbopump prompted a contract to Pratt & Whitney for an alternative design, the Alternative Turbopump Development–High-Pressure Fuel Turbopump (ATD-HPFTP). Childs was employed as a consultant and once again used a computer simulation to predict that the pump would be unstable in subsynchronous whirl, due to the high ΔP and swirl around the turbine interstage seals [60]. The simulation also predicted a serious critical speed problem (unbalance response). Childs recommended that the bearing supports be softened to solve the critical speed problem. A swirl brake was designed and installed to raise the threshold speed of instability, but the soft support recommendation was ignored. Initial tests showed large synchronous vibration that caused rubbing at the seals. The stiff support paradigm lives on.

Noncontacting Damper Seals

The remarkable a priori correct prediction (twice) by Childs that rocket engine turbopumps would be rotordynamically unstable because of seal forces, and the expensive experiences with the Ekofisk and Kaybob compressors, led to the development of noncontacting seals specially designed to suppress whirl and vibration in turbomachinery.

Two main types of damper seals were developed: (1) honeycomb or hole pattern seals, researched and developed by Childs, and (2) pocket damper seals, patented by the first author and Shultz [61] under the trade name TAMSEAL. Detailed description of the invention and development of pocket damper seals can be found in Chapter 6 and in [62] and [63].

Benkert and Wachter [64] conducted experiments in the late 1970s showing that the gas swirl in conventional labyrinth seals, and the resulting pressure distribution around the seal, can produce a destabilizing follower force. This can be modeled mathematically using cross-coupled stiffness coefficients with opposite signs. It produces negative *effective* damping. They measured direct and cross-coupled stiffness coefficients for a wide range of see-through and interlocking labyrinth geometries. Their test procedure involved displacing the rotor statically in the X direction a distance and measuring the resulting circumferential pressure distribution in labyrinth cavities. Thus, they could not measure damping (with no vibratory velocity). Integrating the circumferential pressures yields a reaction force component in the direction of displacement, (direct stiffness), and a component perpendicular to the displacement (cross-coupled stiffness). With the insight that fluid swirl caused the destabilizing forces,

Benckert and Wachter developed swirl brakes to destroy or reduce the preswirl entering a labyrinth. Swirl-brake technology was rapidly implemented into compressors of several German and Swiss manufacturers, substantially improving their stability characteristics.

SHAFT DIFFERENTIAL HEATING (THE MORTON EFFECT)

The last historically significant development in the history of rotordynamics up to 2009 has been the recent rapid progress in understanding the Morton effect. This is another type of rotordynamic instability of an entirely different nature than aerodynamically induced subsynchronous instabilities, or instabilities due to cross-coupled stiffness in bearings or seals. The Morton Effect is a thermally driven phenomenon which relates to synchronous vibration as opposed to subsynchronous. It arises when a source of friction heats one side of a spinning rotor in a way that is proportional to vibration amplitude. Under the right conditions, a thermal bow can result which can change the state of balance of the rotor. If this change in balance leads to increased vibration, then thermal runaway is possible such that vibration grows continuously, eventually leading to a forced shutdown. This phenomenon is known by different names depending on the source of friction. The earliest identification was in 1926 by Newkirk [65], where the source of friction was rubbing between the rotor and housing. The requisite rubbing can take different forms such as in labyrinth seals, oil seal rings, or slip ring assemblies. A case of rub induced instability is sometimes called a case of the "Newkirk Effect". Another form of friction is viscous shearing of oil in a sleeve or tilting pad bearing. This is often referred to as the Morton Effect in reference to the person credited with its earliest identification [66]. An extensive survey of literature on this topic is presented by De Jongh [67].

Both Newkirk Effect and Morton Effect are forms of Shaft Differential Heating (SDH). The name "synchronous spiral instability" is another moniker for this phenomenon. What is often experienced in field cases is synchronous vibration vectors, when displayed on a polar plot, that spiral around and around with increasing amplitude. The spiral is due to a high temperature region (i.e., hot spot) on the rotor surface traveling around the rotor. Figure 7-14 shows an example of an unstable case of the Morton Effect. In this example a compressor has reached full speed, and while holding steady at that speed, vibration embarks on a divergent spiral, eventually leading to a forced shutdown. When IX vibration amplitude is plotted versus time, amplitude is seen to cycle up and down while growing larger as illustrated in Figure 7-15. These two examples show that divergence can range from rather slow to quite rapid. The time per cycle is normally on the order of minutes or longer. This corresponds to

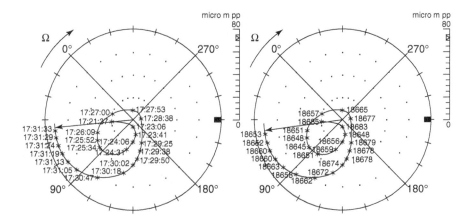

Figure 7-14 From [68] showing unstable divergent behavior in a compressor at 18,600 rpm with a cycle time of about 6 minutes.

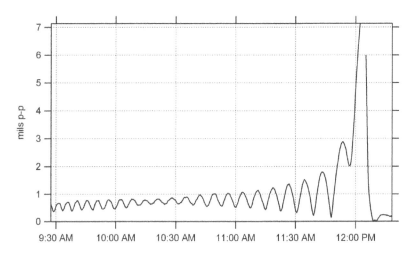

Figure 7-15 Morton effect instability as observed in a generator with a cycle time of about 8 minutes. Speed held constant until amplitude exceeded 7 mils p-p, forcing shutdown.

the amount of time required for the hot spot to travel 360 degrees around the rotor.

The resolution to a case of the Morton Effect could range from simply changing bearing oil inlet temperature or flow rate, to redesigning the shaft to elevate its critical speeds. Another remedy that has been successfully employed is to mount a thermal barrier sleeve on the shaft within the bearing [67]. The sleeve can potentially reduce the temperature difference, or it may reduce how much the rotor bows, or both.

With the Newkirk effect the hot spot can be anyplace on the rotor where rubbing can happen. With the Morton effect the hot spot must be within a bearing. For this reason unstable cases of the Morton effect are essentially limited to high speed flexible rotors that have relatively large overhung masses [69]. This is because rotor bow happening within the confines of a bearing throws overhung weight out of balance, while weight between the bearings is largely unaffected. It has been noted that as high speed overhung rotor designs become more common, cases of the Morton Effect have become more common. The Morton Effect should exist to some extent in all journal bearings, but typically would have no effect on rotors which do not have a large overhung mass.

A rigorous analysis to predict Morton effect requires a transient calculation of coupled thermal, hydrodynamic and rotordynamic effects. This is an active area of research.

REFERENCES

[1] Gunter, E. J. Jr. Dynamic stability of rotor–bearing systems. NASA SP-113, 1966.

[2] Bishop. R. E. D. The vibration of rotating shafts. *Journal Mechanical Engineering Science* 1, No.1 (1959).

[3] Meher-Homji, C. B. The historical evolution of turbomachinery. Proceedings of the 29th Turbomachinery Symposium, Houston, Texas, September 18–21, 2000, pp. 281–310.

[4] Dimarogonas, A. *Vibration for Engineers*, 2nd ed. Englewood Cliffs, NJ: Prentice-Hall, 1995, pp. xvii, 10–13, 535, 539, 573.

[5] Childs, D. W. Rotordynamics of turbomachinery, looking back, looking forward. Proceedings of the Sixth International Conference on Rotordynamics, September 30 - Sydney Australia, September 30–October 4, 2000.

[6] Rankine, W. A. On the centrifugal force of rotating shafts. *Engineer (London)* 27: 249 (1869).

[7] Meriam, J. L., and Kraige, L. G. *Engineering Mechanics*, Vol. 2: *Dynamics*. New York: Wiley 1986, pp. 223–224.

[8] Dunkerley, S. On the whirling and vibration of shafts. *Philosophical Transactions A* 185: 279 (1894).

[9] Strutt, J. W. *The Theory of Sound (Rayleigh)*. New York: Dover, 1945.

[10] Stodola, A. *Steam Turbines*. London: Archibald Constable, 1905, pp. 177–180. English translation by L.C. Loewenstein, Van Nostrand.

[11] Stodola, A. *Steam Turbines*. London: Archibald Constable, 1905, pp. 172–198 and 351–370. English translation by L.C. Loewenstein, Van Nostrand.

[12] Jeffcott, H. H. The lateral vibration of loaded shafts in the neighbourhood of a whirling speed—the effect of want of balance. *Philosophical Magazine*, Series 6, 37: 304 (1919).

[13] Newkirk, B. L. Shaft whipping. *General Electric Review* 27: 169–178 (1924).

[14] Kimball, A. L. Internal friction theory of shaft whirling. *General Electric Review* 27: 244–251 (1924).

[15] Kimball, A. L. Measurement of internal friction in a revolving deflected shaft. *General Electric Review* 28: 554 (1925).

[16] Smith, D. M. The motion of a rotor carried by a flexible shaft in flexible bearings. *Proceedings of the Royal Society of London, Series A* 142: 92–118 (1933).

[17] Thearle, E. L. Dynamics balancing of rotating machinery in the field. *Journal of Applied Mechanics* 56: 745–753 (1934).

[18] Holzer, H., *Die Berechnung der Drehschwin gungen*, Julius Springer, Berlin (1921).

[19] Reynolds, O. On the theory of lubrication and its application to Mr. Towers' experiments. *Philosophical Transactions of the Royal Society of London* 177: 157–234 (1886).

[20] Harrison, W. J. The hydrodynamical theory of lubrication with special reference to air as a lubricant. *Transactions of the Cambridge Philosophical Society* 22: 39 (1913).

[21] Robertson, D. Whirling of a journal in a sleeve bearing. *Philosophical Magazine* Series 7, 15: 113–130 (1933).

[22] Newkirk, B. L., and Taylor, H. D. Shaft whipping due to oil action in journal bearing. *General Electric Review* 28: 559–568 (1925).

[23] Richardson, A. The evolution of the parsons steam turbine. Offices of Engineering, 35 and 36, Bedford Street, Strand, W.C., London, 1911.

[24] Foote, W. R., Poritsky, H., and Slade, J. J., Jr. Critical speeds of a rotor with unequal shaft flexibility. *Journal of Applied Mathematics*, A-77–A-84 (June 1943).

[25] Prohl, M. A. A general method for calculating critical speeds of flexible rotors. *Transactions of the ASME* 67 (1945); *Journal of Applied Mechanics* 12: A-142–A-148.

[26] Myklestad, N. O. A new method of calculating natural modes of uncoupled bending vibration of airplane wings and other types of beams. *Journal of Aeronautical Sciences* 11 153–162 (April 1944).

[27] Greene, R. B. Gyroscopic effects on the critical speeds of flexible rotors. *Journal of Applied Mechanics, Transactions* 70: 369–376 (1948).

[28] Poritsky, H., "Contribution to the Theory of Oil Whip", Transactions of the ASME, August 1953, pp. 1153–1161.

[29] Simons, E. M., "The Hydrodynamic Lubrication of Cyclically Loaded Bearings", Transactions of the ASME, Vol. 72, 1950, page 805.

[30] McHugh, J. D., "Principles of Turbomachinery Bearings", Proceedings of the 8th Turbomachinery Symposium, Texas A&M University, page 143.

[31] Fuller, D. D., Theory and practice of Lubrication for Engineers, 2nd Edition, John Wiley & Sons, page 240.

[32] Thomas, H. J., "Unstable Oscillations of Turbine Rotors Due to Steam Leakage in the Clearance of the Sealing Glands and the Buckets", Bulletin Scientifique, A.J.M. 71, 1958, pp. 1039–1063.

[33] "Design and Application of Squeeze Film Dampers in Rotating Machinery", Proceedings of the 25th Turbomachinery Symposium, Houston, Texas, September 17-19, 1996, pp. 169–188.

[34] Friedericy, J. A., Eppink, R. T., Liu, Y. N., and Cetiner, A., "An Investigation of the Behavior of Floating Ring Dampers and the Dynamics of Hypercritical Shafts on Flexible Supports", USAAML Technical Report 65-34, UVA Report No. CE-3340-104-65U, 1965.

[35] Cooper, S. Preliminary investigation of oil films for the control of vibration. Paper 28, Lubrication and Wear Convention, Institution of Mechanical Engineers, 1963.

[36] White, D. C. The dynamics of a rigid rotor supported on squeeze film bearings. A72-2212708.28, Proceedings of the Conference on Vibrations in Rotating Systems, London, February 14–15, 1972 (Institution of Mechanical Engineers), pp. 213–229.

[37] Goodman, T. P. A least squares method for computing balance corrections. *Journal of Engineering for Industry (Transactions of the ASME), Series* B 86: 273–279 (1964).

[38] Lund, J. W., and Tonnesen, J. Analysis and experiments on multiplane balancing of a flexible rotor. *Journal of Engineering for Industry (Transactions of the ASME)*, 94: 2333, Appendix A (1972).

[39] Muszynska, A., and Bently, D. *Through the Eyes of Others*, 2nd ed. Minden, NV: Bird Rock, 1996.

[40] Thomas, H. A. Precision measuring instrument for small motion of rigid bodies. *The Engineer*, 138–140 (February 1923).

[41] Alford, J. S. Protecting turbomachinery from self-excited rotor whirl. *ASME Journal of Engineering for Power*, 333–344 (October 1965).

[42] Vance, J. M., and Laudadio, F. J. Experimental measurement of Alford's force in axial flow turbomachinery. *ASME Journal of Engineering for Gas Turbines and Power*, 106: 585–590 (1984).

[43] Storace, A. F., et al. Unsteady flow and whirl-inducing forces in axial-flow compressors, part I: experiment. Paper No. 2000-GT-0565, Proceedings of ASME Turbo Expo 2000, Munich, Germany, May 8–11, 2000.

[44] Lund, J. W. *Self-excited Stationary Whirl Orbits of a Journal in a Sleeve Bearing* Thesis, Rensselaer Polytechnic Institute, 1966.

[45] Lomakin, A. A. Calculation of critical speed and securing of dynamic stability of the rotor of hydraulic high pressure machines with reference to forces arising in the seal gaps (in Russian). *Energomashinostroenie* 4(4): 1–5 (1958).

[46] Black, H. F. Effects of hydraulic forces in annular pressure seals on the vibrations of centrifugal pump rotors. *Journal of Mechanical Engineering Science* 11: 206–213 (1969).

[47] Fowlie, D. W. and Miles, D. D. Vibration problems with high pressure centrifugal compressors. ASME Paper 75-Pet-28, Petroleum Mechanical Engineering Conference, Tulsa, Oklahoma, September 21–25, 1975.

[48] Doyle, H. E. Field experiences with rotordynamic instability in high-performance turbomachinery. Rotordynamic Instability Problems in High-Performance Turbomachinery, NASA CP 2133, Texas A&M University Workshop, May 12–14, 1980, pp. 3–4.

[49] Wachel, J. C. Rotordynamic instability field problems. Rotordynamic Instability Problems in High-Performance Turbomachinery, NASA CP 2250, Texas A&M University Workshop, May 10–12, 1982, pp. 3–6.

[50] Lund, J. W. Stability and damped critical speeds of a flexible rotor in fluid-film bearings. *Journal of Engineering for Industry* 96: 682–693 (1974).

[51] Vance, J. M. Torquewhirl: a theory to explain nonsynchronous whirling failures of rotors with high load torque. *Journal of Engineering for Power*, 288–293 (April 1978).

[52] Ek, M. C. Dynamics problems in high performance rocket engine turbomachinery. Workshop on Rotordynamics Technology for Advanced Turbopumps, NASA- Lewis Research Center, Cleveland, February 23, 1981.

[53] Ek, M. C. Solving subsynchronous whirl in the high-pressure hydrogen turbomachinery of the SSME. *Journal of Spacecraft and Rockets* 17: 208–218 (1980).

[54] Biggs, R. E. Space shuttle main engine, the first ten years, history of liquid rocket engine development in the United States, 1955–1980. *American Astronautical Society History Series*, Vol. 13, Part 3, Chapter 4, pp. 69–122.

[55] Childs, D. W. The space shuttle main engine high-pressure fuel turbopump rotordynamic instability problem. ASME Paper No. 77-GT-49, Gas Turbine Conference, Philadelphia, PA, March 27–31, 1977.

[56] Childs, D. W. SSME turbopump technology improvements via transient rotordynamic analysis. Contract Report NASA 31233, The University of Louisville/Speed Scientific School, December 1975.

[57] Black, H. F. Calculation of forced whirling and stability of centrifugal pump rotor systems. *Journal of Engineering for Industry, Transactions of the ASME* (August 1974).

[58] Walton, J., Artiles, A., Lund, J., Dill, J., and Zorzi, E. "Internal rotor friction instability. MTI 88TR39, NASA Contract NAS8-35601, February 1990.

[59] Vance, J. M., and Ying, D. Effect of interference fits on threshold speeds of rotordynamic instability. Paper No. 2007, Proceedings of the International Symposium on Stability Control of Rotating Machinery, South Lake Tahoe, California, August 20–24, 2001.

[60] Childs, D. W. *Turbomachinery Rotordynamics*. New York: Wiley, 1993, p. 434.

[61] Vance, J. M., and Shultz, R. R. A new damper seal for turbomachinery. Proceedings of the 14th Vibration and Noise Conference, Albuquerque, New Mexico, ASME DE-Vol. 60, 1993, pp. 139–148.

[62] Childs, D. W., and Vance, J. M. Annular gas seals and rotordynamics of compressors and turbines. Proceedings of the 26th Turbomachinery Symposium, Houston, Texas, Texas A&M University, September 1997.

[63] Ertas, B. H. Rotordynamic force coefficients of pocket damper seals. Thesis, Texas A&M University, August 2005.

[64] Benckert, H., and Wachter, J. Flow induced spring constants of labyrinth seals. IMechE, Proceedings of the Second International Conference, Vibrations in Rotating Machinery, Cambridge, England, 1980, pp. 53–63.

[65] Newkirk, B. L., 1926, "Shaft Rubbing," Mechanical Engineering, 48, pp. 830–832

[66] Keogh, P. S., and Morton, P. G., 1993, "Journal Bearing Differential Heating Evaluation with Influence on Rotor Dynamic Behaviour," Proceedings Roy. Soc. (London), Series A, Vol. 441, pp. 527–548.

[67] Jongh, F. M., 2008, "The Synchronous Rotor Instability Phenomenon –Morton Effect," Proceedings of the 37th Turbomachinery Symposium, Turbomachinery Laboratory, Texas A&M University, College Station, Texas.

[68] Schmied, J. S., Pozivil, J., and Walch, J., 2008, "Hot Spots in Turboexpander Bearings: Case History, Stability Analysis, Measurements and Operational Experience." Proceedings of ASME Turbo Expo 2008: Power for Land, Sea and Air, Berlin, Germany, June 9-13, 2008. ASME Paper Number GT2008-51179.

[69] Murphy, B. T. and Lorenz, J. A., "Simplified Morton Effect Analysis for Synchronous Spiral Instability," Proceedings of the ASME Power 2009 Conference, Alburquerque, New Mexico, July 21-23, 2009.

INDEX